VOLUME ONE HUNDRED

ADVANCES IN PROTEIN CHEMISTRY AND STRUCTURAL BIOLOGY

Combined Quantum Mechanical and Molecular Mechanical Modelling of Biomolecular Interactions

VOLUME ONE HUNDRED

ADVANCES IN PROTEIN CHEMISTRY AND STRUCTURAL BIOLOGY

Combined Quantum Mechanical and Molecular Mechanical Modelling of Biomolecular Interactions

Edited by

TATYANA KARABENCHEVA-CHRISTOVA

Department of Applied Sciences
Northumbria University
Newcastle, United Kingdom

ELSEVIER

AMSTERDAM • BOSTON • HEIDELBERG • LONDON
NEW YORK • OXFORD • PARIS • SAN DIEGO
SAN FRANCISCO • SINGAPORE • SYDNEY • TOKYO
Academic Press is an imprint of Elsevier

Academic Press is an imprint of Elsevier
225 Wyman Street, Waltham, MA 02451, USA
525 B Street, Suite 1800, San Diego, CA 92101-4495, USA
125 London Wall, London, EC2Y 5AS, UK
The Boulevard, Langford Lane, Kidlington, Oxford OX5 1GB, UK

First edition 2015

ISBN: 978-0-12-802003-6
ISSN: 1876-1623

For information on all Academic Press publications
visit our website at store.elsevier.com

CONTENTS

CONTRIBUTORS

Victoria Erica J. Berryman
Department of Chemistry, Dalhousie University, Halifax, Nova Scotia, Canada

Tomasz Borowski
Jerzy Haber Institute of Catalysis and Surface Chemistry, Polish Academy of Sciences, Kraków, Poland

Russell J. Boyd
Department of Chemistry, Dalhousie University, Halifax, Nova Scotia, Canada

Eric André C. Bushnell
Department of Chemistry, Dalhousie University, Halifax, Nova Scotia, and Department of Chemistry, Brandon University, Brandon, Manitoba, Canada

Ganglong Cui
Key Laboratory of Theoretical and Computational Photochemistry, Ministry of Education, College of Chemistry, Beijing Normal University, Beijing, China

Sam P. de Visser
Manchester Institute of Biotechnology and School of Chemical Engineering and Analytical Science, The University of Manchester, Manchester M1 7DN, Manchester, United Kingdom

Victoria Gisel Dumas
Instituto de Química Física de los Materiales Medio Ambiente y Energía (INQUIMAE), UBA-CONICET, and Departamento de Química Biológica, Facultad de Ciencias Exactas y Naturales, Universidad de Buenos Aires, Ciudad Universitaria, Buenos Aires, Argentina

Wei-Hai Fang
Key Laboratory of Theoretical and Computational Photochemistry, Ministry of Education, College of Chemistry, Beijing Normal University, Beijing, China

James W. Gauld
Department of Chemistry and Biochemistry, University of Windsor, Windsor, Ontario, Canada

Paulino Gómez-Puertas
Molecular Modelling Group, Center of Molecular Biology "Severo Ochoa" (CSIC-UAM), Madrid, Spain

Hansel Gómez
Institut de Biotecnologia i de Biomedicina (IBB), and Departament de Química, Universitat Autònoma de Barcelona, Bellaterra (Barcelona), Spain

Lihong Liu
Key Laboratory of Theoretical and Computational Photochemistry, Ministry of Education, College of Chemistry, Beijing Normal University, Beijing, China

José M. Lluch
Institut de Biotecnologia i de Biomedicina (IBB), and Departament de Química, Universitat Autònoma de Barcelona, Bellaterra (Barcelona), Spain

Iñigo Marcos-Alcalde
Molecular Modelling Group, Center of Molecular Biology "Severo Ochoa" (CSIC-UAM), Madrid, Spain

Marcelo Adrián Marti
Departamento de Química Biológica, Facultad de Ciencias Exactas y Naturales, Universidad de Buenos Aires, and Instituto de Química Biológica de la Facultad de Ciencias Exactas y Naturales (IQUIBICEN) CONICET, Ciudad Universitaria, Buenos Aires, Argentina

Mariano Martin
Departamento de Química Biológica, Facultad de Ciencias Exactas y Naturales, Universidad de Buenos Aires, Ciudad Universitaria, Buenos Aires, Argentina

Laura Masgrau
Institut de Biotecnologia i de Biomedicina (IBB), Universitat Autònoma de Barcelona, Bellaterra (Barcelona), Spain

Jesús I. Mendieta-Moreno
Departamento de Física Teórica de la Materia Condensada and Condensed Matter Physics Center (IFIMAC), Universidad Autónoma de Madrid, and Molecular Modelling Group, Center of Molecular Biology "Severo Ochoa" (CSIC-UAM), Madrid, Spain

Jesús Mendieta
Molecular Modelling Group, Center of Molecular Biology "Severo Ochoa" (CSIC-UAM), and Biomol-Informatics SL, Campus UAM, Madrid, Spain

Fernanda Mendoza
Institut de Biotecnologia i de Biomedicina (IBB), and Departament de Química, Universitat Autònoma de Barcelona, Bellaterra (Barcelona), Spain

Ankur Omer
Division of Toxicology, CSIR-Central Drug Research Institute, Lucknow, Uttar Pradesh, and Academy of Scientific & Innovative Research (AcSIR), New Delhi, India

José Ortega
Departamento de Física Teórica de la Materia Condensada and Condensed Matter Physics Center (IFIMAC), Universidad Autónoma de Madrid, Madrid, Spain

Matthew Quesne
Jerzy Haber Institute of Catalysis and Surface Chemistry, Polish Academy of Sciences, Kraków, Poland, and Manchester Institute of Biotechnology and Faculty of Life Sciences, The University of Manchester, Manchester, United Kingdom

Claudia Lilián Ramirez
Instituto de Química Física de los Materiales Medio Ambiente y Energía (INQUIMAE), UBA-CONICET, Facultad de Ciencias Exactas y Naturales, Universidad de Buenos Aires, Ciudad Universitaria, Buenos Aires, Argentina

Benjamin P. Roberts
Centre for eResearch, The University of Auckland, Auckland, New Zealand

Juan Manuel Romero
Instituto de Química Física de los Materiales Medio Ambiente y Energía (INQUIMAE), UBA-CONICET, Facultad de Ciencias Exactas y Naturales, Universidad de Buenos Aires, Ciudad Universitaria, Buenos Aires, Argentina

Gustavo M. Seabra
Departamento de Química Fundamental, Universidade Federal de Pernambuco, Cidade Universitária, Recife, Pernambuco, Brazil

Chandrabose Selvaraj
Department of Bioinformatics, Computer Aided Drug Design and Molecular Modeling Lab, Science Block, Alagappa University, Karaikudi, Tamil Nadu, India

Poonam Singh
Division of Toxicology, CSIR-Central Drug Research Institute, Lucknow, Uttar Pradesh, and Academy of Scientific & Innovative Research (AcSIR), New Delhi, India

Sanjeev Kumar Singh
Department of Bioinformatics, Computer Aided Drug Design and Molecular Modeling Lab, Science Block, Alagappa University, Karaikudi, Tamil Nadu, India

Venkatesan Suryanarayanan
Department of Bioinformatics, Computer Aided Drug Design and Molecular Modeling Lab, Science Block, Alagappa University, Karaikudi, Tamil Nadu, India

Maciej Szaleniec
Jerzy Haber Institute of Catalysis and Surface Chemistry, Polish Academy of Sciences, Kraków, Poland

Amy Timmins
Manchester Institute of Biotechnology and School of Chemical Engineering and Analytical Science, The University of Manchester, Manchester M1 7DN, Manchester, United Kingdom

Juan Torras
Department of Chemical Engineering, EEI, Universitat Politècnica de Catalunya, Igualada, Spain

Daniel G. Trabada
Departamento de Física Teórica de la Materia Condensada and Condensed Matter Physics Center (IFIMAC), Universidad Autónoma de Madrid, Madrid, Spain

Samuel B. Trickey
Departments of Physics and of Chemistry, Quantum Theory Project, University of Florida, Gainesville, Florida, USA

PREFACE

For biomolecular systems as big as solvated proteins, the combined quantum mechanics/molecular mechanics (QM/MM) method has proved its role as a potent computational technique. This method synergizes the advantages of both the quantum mechanical (QM) and the molecular mechanical (MM) methods. The QM method is needed for modeling of electronic processes including reaction mechanisms and is applicable to systems as big as a few hundreds of atoms. For biomolecular interactions, however, we need to be able to model systems as big as several thousands of atoms which are doable by applying MM methods. The QM/MM method uses the advantages of both the QM and the MM methods, and hence, its accuracy is due to the QM method and the speedup is due to the MM method. In a QM/MM scheme, a QM method is applied to the species of interest for a chemical transformation and the MM method is used for the treatment of the rest of the protein and the solvent.

Warshel and Levitt published the first paper on the QM/MM method and the two authors together with Martin Karplus were awarded the 2013 Nobel Prize in Chemistry (http://www.nobelprize.org/nobel_prizes/chemistry/laureates/2013/) for the development of this method for modeling of complex biomolecular systems.

This thematic volume of *Advances in Protein Chemistry and Structural Biology* is a collection of some of the most recent advancements in both QM/MM method developments and applications to biomolecular systems. In Chapter 1 by Juan Torras et al., the development and application of a new program—PUPIL—for simultaneous treatment of more than one active site in complex biological systems is reviewed. In Chapter 2 by Marcelo Marti et al., the application of HyDRA—a hybrid differential relaxation algorithm—for calculation of an enzymatic reaction free energy is presented. In Chapter 3 by José Ortega et al. is reviewed the application of a new QM/MM method—FIREBALL/AMBER, synergizing a density functional theory method—FIREBALL—and the molecular dynamics package—AMBER—for studying biomolecular interactions. Chapter 4 by Poonam Singh et al. focuses on QM/MM-based approaches for the prediction of novel drug–target interactions. Chapter 5 by Sam de Visser et al. introduces a review of mechanistic computational studies on metal-containing haloperoxidases and halogenases. Chapter 6 by Russell

Boyd et al. focuses on explanation of the effect of the protein environment and the choice of the quantum mechanics method on the reaction mechanism in QM/MM simulations. Chapter 7 by Maciej Szaleniec et al. represents on comparative study of metalloenzymes by QM and QM/MM methods. Important findings of glycosyltransferase reaction mechanisms modeled by QM/MM methods are reviewed in Chapter 8 by Laura Masgrau et al. The application of QM/MM methods in modeling of the excited states of significant for the biology photoactive proteins is represented in Chapter 9 by Wei-Hai Fang.

I would like to thank all authors of this volume of *Advances in Protein Chemistry and Structural Biology* for their excellent contributions to the volume and for their work. I would like also to thank the *APCSB* Editor-in-Chief Dr. Rossen Donev and Helene Kabes, Mary Ann Zimmerman from Elsevier (Oxford, UK), and Surya Narayanan from Elsevier (Chennai, India) for their work on this volume.

Tatyana Karabencheva-Christova
Northumbria University, Newcastle, United Kingdom

ACKNOWLEDGMENTS

Acknowledgments for Chapter 2, *Advances in Protein Chemistry and Structural Biology*, Volume 98 (*Modifications of Natural Peptides for Nanoparticle and Drug Design*, Andrew P. Jallouk, Rohun U. Palekar, Hua Pan, Paul H. Schlesinger, Samuel A. Wickline).

This work was supported by grants from the National Institutes of Health (NIH R01 HL073646 and U54 HL112303), the Barnes-Jewish Hospital Foundation, and the James R. Hornsby Family Dream Garden Investment Partnership to S.A.W.

CHAPTER ONE

PUPIL: A Software Integration System for Multi-Scale QM/MM-MD Simulations and Its Application to Biomolecular Systems

Juan Torras[*,1], **Benjamin P. Roberts**[†], **Gustavo M. Seabra**[‡], **Samuel B. Trickey**[§,1]

[*]Department of Chemical Engineering, EEI, Universitat Politècnica de Catalunya, Igualada, Spain
[†]Centre for eResearch, The University of Auckland, Auckland, New Zealand
[‡]Departamento de Química Fundamental, Universidade Federal de Pernambuco, Cidade Universitária, Recife, Pernambuco, Brazil
[§]Departments of Physics and of Chemistry, Quantum Theory Project, University of Florida, Gainesville, Florida, USA
[1]Corresponding authors: e-mail address: joan.torras@upc.edu; trickey@qtp.ufl.edu

Contents

Abstract

PUPIL (Program for User Package Interfacing and Linking) implements a distinctive multi-scale approach to hybrid quantum mechanical/molecular mechanical molecular dynamics (QM/MM-MD) simulations. Originally developed to interface different external programs for multi-scale simulation with applications in the materials sciences, PUPIL is finding increasing use in the study of complex biological systems. Advanced MD techniques from the external packages can be applied readily to a hybrid QM/MM treatment in which the forces and energy for the QM region can be computed by any of the QM

ISSN 1876-1623
http://dx.doi.org/10.1016/bs.apcsb.2015.06.002

methods available in any of the other external packages. Here, we give a survey of PUPIL design philosophy, main features, and key implementation decisions, with an orientation to biomolecular simulation. We discuss recently implemented features which enable highly realistic simulations of complex biological systems which have more than one active site that must be treated concurrently. Examples are given.

1. INTRODUCTION

One of the most remarkable but under-appreciated insights of contemporary molecular and condensed-phase research is that the atomic constituents of molecules behave in many ways according to the classical mechanics of Newton. A related but also remarkable recognition is that molecules themselves often behave in an essentially classical way. Thus one has, for a single example, the concept of docking, which operationally is purely classical mechanics. The atomic-level insight had been reached by condensed matter and materials physicists (Alder & Wainwright, 1959; Rahman, 1964) over 50 years ago but in the setting of drastically simpler systems than those found in biology. They converted the insight into a powerful computational tool by recognizing, in essence, that classical statistical mechanics for the microcanonical ensemble amounts operationally to simply moving members of a population of interacting particles via Newton's Second Law for a substantial time interval and averaging the resultant energetics to yield the system thermodynamics. (This assumes ergodicity, one of several subtleties we leave aside for the purpose of broad perspective.)

These ideas were anticipated, of course, by molecular vibration theory and analysis on the chemistry side and the corresponding "lattice dynamics" or phonon treatment in the physics of ordered solids. Both start with a "ball and stick" model of the system, with the "balls" representing atoms and the "sticks" representing bonds (and, in a scale model, bond lengths). Such models trace to August Wilhelm von Hofmann (Meinel, 1992). Vibrational analyses then replace the sticks with suitably chosen springs, corrected, if need be, by low-order anharmonicities. Common to both vibrational analyses and molecular dynamics (MD) is the assumption of a potential energy surface or force field. Given the existence of bond-length data, it is an obvious step to calibrate the force field so as to recover the bond lengths at the energy minimum.

In a vernacular way, the summary just given comprises molecular mechanics (MM). MM is the treatment of molecular systems by methods of classical mechanics. Statics, i.e., rigid systems, may be enough to yield valuable insight for the huge (by the standards of materials physics) molecules of biological importance. Elementary docking again provides an example: can two rigid structures be fit together in a snug way? MD adds the classical dynamics to get at the thermodynamics.

Computational approaches to the description of chemical systems in fact fall into two broad categories. The essential differences lie in assumptions about chemical interactions and bonding. MM is one category. MM schemes attempt to describe chemical systems via simple mathematical models for the forces between classical objects. Common aspects of these models (the force fields) include (1) description of bonds themselves, as well as the angles between bonds to the same atom, as simple harmonic oscillators; (2) inclusion of short-range electrostatic interactions using the Coulomb inverse-square law directly; (3) inclusion of long-range electrostatic (Coulomb) interactions using a particle-mesh Ewald summation (Darden, York, & Pedersen, 1993); and (4) description of van der Waals (dispersion) interactions via one of several Lennard-Jones type potentials.

All MM approaches, whether static or MD, are useful primarily for simulating large chemical systems containing 10^3 to 10^5 or more particles, because the computational tasks are relatively simple: calculation of the forces and temporally discretized solution to Newton's equation. However, the descriptions of chemical systems are of limited accuracy. All of the chemistry is in the force field, so it determines the quality of MM calculations. For simple liquids such as Ar, an equally simple Lennard-Jones form goes a long way toward providing realistic thermodynamics from MD. But for biomolecular systems, construction of reliable potentials has proven to be a scientific craft of its own. See, for example, the literature of AMBER (Case et al., 2006) and CHARMM (Brooks et al., 2009). Moreover, the mathematical models used are radically simplified. At most the chemical system composition is that of indivisible atoms. In a united-atom or coarse-grained simulation, composite particles (e.g., rigid molecules) are used.

A pure MM description of a system cannot capture perturbations in electron density reliably and accurately, much less give a uniformly accurate description of the vitally important chemical processes of bond breaking and formation, i.e., chemical reactivity. Moreover, a system can only be described using an MM model if the definition of the system includes all necessary variables (such as bonds, angles, and so forth) and the chosen force field contains the associated parameters. Parameters in this context are coefficients in the force-field equations whose precise values vary depending upon the particular types of atoms involved in the bond, angle, torsion, or nonbonding pair of atoms. If any variable is missing, the MM model is somewhere between intrinsically inaccurate and completely irrelevant. If any variable is defined in the topology but its corresponding parameters are missing from the force field, the MM model simply cannot be solved.

Chemical reactivity is critical in all condensed-phase processes, including biological ones. As illustrated by the structure of the periodic table, chemical reactivity intrinsically arises from quantum mechanics (QM), which identifies the second category of approach. A QM approach to computing the properties of a chemical system typically describes the system in terms of molecular orbitals, electron densities, and, sometimes, low-order density matrices. In addition to handling chemical processes such as bond breaking and formation, many (but not all) QM methods are able to provide insight into the potential energy surface of the chemical system without a need for prior parameterization. QM methods are, in fact, approximations, because of the intractability of the many-electron Schrödinger equation. The approximations range in sophistication and computational expense. There are comparatively inexpensive semi-empirical approaches, such as the various NDO models (see, e.g., Pople, Santry, & Segal, 1965; Pople & Segal, 1965) and the AM1 (Dewar, Zoebisch, Healy, & Stewart, 1985) and PM3 (Stewart, 1989) families of semi-empirical Hamiltonians, all of which involve some degree of parameterization. And there are very accurate but computationally extremely intensive methods such as coupled cluster (Bartlett & Musiał, 2007) and full configuration interaction (Shavitt, 1998).

Within any given formal methodology, computation requires additional technical approximation. For example, in any QM scheme formulated in terms of molecular orbitals, the scientist must choose one or more basis sets (note that this is true even of grid-based methods, for which the basis is Dirac delta functions at the grid points), that is, sets of functions of linear combinations of which give the molecular orbitals. While the computational difficulty of a QM calculation, and the way in which that difficulty scales with number of particles, both vary from method to method, in all cases QM calculations are far more demanding than a MM calculation on a similarly sized system. The phrase "number of particles" itself is deceptive. For MM it is the number of classical objects, hence, at most the number of atoms. In the QM case, it is at least the number of electrons, which is roughly an order of magnitude larger. Added to that is the fact that the computational costs of QM methods for electrons typically scale as some power of the number of electrons, whereas MM cost scaling is roughly $N_p \ln N_p$ with N_p the particle number.

Various approaches have been developed over the years to preserve the advantages of QM calculations (e.g., proper treatment of bond breaking and formation and of electron distribution) while reducing the disadvantages of high computational cost and corresponding intractability of large chemical systems. Among them there are the fragmentation methods that depend upon some scheme for partitioning the system into distinct fragments and obtaining the total properties of the system through aggregation of the fragment properties (Gordon, Fedorov, Pruitt, & Slipchenko, 2012). A clear example is the divide-and-conquer approach as originally formulated by Yang (1991). It treats a system as a set of subsystems that can be solved largely independently of one another. Thus, the density of the system of interest is divided into the sum of the densities of the subsystems, using an efficient one-electron density matrix approach. Consequently, this approach ignores many unimportant interactions and significantly reduces the computational expense.

An alternative approach is hybrid quantum mechanics/molecular mechanics (QM/MM) or multi-scale simulation, which was initially proposed in a 1976 paper by Warshel and Levitt (Warshel & Levitt, 1976). In a QM/MM simulation, the researcher chooses a small region of the system that is of particular chemical importance. That region is treated using QM, while the remainder of the system is treated via MM. The approach, while necessarily less accurate than representing the whole-system quantum mechanically, offers a good balance of physical accuracy and relatively low computational cost (Lin & Truhlar, 2007; Senn & Thiel, 2009). Implicitly, we have introduced another assumption in this discussion, namely, that QM forces from the subsystem electrons determine the classical forces from within that region upon the subsystem nuclei. Thus, we have invoked the Born-Oppenheimer (BO) approximation for QM/MM and QM/MM-MD (Barnett & Landman, 1993). We do not consider beyond BO methodology here. Since we assume the BO approximation, in the discussion that follows the terms "QM package" or "QM code" means an electronic structure code.

An aside on terminology and associated notation may be helpful. QM/MM as used here denotes the separation of the system into subsystems (regions) according to the way in which the forces are generated within that region. But the very use of forces itself means that *all* of the nuclei (or all of the more coarse-grained particles) nevertheless are positioned or moved according to classical mechanics. Hence, the overall approach is still MM and the notation and terminology has two meanings depending on context.

Observe that the materials physics community typically uses "multi-scale" instead of "QM/MM." The early PUPIL papers use that terminology. Identifying the source of the forces introduces a bit of cumbersomeness when discussing MD. Here, we use QM/MM-MD to denote MD with forces from a QM/MM decomposition. Note that for QM force driving MD without any QM/MM separation, the literature has several different names: Born-Oppenheimer MD (Barnett & Landman, 1993), *ab initio* MD (Marx & Hutter, 2009), and even quantum MD (Horner, Lambert, Kress, & Collins, 2009) are common.

Over time, members of the scientific community have released many codes to carry out MM or QM calculations. Notable MM programs include AMBER (Case et al., 2006), CHARMM (Brooks et al., 2009), NAMD (Phillips et al., 2005), DL_POLY (Smith & Forester, 1996), and so forth. Prominent molecular QM packages include GAUSSIAN (Frisch et al., 2009), GAMESS (Schmidt et al., 1993), Jaguar (Bochevarov et al., 2013), Q-CHEM (Shao et al., 2006), NWChem (Valiev et al., 2010), deMon2k (Köster et al., 2011), Siesta (Soler et al., 2002), etc.

One computational approach to QM/MM is to implement both functionalities in the same package. This approach is used by AMBER (which has evolved from pure MM to contain some limited native QM functionality) and by GAUSSIAN and deMon2k (Salahub et al., 2015), both of which started as pure QM codes but now have some MM functionality. However, this approach seems, more often than not, to be of limited utility, as those who develop and maintain the software put most of their effort into the program's "strong suit." A tendency to bias toward a development group's strength is, of course, completely understandable. The equally understandable consequence is that the implementation of the other component is restricted as to available techniques and the size and complexity of systems that can be considered.

An alternative is for an MM program and a QM program to interface directly. This approach allows the MM program to access much of the functionality offered by the QM program, and so forth. Commonly, this strategy is implemented by making one code into a library for the other, yielding a monolithic package upon compilation. But doing so results in an intimate linking of multiple codes developed by distinct groups. The result is a distinct challenge for both maintenance and enhancement. Changes in one of the codes very often cause changes deep in the others to be mandatory if generation of the monolithic package is to remain supported.

A third (and very distinct) option, the philosophy used by PUPIL, is for the MM program and the QM program each to communicate with a linker program. All the codes, whether MM or QM, maintain their own architectural and developmental autonomy to the greatest degree possible. In this way, theoretically, only one interface needs to be maintained by each MM program and each QM program, while the development effort can be focused on the linker program. The philosophy insists upon minimal if any modification to either a QM or MM package. And it presumes that any MM package supported by PUPIL can be utilized in combination with any QM package supported by PUPIL. Recent changes to the ChemShell code (Metz, Kästner, Sokol, Keal, & Sherwood, 2014) are somewhat analogous in that they have included a number of interfaces to QM codes.

PUPIL began in the materials physics simulation community, so the notation and some of the motivating text in the original papers (Torras, Deumens, & Trickey, 2006; Torras et al., 2007) may be a bit unfamiliar. Thus it is fitting, before going to PUPIL itself, to set formulations and notation for MM, MD, and QM/MM and QM/MD.

2. QM/MM-MD METHODOLOGY

Most current QM/MM schemes, including the one used in PUPIL, are based on the approach developed by the 2013 Nobel Prize recipients Arieh Warshel, Michael Levitt, and Martin Karplus (Field, Bash, & Karplus, 1990; Warshel & Levitt, 1976). As noted already, the simplest version involves partitioning the system into two regions ("inner" and "outer" regions) to be treated at different levels of approximation. The inner region contains only a small number of atoms that are the chemically relevant part of the system. It is treated with a QM method for the forces from that region upon the nuclei. The outer region, the remainder of the system, is treated via MM. At this point, there are two main schemes to consider for the QM/MM energy expressions: the subtractive and the additive QM/MM schemes. The subtractive QM/MM scheme calculates the entire system at the MM level of approximation, whereas the inner system is calculated at the QM level. Subtraction of the energy of the MM calculation for the inner QM region to avoid double counting then yields the final energy expression. This kind of scheme was initially implemented by Morokuma and coworkers (Maseras & Morokuma, 1995) and later extended to include calculations using electrostatic embedding (Chung, Hirao, Li, & Morokuma, 2012). However, in such an interpolation scheme the QM/MM interactions are handled entirely at the MM level. On the other hand, the additive QM/MM scheme utilizes a similar system partition but requires an explicit treatment of the QM/MM coupling terms. Given this partitioning, the potential energy for the system can be written as

$$E = E^{\mathrm{QM}} + E^{\mathrm{MM}} + E^{\mathrm{QM/MM}} \tag{1}$$

Here, E^{QM} is the sum of electronic energies and internuclear Coulomb repulsion energies for the QM part of the system, perturbed by the presence of the atoms in the MM region, E^{MM} is the potential energy for the MM part of the system, and $E^{\mathrm{QM/MM}}$ describes the interaction energy between the two regions. Typically, it contains terms for electrostatic, van der Waals, and bonded interactions across region boundaries

$$E^{\mathrm{QM/MM}} = E^{\mathrm{QM/MM}}_{\mathrm{vdW}} + E^{\mathrm{QM/MM}}_{\mathrm{electr}} + E^{\mathrm{QM/MM}}_{\mathrm{bond}}, \tag{2}$$

In the PUPIL implementation (Torras et al., 2006, 2007), the van der Waals interactions between the QM and MM atoms are calculated as usual by the MM program, utilizing standard parameters for whatever force field is adopted. It has been shown that considering this interaction as purely classical does not introduce significant errors in the calculation (Riccardi, Li, & Cui, 2004). For instance, in the AMBER–PUPIL interface (which we describe below), the van der Waals interactions are calculated using a 12–6 Lennard-Jones potential:

$$E^{\mathrm{QM/MM}}_{\mathrm{vdW}} = \sum_{\alpha}^{\mathrm{QM}} \sum_{i}^{\mathrm{MM}} \left[\frac{A_{\alpha i}}{R^{12}_{\alpha i}} - \frac{B_{\alpha i}}{R^{6}_{\alpha i}} \right], \tag{3}$$

where Greek letters label atoms in the QM region and Roman letters label those in the MM region, $R_{\alpha i}$ is the distance between atoms α and i, and A and B are standard Lennard-Jones parameters for the interaction between α and i.

The calculation of the electrostatic interaction between QM and MM atoms depends on the resources and information made available by the QM program of choice, but usually can be divided in two parts: the influence of the atoms in the MM region on the atoms in the QM region and vice versa,

$$E^{\mathrm{QM/MM}}_{\mathrm{elect}} = E^{\mathrm{QM}\leftarrow(\mathrm{MM})} + E^{\mathrm{MM}\leftarrow(\mathrm{QM})}, \tag{4}$$

Whenever possible, PUPIL makes use of electrostatic embedding for the calculation of the $E^{\mathrm{QM}\leftarrow(\mathrm{MM})}$ term, i.e., the atoms in the MM region are passed to the QM program as effective point charges fixed at their respective coordinates, with the classical force- values used for the charges. As a result, this term usually is already accounted for in E^{QM} and the forces acting on the QM atoms. The $E^{\mathrm{MM}\leftarrow(\mathrm{QM})}$ term usually is not calculated, since the object of interest in a MD calculation is actually the force that arises from that interaction, which can be calculated directly. As some QM packages do not calculate the forces exerted on the atoms of the MM region due to the interaction with the atoms in the QM region, calculation of this contribution depends on the specific QM program used. GAUSSIAN, for example, can provide the electric field $\vec{E}$ at the locations of the point charges, which can be used to calculate the forces (Roberts et al., 2012):

$$\vec{F}_i^{\mathrm{QM}} = q_i \vec{E}, \tag{5}$$

where $\vec{F}_i^{\mathrm{QM}}$ is the force acting on atom i of classical charge q_i in the MM region due to the interaction with the QM density. Also, recent program versions of NWChem and deMon2k are able to calculate the forces exerted on the atoms of the MM region due to the interaction with the atoms in the QM region as a normal program output. For other programs and older versions of NWChem and deMon2k, this force contribution can be obtained by projecting the electronic density of the QM system on a grid, then calculating the interaction between the classical charges and each point of the grid,

$$\vec{F}_i^{\mathrm{QM}} = \sum_j^N \vec{r}_{ij} \frac{q_i}{|r_{ij}|^3} \mathrm{d}q_j, \tag{6}$$

where N is the number of points in the grid and $\mathrm{d}q_j = \rho_j \mathrm{d}x\mathrm{d}y\mathrm{d}z$.

Note well that special treatment must be used whenever there are covalent bonds that cross the QM/MM boundaries, such that cutting the molecule at the boundary would leave both the MM and QM regions with incomplete valences. This can be the case in biomolecular simulations, for instance, wherein only a limited set of the residues, e.g., the residues in the active site, are to be included in the QM region. To deal with this issue, PUPIL uses the link-atom method (Singh & Kollman, 1986). A nonphysical quantum atom (the *link atom*) is introduced in the QM region along the covalent bond between the MM and QM region, at the appropriate distance from the QM atom, completing its valence. To avoid improperly high electrostatic interactions, the MM atom is not included among the point charges around the QM region. In the case of the MM region, all force-field terms that include at least one MM atom are calculated, and all terms involving QM atoms exclusively are omitted. There are no MM terms including the link atom. Once the QM gradient (i.e., the force on the link atom treated as a QM atom) is available, it is redistributed to the QM and MM atoms that form the linked pair. Redistribution is via a chain-rule recipe from the energy gradient with respect to the link-atom coordinate to gradients with respect to QM and MM atom coordinates. A detailed discussion can be found in Field, Albe, Bret, Proust-De Martin, and Thomas (2000) and Walker, Crowley, and Case (2008).

3. THE PUPIL FRAMEWORK

PUPIL is an acronym for "Program for User Package Interfacing and Linking." Its original design (Torras et al., 2006) was motivated by a materials physics problem, namely, hydrolytic weakening: in the presence of water, a ceramic under tension fractures much more readily when wet than when dry. The design process involved serious examination of the QM and MM software that was deemed potentially relevant. For the QM side that meant quantum chemistry software, with an initial emphasis on semi-empirical methods calibrated to high-level coupled cluster calculations on model systems (Mallik, Runge, Dufty, & Cheng, 2007). It was clear even at that stage, however, that the design had to accommodate more sophisticated QM software without fundamental alteration. For the MM side, the design considerations were most strongly influenced by what seemed to be the dominant MD code for materials at the time, DL_POLY (Todorov, Smith, Trachenko, & Dove, 2006). Little if any consideration was given to MM in the static sense but that has not turned out to be a limitation. A third major category of functionality was "domain identification" (DI). The DI concept is to provide automated identification of the chemically active region within which the QM forces are necessary. Even today, that identification almost always is done by the software user (identification "by hand") rather than by the software itself, but the design anticipated automation. The fourth major category of functionality was user support via an easy-to-use interface.

As we have summarized already, at the time of the PUPIL initial design, there were two main ways to do a QM/MM simulation which exploited the capabilities of existing codes. One was to merge them into a single code of some sort. Typically, this takes considerable rewriting of the component codes, as well as writing of new data-interchange and control-interchange code, with the architectural outcome being that one component code becomes the manager of the other. In addition to the labor involved, there is another serious problem. Much of that work must be redone each time there is a new major release of any of the component codes. Almost universally, the modifications are too deeply entangled in the internals of the component codes to allow easy updates. Those barriers are part of the motivation for an approach that is common now, namely, to construct the simulation via scripting: the component codes are invoked and controlled via the script and data are moved, reformatted, combined, and processed via the script. For a skillful user, the benefit is that a novel simulation can be assembled quickly, but the prerequisite is intimate familiarity with the input, output, and control structures of each component code. The disadvantages are a very high barrier to less-experienced users, replication of effort, and error proneness owing to lack of systematic protocols (an almost inescapable side effect of the flexibility provided by scripting).

PUPIL therefore is a software environment mainly designed with three aims. The first was (and is) to bring to the user a general open-source tool to perform QM/MM (multi-scale) simulations within materials physics, chemistry, and biochemistry by exploiting existing QM and MM codes, the "user packages." (Notice that this "packages" terminology implicitly acknowledges that there may be some MM capability in a QM code and conversely. PUPIL design does not care; its focus is on interoperation.) The second was (and is) to provide a means for developers to contribute new and improved capabilities easily, either by adding support for new user packages or by adding new common and interesting functionalities. The third aim was (and is) to achieve the first two as generically as possible, that is to say, with as minimal as practicable intrusion into the user packages as possible. The extent to which PUPIL has achieved those design aims is illustrated at least in part by the fact its range of usage now goes outside materials physics and chemistry to more general QM/MM-MD simulations on structures as demanding as complex enzymes.

The design philosophy of PUPIL is to provide an environment wherein all the common capabilities needed for the multi-scale simulation are collected into PUPIL itself. Thus, developers of user packages can link their applications to PUPIL to perform simulations interacting with other software in which data and simulation control are transferred from one user package to another in a straightforward manner. PUPIL itself acts as the Supervisor program. It coordinates execution and communication between the user packages, each of which provides a calculation unit (CU). The Supervisor is implemented as a distributed program with one Manager and several Workers, one Worker for each CU. The Manager and Workers communicate through the network using the client–server model design.

3.1 Features

All QM/MM-MD simulations performed within the PUPIL framework involve the execution of at least two user packages, an MD engine (e.g., AMBER, DL_POLY) and a QM engine (e.g., NWChem, deMon2k, etc.). We leave DI aside for the moment. These programs are executed, coordinated, and managed by the PUPIL Manager. A user therefore may choose, in mix and match fashion, among any of the possible combinations of codes currently interfaced with PUPIL. The list of QM and MM codes currently interfaced to PUPIL is given in Table 1. Each code has a specific interface in PUPIL, written in Java (details below), to allow for communication and exchange of information to support all features in the external codes. The communication between packages is basic enough, i.e., coordinates, atom types, atomic charges, forces, and energy, that new additional features in a user package almost always are incorporated immediately by the general features of the PUPIL Manager. (The exception would be some scientific method or concept not previously supported at all.)

Table 1 External QM and MM Codes that Currently Interfaced to PUPIL

	Electrostatic Embedding	Start–Stop Behavior	Cyclic Behavior	Tightly Coupled Interface	MPI Execution
QM codes					
deMon2k	X	X			X
GAUSSIAN09	X	X			–[a]
NWChem	X	X			X
Siesta		X	X	X	X
MNDO		X			-
MM codes					
AMBER14	X			X	
DL_POLY 2	X			X	

[a]Conventional parallel execution using threads and LINDA software.

PUPIL functionalities and capabilities are distributed among the QM Engine, MD engine, and the PUPIL Manager. The user must be familiar with the user packages selected for the particular QM/MM-MD simulation at the level of knowledge of input and output file formats and contents. A set of input file templates from each external engine must be supplied by the user with the usual information associated with each engine. For example, this would consist of system coordinates, MM atom particle types, and force field for the MD engine, and QM atom types, QM approximation ("level of theory" to use common but unhappy terminology), and convergence directives for the QM engine. However, information related to the QM/MM-MD simulation and the coupling between QM and MM calculations is supplied through the PUPIL Graphical User Interface (GUI). It yields an output file containing all the information necessary to assist the PUPIL Manager to conduct the whole simulation. Neither engine knows explicitly about the other.

As laid out in Section 2, all QM/MM-MD calculations are performed within the additive QM/MM scheme of energy partition between an "inner" (QM) and an "outer" (MM) region plus a coupling term between them. Two different QM/MM coupling schemes are allowed, mechanical and electrostatic embedding. In the former scheme, the QM calculations are performed in the inner region in the absence of the outer region, with the interaction between the outer and inner regions treated at the MM level (both bonded and nonbonded interactions) of approximation. In the latter scheme, the QM Hamiltonian includes classical partial charges from the MM description as point charges which thereby polarize the QM region. Similarly, the forces on the classical partial charges due to the interaction with the electronic density of the QM region are also included. That, in turn, induces polarization of the MM region by the QM region along an MD trajectory.

QM packages linked to PUPIL can have two different behaviors depending upon how they are invoked by the PUPIL Manager: Cyclic or Start–Stop. In Cyclic mode, a CU is started by the Manager, and all the actions involved in any individual QM/MM-MD simulation step (e.g., data request, data insertion, computation, and return of results to the PUPIL Manager) are performed at that step without restarting the QM package. Start–Stop mode so far has been used for QM CUs. In that mode, the QM CU is started by the PUPIL Manager. Upon completion for that step, the QM CU terminates and its output files are parsed to get the information back to the Manager for use in the MD CU. Thus, a new instance of the QM CU is started and executed at each force evaluation.

The advantage of Start–Stop mode lies in the ease with which PUPIL can link packages without requiring any source-code modification or recompilation. It is the only route available to supporting closed-source codes. The disadvantage is loss of speed and flexibility. Some of the CUs that are more tightly coupled to PUPIL require minimal source-code modification and linking with the PUPIL libraries to be used with the PUPIL interface. Specifically, those QM CUs which operate in Cyclic mode as well as all MD user packages so far have this kind of link to the PUPIL Manager.

We have already remarked on treatment of the QM-MM region boundary by the link-atom approach to saturate the dangling bond of any QM atom left over from a broken QM–MM bond. This link atom is usually taken to be a hydrogen atom. However, the user also may use any of the quantum atoms allowed by the external QM engine to saturate the free valence of the QM atom. An example is pseudoatoms with a parameterized effective core potential which can be adjusted to mimic the properties of the original chemical bond being cut (Mallik, Taylor, Runge, Dufty, & Cheng, 2006).

3.1.1 High Performing Computing

Distinct (so far as we know) from other implementations, PUPIL treats the MD, QM, and DI codes at the same level, so that PUPIL can control their execution on the same footing. The necessary resources, i.e., processors, and the information to communicate and control the external code execution are stored and coordinated by the PUPIL Manager. In this way, PUPIL is capable of dynamically starting and stopping external MPI codes on demand, with communication among external codes conducted within the CORBA (Common Object Request Broker Architecture) protocol by means of the client/server paradigm (Torras et al., 2007). See details below. The great advantage of this architecture lies in the ease with which workloads can be distributed across multiple computing resources. The recent addition to the PUPIL Manager of the capability to handle a fixed number of multiple independent active zones (QM regions) during the whole simulation (Torras, 2015) is made possible by this architecture. Another beneficial aspect is the capability to assign different computational resource to the different CUs depending on their computational time scaling (thereby managing load balancing). Indeed, running two or more separate binaries in a high-performance computing (HPC) environment can be optimized efficiently by balancing the resources assigned to each parallelized external code involved.

This approach overall is a generalization of the original "inner" and "outer" region paradigm of Warshel, Levitt, and Karplus. In principle, the DI user package would decide the number, type, and extent of active zones on the fly during the simulation. Although the current PUPIL implementation does not support that advanced feature, it is conceptually possible. It would require capabilities of a much more sophisticated DI code than presently exists. The crucial PUPIL property to be emphasized is that such a DI code again would not need to know, nor would it know, about the QM and MM user packages. PUPIL simply would provide the sophisticated DI with the data (from the QM and MM package results) needed to determine the boundaries and properties of each active region, with those results then relayed back to the appropriate user packages by PUPIL.

The most time-consuming process, hence the major bottleneck, in QM/MM-MD simulations is the force calculation by the external QM engine. Typically, 80–95% of the total time invested in one QM/MM-MD step is consumed by the QM calculation. The next largest time cost is from building the quantum zone embedding (Torras et al., 2006, 2007). Thus, major effort must be given to the parallelization of the QM code calculation. PUPIL is able to deal with parallelized code for a user package, i.e., the QM code. To facilitate the execution within an MPI environment of any external code, the PUPIL Manager takes advantage of its capability to dynamically start and stop processes to assign specific resources to each CU. Thus, prior to starting any parallel Worker (QM or MD), the parallel environment must be initiated in accordance with local hardware and software cluster characteristics and policies, e.g., OpenMPI, MPICH2, etc. In fact, an automatic startup shell script is generated from the PUPIL core following a user-provided shell script template which incorporates those local cluster characteristics and policies. An example would be the MPI environment commands to get the Worker running in the local hardware environment and the execution syntax for the corresponding CU. Though this is a platform-dependent solution, experience suggests that just a few templates can cope with most MPI environments.

The PUPIL core is implemented in Java. To provide good performance in building the quantum zone embedding, specific parts of the PUPIL core (for example, application of embedding rules) are executed in parallel using Java threads. Also, the most computationally demanding coupling terms are calculated through a parallel execution using the Java Native Interface (JNI) combined with native C code.

3.2 User Interface

All simulations are done in three steps. Initially, the PUPIL GUI supports preparation of input data. Second, the PUPIL Manager uses information prepared by the GUI to start the simulation and all the externally linked CUs. Finally, output files from the Manager and each of the CUs are analyzed by the user.

The main functionality of the GUI is collection of general simulation information along with the required input files for the CUs to be used. Thus, the user must already be familiar with the external interfaces offered and input data formats required by each CU. Recall the mention above about the user needing to supply input file templates. The GUI helps the user to generate all necessary information to conduct the QM/MM-MD simulation. Thus, the input file templates are preprocessed and parsed to extract information. This information is used during the simulation to coordinate data exchange between CUs. At the end, the GUI saves all collected information in a structured data file (XML) that is then supplied to the Supervisor as an input file at simulation time. (As a remark about the limitations of design, in spite of our efforts, the GUI design was subconsciously biased to materials systems, which typically have a small number of different atoms compared to the number found in biomolecules. An unintended consequence is a bit of cumbersomeness.)

3.2.1 QM Program and Method Selection

Obviously, the user must specify a CU for each of the three main factors involved in any QM/MM-MD simulation, viz., the force generator (QM engine), molecular dynamics (MD engine), and domain identification (DI engine) method. The GUI helps in the selection. Thus, several QM engines are available from which to choose (see Table 1) to provide the energetics and forces in the QM region. The GUI also enables specification of the common parameters for each QM engine involved in the simulation, such as the use of periodic boundary conditions (PBC), selection of electrostatic embedding, the use of link-atom pairs when there are QM/MM chemical bonds crossing the QM region, and the use of long-range electrostatics in the QM/MM coupling term.

3.2.2 QM Region Selection Rules

In addition to selection of the QM engine itself, an obviously important step is selection of the QM region and its environment. The DI Worker is a module for control of the QM/MM partitioning, that is, setting of the inner

(QM) region and outer (MM) regions. Currently, two kinds of Domain Identification are supported: Manual Region Specification and specification through an external program. Manual Region Specification is determination of the QM region by user choice ("by hand"), along with the link pairs connecting the quantum and classical regions, and the embedding particles used as point charges. The selection is made by a user-friendly interface that enables specification of rules to define the different layers that will comprise the QM region.

In the design of the quantum zone embedding, it is allowable to choose not only the QM region but also some additional embedding regions. One may distinguish among three main regions, namely, the quantum, classical, and static-charge regions. All system particles involved in the simulation must be assigned to one of those three regions. There are four different basic categories that allow the user to define all atoms/residues belonging to one of the three regions easily: direct atom/residue type assignment, fixed link pairs during the whole simulation, variable link pairs during simulation (distance-based assignment), and neighboring-residue type assignment. Direct assignment is the normal method of choice for the typical QM/MM-MD simulation in which neither the QM region nor its embedding region changes shape or chemical composition during the whole simulation. In contrast, the neighboring rules are designed mainly for those simulations with a variable quantum region.

We have already mentioned a more sophisticated capability designed into PUPIL, domain identification through an external program. This option allows specification of a complex QM region via a user package as another CU which interacts with the PUPIL simulation Manager analogously with the external MD and external QM programs. This functionality is useful when specification of the quantum region by means of the usual manual region specification rules is complicated, e.g., assignment of multiple QM regions, and the design of variable quantum regions on the fly depending upon some physical or chemical property of the system. Thus, the external DI should interact with the Simulation Manager by exchanging information relevant to the QM region, whereas the embedding region can be managed through the usual neighboring rules.

3.3 Technical Details

We noted above that the PUPIL Manager, GUI, and substantial portions of the Workers are implemented in Java. The main advantages of using Java are fast implementation, easy maintenance, software reuse, and multi-platform support. Though there is some platform-dependent code, it is mainly localized in the wrapper interface between the PUPIL system and the CUs which are tightly coupled with PUPIL, e.g., MD engines and some QM engines. Most of the CUs are written in FORTRAN, though some are in other languages. Therefore, a wrapper written in C was built through the JNI as a natural bridge between both languages (Liang, 1999). All the wrappers have been merged in a single C library. However, additional code modifications to the CUs which are tightly coupled to PUPIL become a simple packing and unpacking of data to be exchanged with the PUPIL system on the QM engines, plus some additional routines to hold the QM/MM coupling in MD engines.

Figure 1 shows the general behavior and exchange of information of the whole PUPIL framework (Torras et al., 2007). The Manager is the main application to execute the user directives previously assigned using the GUI. The majority of its code has generic behavior, resulting in significant software reuse for support of all the CU Workers. Generally speaking, QM/MM-MD simulations are performed as a distributed execution that runs with a main application and several Java Virtual Machines, one for each CU Worker. The Manager prepares and starts the simulation environment,

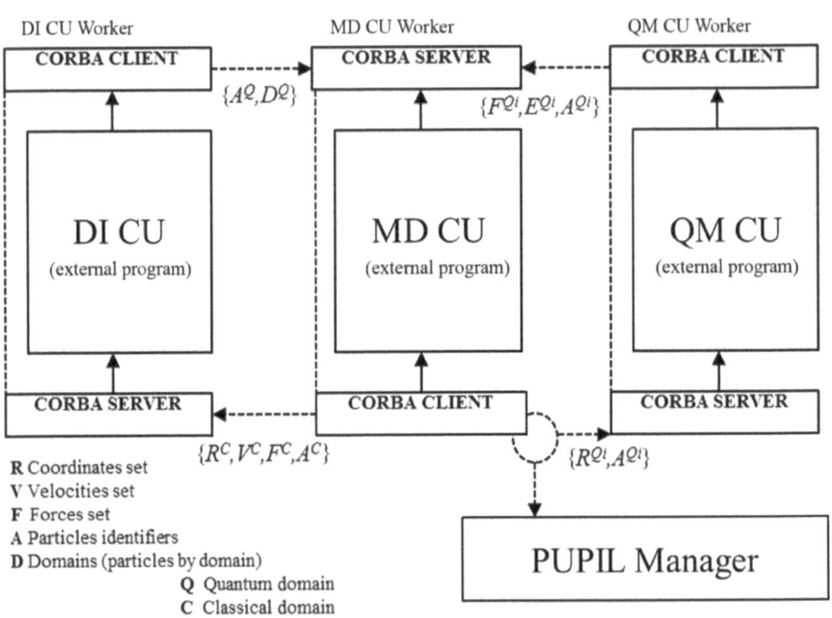

Figure 1 Distributed Supervisor processes at each Worker and their associated CORBA clients and servers.

logs all the distributed processes, performs error control, and concludes the simulation. Each CU (MD, DI, and QM engines) has its own Worker.

The data flow at a given MD step of the QM/MM-MD simulation is this: the MD Worker receives atom types, coordinates, and velocities from the MD engine. Then, the MD Worker decides about the procedure to identify the quantum region (whether to use an external DI or not), and freezes the MD engine. Prior to submitting the information to the QM Worker, the MD Worker adjusts the quantum region by adding the required embedding particles (electrostatic embedding or link-pair atoms). Upon receiving the set of forces associated with the QM region from the QM Worker, the MD Worker puts those forces into the MD engine and releases it to proceed with the subsequent MD step.

All the communications among the distributed processes are done through the CORBA protocol. Each Worker (process) has at least one CORBA server associated with it, along with several CORBA clients depending upon the other servers with which it is communicating (see Fig. 1). Thus, all the client–server communications are performed from Java code and the communication between them and their associated CU, i.e., MD, QM, and DI engines, is through the wrapper interface described above.

4. BIOMOLECULAR APPLICATIONS

A particular advantage of the interfacing and linking approach of PUPIL is that, during each computation of the system's Hessian matrix (typically in the MM Worker), the forces on the particles are modified in place according to the results of a QM calculation. This approach means that, in general, any scientific method implemented in the MM calculation unit that involves computation of forces on a per-particle basis can support a QM/MM treatment with PUPIL, even if any native QM/MM implementation in that particular user package does not support that particular method. (Of course, it helps if the user package is designed in such a way that the forces are computed in one single subroutine, which then is easy to modify to support outgoing connections to the PUPIL Manager.)

One such scientific method is energy minimization (alternatively known as geometry optimization, Fig. 2). While this is a fundamental operation in MM, and widely considered necessary before commencing a MD simulation, it often is done only part-way, so as to eliminate egregious close contacts between particles and other grossly unfavorable structural features.

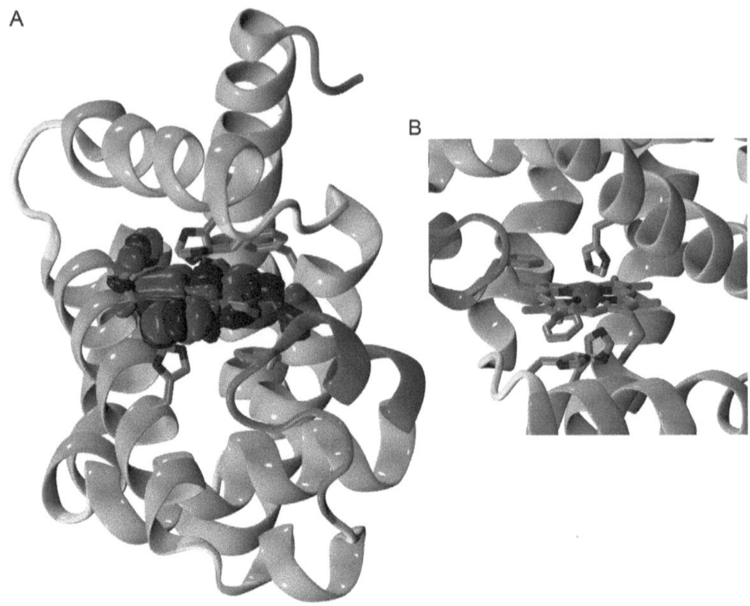

Figure 2 Minimization of the heme group (QM region) within the myoglobin protein (MM region). (A) Highest occupied molecular orbital (HOMO) and (B) the quantum region are also shown.

A common approach in the MD Worker, therefore, is to use a conjugate-gradients algorithm, which comes with the distinct disadvantage (in a QM context) that it requires many force evaluations. An alternative, if offered by the MD Worker, is to optimize the structure using a quasi-Newtonian method such as the limited-memory Broyden–Fletcher–Goldfarb–Shanno (L-BFGS) method (Liu & Nocedal, 1989). Although each step may be more computationally intensive (once forces are evaluated), L-BFGS requires fewer steps and thus fewer force evaluations overall. This approach to energy minimization is thus very useful in the context of a QM/MM calculation.

Furthermore, PUPIL has the potential to allow for access to a multitude of advanced MD simulation techniques, again depending on the battery of methods implemented in the user package. Here, we describe a few examples. The technique of replica-exchange MD (REMD) (Sugita & Okamoto, 1999) permits a more comprehensive sampling of phase space in shorter timescales than would be possible via conventional MD. This sampling is achieved by running different simulations of the same physical system at different temperatures, and at chosen times offering those different simulations (replicas) an opportunity to exchange positions and momenta, the latter being scaled according to the difference in temperature. Whether an exchange actually occurs is determined by a probability function such as the Metropolis criterion. Zhang *et al.* have shown that at high temperatures ($T > 300$ K), REMD offers an excellent alternative to long-timescale MD simulations (Zhang, Wu, & Duan, 2005).

An alternative, if one wishes to bias the system toward a particular, known configuration, is to use targeted MD. In it, the potential energy is biased by the addition of a constraint force on each of the particles such that the particle is pulled toward the position it would hold in a final configuration that is specified by the operator. At each step, the constraint force is computed from the vector difference between the final configuration and the current configuration, along with an appropriately chosen Lagrange parameter (Schlitter, Engels, & Krüger, 1994; Schlitter, Engels, Krüger, Jacoby, & Wollmer, 1993). Targeted MD is a particularly useful technique for simulating gross changes in the structure of a biomolecule that might not be expected to occur on the timescale of a conventional, unbiased MD simulation. To the extent that the constraint forces are expressed mathematically as modest corrections at each step to an otherwise ordinary force matrix, targeted MD is eminently compatible with the PUPIL approach to force evaluation.

A particularly important application of QM/MM techniques is modeling chemical reactions involving bond breaking and formation. Scientists simulating these reactions commonly use the potential of mean force (PMF) technique, which is implemented in various packages. PUPIL allows the investigator to construct PMF profiles of reaction coordinates (that is, the bonds to be broken or formed) using QM methods or basis sets that are not native to the MD program. One particular method for computing a PMF is umbrella sampling, by which a bond length is constrained to be near a specified value (or, more commonly, a series of specified values) by a harmonic potential. The statistical distribution of actual bond lengths arising from the sum of the harmonic potential and the underlying potential energy equation for the system (as specified by the QM and MM approaches used) can be analyzed using the weighted histogram analysis method (WHAM) (Kumar, Rosenberg, Bouzida, Swendsen, & Kollman, 1992). Thus, the scientist can derive an equilibrium constant and free energy profile for the reaction. An alternative approach to umbrella sampling and WHAM is to use steered molecular dynamics (SMD) and Jarzynski's equality, as described by Park and Schulten (2004).

Although initially developed with materials simulations in mind, PUPIL is general enough to be applied in any field of molecular simulation.

It requires only a comparatively short development of an interface between PUPIL and the desired programs. On account of the software architecture and modular construction of PUPIL, that development typically is limited to some wrappers and file parsers. Most of the QM/MM interface is already implemented and thus can be reused (Torras et al., 2006, 2007). For example, an interface with the biomolecular simulations package AMBER9 (Case et al., 2006) was developed in 2008 (Torras, Seabra, Deumens, Trickey, & Roitberg, 2008), together with an interface to the GAUSSIAN QM package (Frisch et al., 2009). This interface was used to study the decomposition of Angeli's salt in explicit solvent (Torras, Seabra, & Roitberg, 2009). Angeli's salt, $Na_2N_2O_3$, has unique cardiovascular effects, associated with its ability to yield HNO upon dissociation under physiological conditions. Its dissociation had been studied earlier, using a polarized continuum model to represent the solvent (Dutton, Fukuto, & Houk, 2004). The use of the AMBER–PUPIL–GAUSSIAN interface allowed the study of the reaction by use of the Multiple Steered MD capabilities built in AMBER with the Jarzynski relationship to calculate the free energy of the process, and by use of GAUSSIAN to calculate the energy of the QM region at UB3LYP and UMP2 levels of density functional theory (DFT) approximation with 6-311 + G(d) basis sets. The explicit inclusion of the solvent molecules allowed a more precise determination of the free energy barrier of decomposition, thereby giving evidence of the importance of explicit consideration of the solvent molecules.

Later, a PUPIL interface to the quantum chemistry program NWChem was developed and used to analyze the conformational preferences of proline (Pro) analogues containing a fused benzene ring, which reduces the molecule flexibility (Warren et al., 2010). Proline is the only proteinogenic amino acid that is naturally conformationally constrained. This constraint is highly significant in protein structure and has stimulated the search for Pro analogues with tailored properties. The incorporation of functional groups from different amino acids is particularly interesting. In that study, the authors concentrated on indoline-2-carboxylic acid (Inc) and its methylated derivative, resulting of a fusion of a benzene ring, present in phenylalanine (Phe), to the pyrrolidine bond linking the γ and δ carbons in Pro, and consequently can be considered either a Pro or a Phe analogue, a combination with important applications in drug design. In an attempt to understand the effects of the additional benzene ring in the conformational preferences of Pro, the authors used DFT calculations in vacuum and estimated the effects of the solvent environment by use of an implicit (SCRF) solvent and explicit solvent by hybrid QM/MM-MD calculations using the AMBER–PUPIL–NWChem interface. The authors noted that the DFT calculations in vacuum overestimated the stability of the structures with *cis* distribution around the ω angle even though, experimentally, only the *trans* arrangement has been detected for the derivative. Inclusion of solvent effects by means of PCM/SCRF calculations did decrease the free energy difference between *cis* and *trans* structures, but the *cis* is still overstabilized. Only after explicit solvent molecules were considered by means of the QM/MM-MD interface was the *trans* disposition predicted to be considerably lower in energy than the *cis*. The authors noted, however, that now the *trans* arrangement was likely overstabilized.

More recently, the same interface has been used to study the characteristics of bioactive platforms based on biocomposites of poly(3,4-ethylenedioxythiophene) (PEDOT) and collagen (CLG), named P(EDOT:CLG), where the presence of the collagen protein affects both the morphology and electrochemical activity of PEDOT (Soto-Delgado, Torras, del Valle, Estrany, & Aleman, 2015). The specific interactions between PEDOT and CLG were studied quantum mechanically with MP2/6-31+G(d,p) methodology both in vacuum and in solution The solvent presence was represented implicitly using PCM/SCRF and explicitly via the AMBER–PUPIL–NWChem interface at the UB3LYP/6-31+G(d,p) level of approximation, using chloroform or water as solvent. In the calculations, the PEDOT was modeled by the monomer, EDOT, while the CLG was represented by proline or L-hydroxyproline, each terminated by an acetyl and *N*-methylamide to yield Ac-L-Pro-NMe and Ac-L-Hyp-NMe. The structures derived from the QM/MM-MD calculations were in good agreement with the ones obtained with the implicit solvent models. The same specific interactions for EDOT/Ac-L-Pro-NMe complexes in chloroform and water solutions were found using implicit solvent with PCM/SCRF or explicit solvent with QM/MM-MD. On the other hand, the QM/MM-MD method reveals three different types of specific interactions between the components in EDOT/Ac-L-Hyp-NMe, which turned out to be the combination of the two modes predicted by the implicit solvent model, which were found to be practically isoenergetic.

5. RECENT DEVELOPMENTS

One of the major challenges of *in silico* simulations on complex biological systems is to treat several chemically active zones concurrently because their distinct evolution is linked critically to the global system behavior. Very recently, the capability for handling such multiple, disjoint QM zones in QM/MM-MD simulations has been developed within the PUPIL framework (Torras, 2015). This new capability will allow simulational treatment of complex proteins such as those that contain multiple metallic centers, e.g., the ferritin cage (see Fig. 3), ubiquinone oxidoreductase and Laccase, among others. In the first case, the ferritin cage holds several metallic ions within its structure which have been shown to be important in protein–protein interactions via formation of metal-induced self-assembly cages (Liu & Theil, 2005). The complex in NADH: ubiquinone oxidoreductase plays a major role in the respiratory electron transport chain from the NADH to ubiquinone across the membrane, which is necessary for ATP synthesis (Hayashi & Stuchebrukhov, 2010). A dynamical treatment of independent active zones to deal with distinct electron tunneling pathways between neighboring Fe/S clusters is indicated. And in the study of metalloenzymes such as Laccase, which has several active metal sites, the new methodology should be especially useful to characterize synergies among those sites (Piontek, Antorini, & Choinowski, 2002).

5.1 Working with Multiple Active Zones

To describe the multiple active zone scheme for QM/MM-MD calculations (hereafter *maz*-QM/MM-MD approach), we return briefly to the general QM/MM approach. The entire system (*S*) is partitioned into an inner region (*I*) that is treated by QM and the outer region (*O*) described by a force

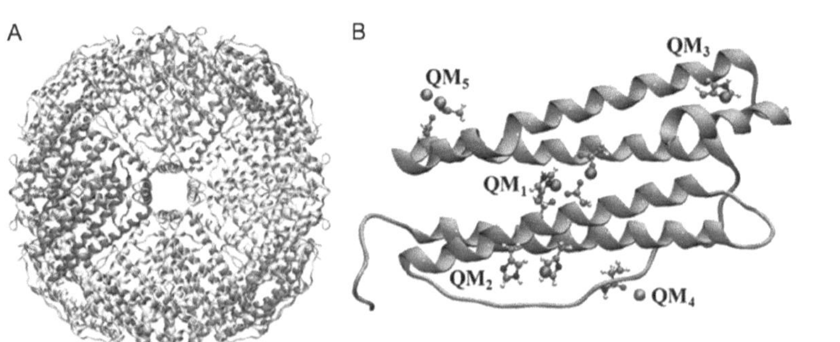

Figure 3 Cu-Ferritin cage (A) with a selected building block monomer (green, dark gray in the print version). (B) Detailed location of the active zones in a monomer of the Cu-Ferritin cage.

field. The energy partition of the two main regions is modeled via the additive QM/MM scheme,

$$E(S) = E_{\text{QM}}(I) + E_{\text{MM}}(O) + E_{\text{QM/MM}}(I, O) \tag{7}$$

In the *maz*-QM/MM-MD extension, the QM region is defined as the sum of several disjoint QM subregions (or active zones, AZs). The energy and its gradients (forces) follow from the general QM/MM approach. Thus, any simulation particle (nucleus or more coarse-grained) within an AZ is subject to QM forces from the electrons in that subregion. But the interactions with the other AZs are treated the same as with the MM region, namely, as forces from point charges in those remote subregions. At present, those other AZ point charges are calculated as Mulliken charges, but that is a choice, not an essential design property. This procedure is similar to the one previously proposed by Kiyota, Hasegawa, Fujimoto, Swerts, and Nakatsuji (2009) but is more general. The energy partition of the QM region then is formulated as follows,

$$E_{\text{QM}}(I) = \sum_{A} E_{\text{QM}}(I_A) + \frac{1}{2}\sum_{A}\sum_{B \neq A} E_{\text{QM/QM}}(I_A, I_B) \tag{8}$$

The coupling term between two disjoint AZs (the only type allowed; see below) has contributions only from the van der Waals and electrostatic interactions between the QM atoms of those two subregions.

$$E_{\text{QM/QM}}(I_A, I_B) = E^{\text{vdw}}(I_A, I_B) + E^{\text{el}}(I_A, I_B) \tag{9}$$

All AZs are treated as in the ordinary QM/MM method except for incorporating the electrostatic interactions between different AZs by means of the electrostatic-embedding scheme. Different QM/MM calculations therefore are performed concurrently, one for each AZ. As a result the conventional MM region polarizes each AZ in addition to the polarization from the different sets of point charges, each of which sets represents one of the other AZs. Thus, the electrostatic interaction between different subregions, $E^{\text{el}}(I_A, I_B)$, is approximated by the interaction between the electron density of one AZ with a set of charges of the other (Q_B) to simulate the charge polarization of the remote quantum subregion instead of using its electronic density and associated multipoles:

$$E^{\text{el}}(I_A : \rho_A, I_B : \rho_B) \cong \frac{1}{2}\left(E^{\text{el}}(I_A : \rho_A, I_B : Q_B) + E^{\text{el}}(I_A : Q_A, I_B : \rho_B)\right) \tag{10}$$

Observe that this expression is symmetrized between subregions so as not to introduce a violation of Newton's Third Principle.

The potential energy of the whole system can be written as the energy of N independent QM subregions plus their corresponding QM/QM and QM/MM coupling terms:

$$E(S) = E_{\mathrm{MM}}(O) + \sum_{A}^{N} \left[E_{\mathrm{QM}}(I_A) + E_{\mathrm{QM/MM}}(I_A, O) \right] + \sum_{A}^{N} \sum_{B>A}^{N} \frac{1}{2} \left[E_{\mathrm{QM/QM}}(I_A : \rho_A, I_B : Q_B) + E_{\mathrm{QM/QM}}(I_B : \rho_B, I_A : Q_A) \right] \tag{11}$$

The whole approach is under the hypothesis that all the AZs are sufficiently separated that their charge distributions are essentially non-overlapping. Otherwise, this approach is not valid. Thus, only disjoint active zones are allowed. However, *maz*-QM/MM-MD opens the opportunity either to merge different AZs as they approach or to split an AZ if some part of it drifts away the rest. These opportunities are not yet implemented however.

This new methodology, implemented within PUPIL, has been demonstrated with treatment of small molecules in solution and of all five QM regions of the Cu-Ferritin monomer in a unique *maz*-QM/MM-MD simulation, successfully (Fig. 3) (Torras, 2015). Thus, it opens the possibility to perform further studies analyzing the stability of the Cu-4His-ΔC* cage, which holds about 120 AZs with about 50 of them involved on the self-assembly of protein. Indeed, modeling of large biomolecular systems that present an interrelationship between different active sites becomes much more readily accessible than heretofore.

5.2 Treatment of Long-Range Electrostatic Interactions

Long-range electrostatic interactions in conjunction with PBC are extensively used for prediction of condensed system properties. Treatment for long-range electrostatic interactions via PBC based on the QM/MM-Ewald summation methodology was added to the PUPIL framework recently (Torras, 2015). This addition allows the user to choose between a simple electrostatic embedding using all the point charges of the MM particles within the simulation box (real space) and incorporation into real-space interactions those electrostatic interactions with infinitely many images of the simulation box (reciprocal space). Such long-range electrostatics via QM/MM-Ewald summation was described initially by Nam, Gao, and York (2004). The technique involves addition of a periodic correction term to both QM and QM/MM interactions for the usual real-space electrostatic interaction between QM and MM partitioning.

6. CONCLUSIONS

The QM/MM-MD methodology has proven to be a powerful approach to handle large-scale simulations in biology from the dynamics point of view. Its great potential complements classical methods used so far to obtain either a static image by means of a high-level calculation or a structural evolution of biological systems using a low-level calculation. To this context, the PUPIL framework adds a general, flexible, modular, and readily scalable environment for performing QM/MM-MD simulations. PUPIL users can choose and match their preferred QM and MD external packages through a well-established interface. They can add packages systematically and with very substantial software reuse. Besides the basic QM/MM coupling terms among the QM and MM regions, all accessible capabilities in any PUPIL-based QM/MM-MD simulation are limited only by those available in the external packages used. The internal structure of PUPIL is designed to facilitate the management of computing resources, allowing different external packages to be executed concurrently in a parallel environment. Several applications using PUPIL to apply this methodology have been described. Generally, PUPIL applications have ranged from the solid-state to complex biomolecular systems. The continuous evolution and refinement of the QM/MM-MD model allows a more accurate system environment treatment, thus obtaining better observables and opening the possibility to study biochemical reactions from a dynamic point of view. Using the new *maz*-QM/MM-MD approach, the users of PUPIL have access to large and complex biological systems to explore synergies among different active sites.

For further information about the open-source PUPIL project, code download, and new developments, visit the Web site http://pupil.sourceforge.net.

ACKNOWLEDGMENTS

This work has been supported by MINECO and FEDER funds (MAT2012-34498), and by the DIUE of the Generalitat de Catalunya (Research group 2009 SGR 925). S. B. T. was supported under U.S. Dept. of Energy grant DE-SC0002139.

REFERENCES

Alder, B. J., & Wainwright, T. E. (1959). Studies in molecular dynamics. I. General method. *The Journal of Chemical Physics*, *31*(2), 459–466. http://dx.doi.org/10.1063/1.1730376.

Barnett, R. N., & Landman, U. (1993). Born-Oppenheimer molecular-dynamics simulations of finite systems: Structure and dynamics of $(H_2O)_2$. *Physical Review B*, *48*(4), 2081–2097.

Bartlett, R. J., & Musiał, M. (2007). Coupled-cluster theory in quantum chemistry. *Reviews of Modern Physics*, *79*(1), 291–352.

Bochevarov, A. D., Harder, E., Hughes, T. F., Greenwood, J. R., Braden, D. A., Philipp, D. M., et al. (2013). Jaguar: A high-performance quantum chemistry software program with strengths in life and materials sciences. *International Journal of Quantum Chemistry*, *113*(18), 2110–2142. http://dx.doi.org/10.1002/qua.24481.

Brooks, B. R., Brooks, C. L., Mackerell, A. D., Nilsson, L., Petrella, R. J., Roux, B., et al. (2009). CHARMM: The biomolecular simulation program. *Journal of Computational Chemistry*, *30*(10), 1545–1614. http://dx.doi.org/10.1002/jcc.21287.

Case, D. A., Darden, T. A., III, Cheatham, T. E., Simmerling, C. L., Wang, J., Duke, R. E., et al. (2006). *AMBER 9*. San Francisco: University of California.

Chung, L. W., Hirao, H., Li, X., & Morokuma, K. (2012). The ONIOM method: Its foundation and applications to metalloenzymes and photobiology. *Wiley Interdisciplinary Reviews: Computational Molecular Science*, *2*(2), 327–350. http://dx.doi.org/10.1002/wcms.85.

Darden, T., York, D., & Pedersen, L. (1993). Particle mesh Ewald: An N·log(N) method for Ewald sums in large systems. *The Journal of Chemical Physics*, *98*(12), 10089–10092. http://dx.doi.org/10.1063/1.464397.

Dewar, M. J. S., Zoebisch, E. G., Healy, E. F., & Stewart, J. J. P. (1985). Development and use of quantum mechanical molecular models. 76. AM1: A new general purpose quantum mechanical molecular model. *Journal of the American Chemical Society*, *107*(13), 3902–3909. http://dx.doi.org/10.1021/ja00299a024.

Dutton, A. S., Fukuto, J. M., & Houk, K. N. (2004). Mechanisms of HNO and NO production from Angeli's salt: Density functional and CBS-QB3 theory predictions. *Journal of the American Chemical Society*, *126*(12), 3795–3800. http://dx.doi.org/10.1021/ja0391614.

Field, M. J., Albe, M., Bret, C., Proust-De Martin, F., & Thomas, A. (2000). The dynamo library for molecular simulations using hybrid quantum mechanical and molecular mechanical potentials. *Journal of Computational Chemistry*, *21*(12), 1088–1100. http://dx.doi.org/10.1002/1096-987x(200009)21:12<1088::aid-jcc5>3.0.co;2-8.

Field, M. J., Bash, P. A., & Karplus, M. (1990). A combined quantum mechanical and molecular mechanical potential for molecular dynamics simulations. *Journal of Computational Chemistry*, *11*(6), 700–733. http://dx.doi.org/10.1002/jcc.540110605.

Frisch, M. J., Trucks, G. W., Schlegel, H. B., Scuseria, G. E., Robb, M. A., Cheeseman, J. R., et al. (2009). *Gaussian 09, Revision D.01*. Wallingford CT: Gaussian, Inc.

Gordon, M. S., Fedorov, D. G., Pruitt, S. R., & Slipchenko, L. V. (2012). Fragmentation methods: A route to accurate calculations on large systems. *Chemical Reviews*, *112*(1), 632–672. http://dx.doi.org/10.1021/cr200093j.

Hayashi, T., & Stuchebrukhov, A. A. (2010). Electron tunneling in respiratory complex I. *Proceedings of the National Academy of Sciences*, *107*(45), 19157–19162. http://dx.doi.org/10.1073/pnas.1009181107.

Horner, D. A., Lambert, F., Kress, J. D., & Collins, L. A. (2009). Transport properties of lithium hydride from quantum molecular dynamics and orbital-free molecular dynamics. *Physical Review B*, *80*(2), 024305.

Kiyota, Y., Hasegawa, J.-Y., Fujimoto, K., Swerts, B., & Nakatsuji, H. (2009). A multicore QM/MM approach for the geometry optimization of chromophore aggregate in

protein. *Journal of Computational Chemistry, 30*(8), 1351–1359. http://dx.doi.org/10.1002/jcc.21156.
Köster, A. M., Geudtner, G., Calaminici, P., Casida, M. E., Dominguez, V. D., Flores-Moreno, R., et al. (2011). *deMon2k, version 3*. Mexico City: Cinvestav.
Kumar, S., Rosenberg, J. M., Bouzida, D., Swendsen, R. H., & Kollman, P. A. (1992). The weighted histogram analysis method for free-energy calculations on biomolecules. I. The method. *Journal of Computational Chemistry, 13*(8), 1011–1021. http://dx.doi.org/10.1002/jcc.540130812.
Liang, S. (1999). *Java native interface: Programmer's guide and reference* (1st ed.). Boston, MA: Addison-Wesley Longman Publishing Co., Inc.
Lin, H., & Truhlar, D. (2007). QM/MM: What have we learned, where are we, and where do we go from here? *Theoretical Chemistry Accounts, 117*(2), 185–199. http://dx.doi.org/10.1007/s00214-006-0143-z.
Liu, D., & Nocedal, J. (1989). On the limited memory BFGS method for large scale optimization. *Mathematical Programming, 45*(1–3), 503–528. http://dx.doi.org/10.1007/bf01589116.
Liu, X., & Theil, E. C. (2005). Ferritins: Dynamic management of biological iron and oxygen chemistry. *Accounts of Chemical Research, 38*(3), 167–175. http://dx.doi.org/10.1021/ar0302336.
Mallik, A., Runge, K., Dufty, J. W., & Cheng, H.-P. (2007). Multiscale modeling of materials based on force and charge density fidelity. *The Journal of Chemical Physics, 127*(22), 224707. http://dx.doi.org/10.1063/1.2802545.
Mallik, A., Taylor, D., Runge, K., Dufty, J., & Cheng, H. P. (2006). Procedure for building a consistent embedding at the QM–CM interface. *Journal of Computer-Aided Materials Design, 13*(1–3), 45–60. http://dx.doi.org/10.1007/s10820-006-9014-0.
Marx, D., & Hutter, J. (2009). *Ab initio molecular dynamics: Basic theory and advanced methods*. Cambridge: Cambridge University Press.
Maseras, F., & Morokuma, K. (1995). IMOMM: A new integrated ab initio + molecular mechanics geometry optimization scheme of equilibrium structures and transition states. *Journal of Computational Chemistry, 16*(9), 1170–1179. http://dx.doi.org/10.1002/jcc.540160911.
Meinel, C. (1992). August Wilhelm Hofmann—"Reigning Chemist-in-Chief". *Angewandte Chemie, International Edition in English, 31*(10), 1265–1282. http://dx.doi.org/10.1002/anie.199212653.
Metz, S., Kästner, J., Sokol, A. A., Keal, T. W., & Sherwood, P. (2014). ChemShell—A modular software package for QM/MM simulations. *Wiley Interdisciplinary Reviews: Computational Molecular Science, 4*(2), 101–110. http://dx.doi.org/10.1002/wcms.1163.
Nam, K., Gao, J., & York, D. M. (2004). An efficient linear-scaling ewald method for long-range electrostatic interactions in combined QM/MM calculations. *Journal of Chemical Theory and Computation, 1*(1), 2–13. http://dx.doi.org/10.1021/ct049941i.
Park, S., & Schulten, K. (2004). Calculating potentials of mean force from steered molecular dynamics simulations. *The Journal of Chemical Physics, 120*(13), 5946–5961. http://dx.doi.org/10.1063/1.1651473.
Phillips, J. C., Braun, R., Wang, W., Gumbart, J., Tajkhorshid, E., Villa, E., et al. (2005). Scalable molecular dynamics with NAMD. *Journal of Computational Chemistry, 26*(16), 1781–1802. http://dx.doi.org/10.1002/jcc.20289.
Piontek, K., Antorini, M., & Choinowski, T. (2002). Crystal structure of a laccase from the FungusTrametes versicolor at 1.90-Å resolution containing a full complement of coppers. *Journal of Biological Chemistry, 277*(40), 37663–37669.
Pople, J. A., Santry, D. P., & Segal, G. A. (1965). Approximate self-consistent molecular orbital theory. I. Invariant procedures. *The Journal of Chemical Physics, 43*(10), S129–S135. http://dx.doi.org/10.1063/1.1701475.

Pople, J. A., & Segal, G. A. (1965). Approximate self-consistent molecular orbital theory. II. Calculations with complete neglect of differential overlap. *The Journal of Chemical Physics*, *43*(10), S136–S151. http://dx.doi.org/10.1063/1.1701476.

Rahman, A. (1964). Correlations in the motion of atoms in liquid argon. *Physical Review*, *136*, A405. http://dx.doi.org/10.1103/PhysRev.136.A405.

Riccardi, D., Li, G., & Cui, Q. (2004). Importance of van der Waals Interactions in QM/MM simulations. *The Journal of Physical Chemistry B*, *108*(20), 6467–6478. http://dx.doi.org/10.1021/jp037992q.

Roberts, B. P., Seabra, G. M., Roitberg, A. E., Merz, K. M., Deumens, E., Torras, J., et al. (2012). Comment on "a minimal implementation of the AMBER–GAUSSIAN interface for ab initio QM/MM-MD simulation". *Journal of Computational Chemistry*, *33*(19), 1643–1644. http://dx.doi.org/10.1002/jcc.23003.

Salahub, D., Noskov, S., Lev, B., Zhang, R., Ngo, V., Goursot, A., et al. (2015). QM/MM Calculations with deMon2k. *Molecules*, *20*(3), 4780–4812.

Schlitter, J., Engels, M., & Krüger, P. (1994). Targeted molecular dynamics: A new approach for searching pathways of conformational transitions. *Journal of Molecular Graphics*, *12*(2), 84–89. http://dx.doi.org/10.1016/0263-7855(94)80072-3.

Schlitter, J., Engels, M., Krüger, P., Jacoby, E., & Wollmer, A. (1993). Targeted molecular dynamics simulation of conformational change—Application to the T↔R transition in insulin. *Molecular Simulation*, *10*(2–6), 291–308. http://dx.doi.org/10.1080/08927029308022170.

Schmidt, M. W., Baldridge, K. K., Boatz, J. A., Elbert, S. T., Gordon, M. S., Jensen, J. H., et al. (1993). General atomic and molecular electronic structure system. *Journal of Computational Chemistry*, *14*(11), 1347–1363. http://dx.doi.org/10.1002/jcc.540141112.

Senn, H. M., & Thiel, W. (2009). QM/MM methods for biomolecular systems. *Angewandte Chemie International Edition*, *48*(7), 1198–1229. http://dx.doi.org/10.1002/anie.200802019.

Shao, Y., Molnar, L. F., Jung, Y., Kussmann, J., Ochsenfeld, C., Brown, S. T., et al. (2006). Advances in methods and algorithms in a modern quantum chemistry program package. *Physical Chemistry Chemical Physics*, *8*(27), 3172–3191. http://dx.doi.org/10.1039/b517914a.

Shavitt, I. (1998). The history and evolution of configuration interaction. *Molecular Physics*, *94*(1), 3–17. http://dx.doi.org/10.1080/002689798168303.

Singh, U. C., & Kollman, P. A. (1986). A combined ab initio quantum mechanical and molecular mechanical method for carrying out simulations on complex molecular systems: Applications to the CH_3Cl + Cl− exchange reaction and gas phase protonation of polyethers. *Journal of Computational Chemistry*, 7(6), 718–730. http://dx.doi.org/10.1002/jcc.540070604.

Smith, W., & Forester, T. R. (1996). DL_POLY_2.0: A general-purpose parallel molecular dynamics simulation package. *Journal of Molecular Graphics*, *14*(3), 136–141. http://dx.doi.org/10.1016/S0263-7855(96)00043-4.

Soler, J. M., Artacho, E., Gale, J. D., García, A., Junquera, J., Ordejón, P., et al. (2002). The SIESTA method for ab initio order—N materials simulation. *Journal of Physics: Condensed Matter*, *14*(11), 2745.

Soto-Delgado, J., Torras, J., del Valle, L. J., Estrany, F., & Aleman, C. (2015). Examining the compatibility of collagen and a polythiophene derivative for the preparation of bioactive platforms. *RSC Advances*, *5*(12), 9189–9203. http://dx.doi.org/10.1039/c4ra13812k.

Stewart, J. J. P. (1989). Optimization of parameters for semiempirical methods I. Method. *Journal of Computational Chemistry*, *10*(2), 209–220. http://dx.doi.org/10.1002/jcc.540100208.

Sugita, Y., & Okamoto, Y. (1999). Replica-exchange molecular dynamics method for protein folding. *Chemical Physics Letters*, *31*(1–2), 141–151. http://dx.doi.org/10.1016/S0009-2614(99)01123-9.

Todorov, I. T., Smith, W., Trachenko, K., & Dove, M. T. (2006). DL_POLY_3: New dimensions in molecular dynamics simulations via massive parallelism. *Journal of Materials Chemistry*, *16*(20), 1911–1918. http://dx.doi.org/10.1039/b517931a.

Torras, J. (2015). Multiple active zones in hybrid QM/MM molecular dynamics simulations for large biomolecular systems. *Physical Chemistry Chemical Physics*, *17*(15), 9959–9972. http://dx.doi.org/10.1039/c5cp00905g.

Torras, J., Deumens, E., & Trickey, S. B. (2006). Software integration in multi-scale simulations: The PUPIL system. *Journal of Computer-Aided Materials Design*, *13*(1–3), 201–212.

Torras, J., He, Y., Cao, C., Muralidharan, K., Deumens, E., Cheng, H.-P., et al. (2007). PUPIL: A systematic approach to software integration in multi-scale simulations. *Computer Physics Communications*, *177*(3), 265–279. http://dx.doi.org/10.1016/j.cpc.2007.01.009.

Torras, J., Seabra, G. M., Deumens, E., Trickey, S. B., & Roitberg, A. E. (2008). A versatile AMBER-Gaussian QM/MM interface through PUPIL. *Journal of Computational Chemistry*, *29*(10), 1564–1573. http://dx.doi.org/10.1002/jcc.20915.

Torras, J., Seabra, G. M., & Roitberg, A. E. (2009). A multiscale treatment of Angeli's salt decomposition. *Journal of Chemical Theory and Computation*, *5*(1), 37–46. http://dx.doi.org/10.1021/ct800236d.

Valiev, M., Bylaska, E. J., Govind, N., Kowalski, K., Straatsma, T. P., Van Dam, H. J. J., et al. (2010). NWChem: A comprehensive and scalable open-source solution for large scale molecular simulations. *Computer Physics Communications*, *181*(9), 1477–1489. http://dx.doi.org/10.1016/j.cpc.2010.04.018.

Walker, R. C., Crowley, M. F., & Case, D. A. (2008). The implementation of a fast and accurate QM/MM potential method in Amber. *Journal of Computational Chemistry*, *29*(7), 1019–1031. http://dx.doi.org/10.1002/jcc.20857.

Warren, J. G., Revilla-López, G., Alemán, C., Jiménez, A. I., Cativiela, C., & Torras, J. (2010). Conformational preferences of proline analogues with a fused benzene ring. *The Journal of Physical Chemistry B*, *114*(36), 11761–11770. http://dx.doi.org/10.1021/jp105456r.

Warshel, A., & Levitt, M. (1976). Theoretical studies of enzymic reactions: Dielectric, electrostatic and steric stabilization of the carbonium ion in the reaction of lysozyme. *Journal of Molecular Biology*, *103*(2), 227–249. http://dx.doi.org/10.1016/0022-2836(76)90311-9.

Yang, W. (1991). Direct calculation of electron density in density-functional theory. *Physical Review Letters*, *66*(11), 1438–1441.

Zhang, W., Wu, C., & Duan, Y. (2005). Convergence of replica exchange molecular dynamics. *The Journal of Chemical Physics*, *123*(15), 154105. http://dx.doi.org/10.1063/1.2056540.

CHAPTER TWO

Efficient Calculation of Enzyme Reaction Free Energy Profiles Using a Hybrid Differential Relaxation Algorithm: Application to Mycobacterial Zinc Hydrolases

Juan Manuel Romero*[,1], Mariano Martin[†,1], Claudia Lilián Ramirez*, Victoria Gisel Dumas*[,†], Marcelo Adrián Marti[†,‡,2]

*Instituto de Química Física de los Materiales Medio Ambiente y Energía (INQUIMAE), UBA-CONICET, Facultad de Ciencias Exactas y Naturales, Universidad de Buenos Aires, Ciudad Universitaria, Buenos Aires, Argentina

[†]Departamento de Química Biológica, Facultad de Ciencias Exactas y Naturales, Universidad de Buenos Aires, Ciudad Universitaria, Buenos Aires, Argentina

[‡]Instituto de Química Biológica de la Facultad de Ciencias Exactas y Naturales (IQUIBICEN) CONICET, Ciudad Universitaria, Buenos Aires, Argentina

[1]Both authors contributed equally to this work.

[2]Corresponding author: e-mail address: marcelo@qi.fcen.uba.ar

Contents

Advances in Protein Chemistry and Structural Biology, Volume 100
ISSN 1876-1623
http://dx.doi.org/10.1016/bs.apcsb.2015.06.006

Abstract

Determination of the free energy profile for an enzyme reaction mechanism is of primordial relevance, paving the way for our understanding of the enzyme's catalytic power at the molecular level. Although hybrid, mostly DFT-based, QM/MM methods have been extensively applied to this type of studies, achieving accurate and statistically converged results at a moderate computational cost is still an open challenge. Recently, we have shown that accurate results can be achieved in less computational time, combining Jarzynski's relationship with a hybrid differential relaxation algorithm (HyDRA), which allows partial relaxation of the solvent during the nonequilibrium steering of the reaction. In this work, we have applied this strategy to study two mycobacterial zinc hydrolases.

Mycobacterium tuberculosis infections are still a worldwide problem and thus characterization and validation of new drug targets is an intense field of research. Among possible drug targets, recently two essential zinc hydrolases, MshB (Rv1170) and MA-amidase (Rv3717), have been proposed and structurally characterized. Although possible mechanisms have been proposed by analogy to the widely studied human Zn hydrolases, several key issues, particularly those related to Zn coordination sphere and its role in catalysis, remained unanswered.

Our results show that mycobacterial Zn hydrolases share a basic two-step mechanism. First, the attacking water becomes deprotonated by the conserved base and establishes the new C–O bond leading to a tetrahedral intermediate. The intermediate requires moderate reorganization to allow for proton transfer to the amide N and C–N bond breaking to occur in the second step. Zn ion plays a key role in stabilizing the tetrahedral intermediate and balancing the negative charge of the substrate during hydroxide ion attack. Finally, comparative analysis of other Zn hydrolases points to a convergent mechanistic evolution.

1. INTRODUCTION

1.1 Free Energy Profiles of Enzymatic Reactions

Understanding the origins of enzyme catalysis is one of the main challenges in current biochemical research, with potential impact in the fields of protein evolution, protein engineering, and drug development, among others. To uncover the underlying physicochemical reasons of a given protein-catalyzed reaction and relate it with the protein's structure and dynamics at a residue-based level, it is of primordial relevance to determine the corresponding reaction free energy profile (FEP). The FEP, also sometimes referred as the potential of mean force, can not only be directly related to experimentally determined properties, such as reaction rates and equilibrium constants, but also constitutes a proof of concept of the reaction dynamics, defined as the motion of atoms, with the concomitant breaking and forming of chemical bonds during the reactive process (Kamerlin & Warshel, 2010).

To be able to describe a (bio)chemical reaction, the system needs to be treated using quantum mechanics (QM) methods, such as density functional theory (DFT). QM methods, however, are computationally expensive, and describing systems larger than a few hundreds of atoms, with that level of accuracy, is very difficult and an active research field (Ferrer et al., 2011). For enzyme-based reactions, those which require consideration of the whole protein-solvent environment—tens of thousands of atoms, hybrid quantum mechanical/molecular mechanical (QM/MM) schemes are the best choice (Warshel & Levitt, 1976). QM/MM methods allow accurate description of the chemical events that take place in the enzyme active site modeled at the QM level, while treating the rest of the system using a less expensive classical force-field based level of theory. Key to these methods is the coupling between the QM and MM regions, which must properly describe the environment electrostatic as well as the steric effect on the reactive subsystem. Although they have been extensively and successfully used during the last decade to study reaction mechanisms, the configurational sampling required to obtain an accurate FEP is still an open challenge faced by QM/MM methods and their application to enzyme (Kamerlin & Warshel, 2010; Van der Kamp & Mulholland, 2013).

1.1.1 MSMD and Jarzynski's Relationship

In plain molecular dynamics (MD), the system under study is usually unable to cross moderate high barriers ($\gg kT$), such as those presented by enzyme reactions, thus remaining trapped in the initial (reactant) state, unless driven up the hill. There are several enhanced or biased sampling strategies that nonetheless allow determination of FEP with their associated barriers. The most old and possibly common is the umbrella sampling strategy (Leach, 2001). More recently, newer strategies, such as metadynamics (Laio & Parrinello, 2002), adaptive biasing force (Hénin & Chipot, 2004), free energy perturbation (Pohorille, Jarzynski, & Chipot, 2010), and orthogonal space random walk (Zheng, Chen, & Yang, 2008), have also been developed. One of the most easily implemented, with wide applicability, solid statistical thermodynamic background and multiple computational and experimental validations is the multiple steered molecular dynamics

(MSMD) strategy combined with Jarzynski's relationship (JR) (Jarzynski, 1997; Liphardt, Dumont, Smith, Tinoco, & Bustamante, 2002; Park, Khalili-Araghi, Tajkhorshid, & Schulten, 2003). In MSMD (as shown schematically in Fig. 1), the system is driven "multiple" times along the selected reaction coordinate (lambda) under nonequilibrium conditions, by applying an external force onto the reaction coordinate. For each individual trajectory, the work performed by the force is determined ($W_i(\lambda)$). Finally, multiple works are exponentially average in JR (Eq. 1) to obtain the corresponding FEP.

$$G(\lambda) = -\beta^{-1} \ln \left(e^{-\beta W_i(\lambda)} \right) \tag{1}$$

where $G(\lambda)$ represents the FEP as a function of the reaction coordinate, $\beta = 1/k_B T$, where k_B is Boltzmann constant and T is the system temperature. The brackets represent the average of the function within them. In JR, the exponential average is computed on the work values distribution, and the more narrow the distribution, the more accurate the average. The width of the work distribution is directly proportional to the pulling speed. Thus, in the practice, the main drawback with JR is that in order to obtain a well-converged average, and thus an accurate FEP, either very large number of trajectories and/or very low pulling speeds are needed. These facts result in a high computational cost and sometimes in an insurmountable problem that prevents accurate convergence of the FEP (Pohorille et al., 2010; Xiong, Crespo, Marti, Estrin, & Roitberg, 2006).

1.1.2 Hybrid Differential Relaxation Algorithm

To overcome the above-mentioned difficulties and reduce the overall computational cost of MSMD-JR strategy for FEP determination, we have recently developed a hybrid differential relaxation algorithm (HyDRA)

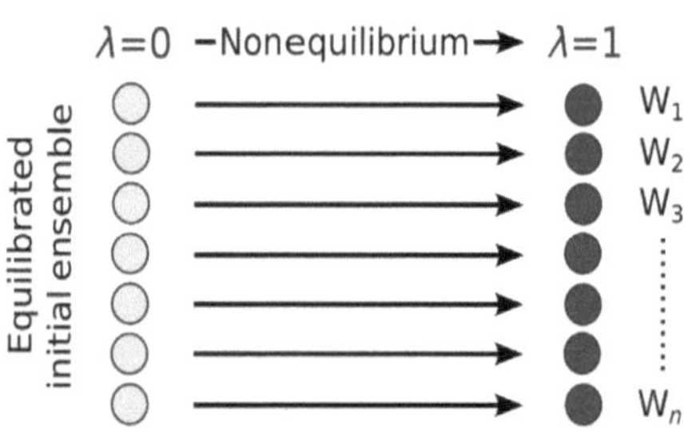

Figure 1 Multiple steered molecular dynamics (MSMD) scheme.

(Ramírez, Zeida, Jara, Roitberg, & Martí, 2014). This scheme allows faster equilibration of the classical environment during the steering process that drives the QM system along the reaction under study. The differential relaxation strategy, inspired in the multiple time step schemes developed earlier, takes advantage of the less expensive calculation of the classical environment and performs multiple pure classical relaxation steps for each QM perturbation step. This allows a better relaxation of the whole system, resulting in closer to equilibrium steering trajectories, with more narrow work distributions, and thus better computed averages. The method, which is implemented in the AMBER (Case et al., 2014) computer simulation package and freely available, therefore, allows obtaining accurate FEPs, with a smaller number of independent trajectories and/or faster pulling speeds, at a smaller computational cost when compared to standard MSMD. In this work, we have applied this strategy to the study of mycobacterial zinc hydrolases reaction mechanism.

1.2 *Mycobacterium tuberculosis* Zinc Hydrolases

Mycobacterium tuberculosis (Mtb) infections are still a worldwide problem, with over a million people dying from its cause annually. Current regular treatment is long and costly, which added to the increasing emergence of multi- and extreme-drug-resistant strains results in an urgent need for new antituberculinic drugs. Therefore, characterization and validation of new drug targets is an intense field of research (Koul, Arnoult, Lounis, Guillemont, & Andries, 2011). Zinc hydrolases are possibly the largest group of Zn containing enzymes; they catalyze the hydrolysis of chemical bonds using an active site water or related nucleophile. Well-known Zn hydrolases are human ones, like carboxy peptidase A, the angiotensin-converting enzyme (ACE) and the histone deacetylases, as well as bacterial thermolysin and metallo-beta-lactamases. They have been extensively studied as drug targets and several inhibitors are available (Hernick & Fierke, 2005). Therefore, Mtb Zn hydrolases have also caught the attention as potential targets for tuberculosis (TB) treatment.

1.2.1 MshB (Rv1170)

Instead of having glutathione for maintaining intracellular redox regulation, like eukaryotes and other bacteria, *Mtb* uses mycothiol (MSH)—a small molecule—as the key reducing agent and main reactive for toxin detoxification (Newton, Buchmeier, & Fahey, 2008). Zn containing MshB catalyzes the third step of MSH biosynthesis: the hydrolysis of GlcNAc-Ins to form 1-D-myo-inosityl-2-amino-2-deoxy-α-D-glucopyranoside and acetate. MshB is an attractive drug target; since it catalyzes the rate-limiting step in MSH biosynthesis (Newton, Av-Gay, & Fahey, 2000), the three-dimensional structure is known (Maynes et al., 2003) and interestingly displays activity with other ions as well (Huang, Kocabas, & Hernick, 2011). MshB consists of one large nine-stranded mixed β-sheet and one small three-stranded antiparallel β-sheet. The metal binding site is comprised of residues from the C terminus of β-strand 1 and the loop linking strand 1 to the helix α_1 (His13 and Asp16) as well as His147 from the N terminus of helix α_5. In addition, in the X-ray structure there are two water molecules that coordinate to the Zn ion, one of which establishes a hydrogen bond with Asp15, the proposed general base catalyst (Newton et al., 2000). Tyr142, His144, and Asp146 have also been suggested to play an important role in the catalytic mechanism, although details are not clear (Huang & Hernick, 2012).

1.2.2 MA-Amidase (Rv3717)

Peptidoglycan hydrolases are key enzymes in bacterial cell wall metabolism and have multiple functions in cell growth, peptidoglycan maturation, and fragment recycling (Wyckoff, Taylor, & Salama, 2012). The Mtb MA-amidase (Rv3717) is a zinc-dependent protein that hydrolyzes *N*-acetylmuramyl-L-alanine, homolog of well-studied enzymes AmiA, AmiB, and AmiC of *E. coli* and CwlB of *B. subtilis*. It consists of a single domain protein which contains the typical features of the Amidase_3 fold: a central six-stranded β-sheet, six surrounding α-helices, and a Zn ion coordinated by two histidines (His35 and His125) and one glutamate (Glu70) in the active site. A conserved amino acid, Glu200, is suggested to occupy an important role in catalysis, working as a general acid/base along the reaction (see below) (Kumar et al., 2013; Prigozhin, Mavrici, Huizar, Vansell, & Alber, 2013). Furthermore, the proposed key role of MA-amidase in bacterial wall metabolism makes it a very attractive drug target.

1.2.3 Zn Hydrolases Reaction Mechanism

Although possible mechanisms have been proposed by analogies to other widely studied Zn hydrolases, several key issues, particularly those related to Zn coordination sphere and its role in catalysis, remain unanswered. Initial studies suggested two main mechanisms which differ in the role assigned to the key active site but noncoordinated acid group (either Asp or Glu). Assuming that no water molecule is present in the active site, the anhydride mechanism designates the carboxylate as the nucleophile, which upon attack on the scissile carbonyl results in an acyl-enzyme intermediate which is subsequently hydrolyzed by water. The alternative mechanism is the hydroxide attack mechanism, which assumes the existence of a water nucleophile, that can (or not) be bound to the Zn ion, which is activated upon proton transfer to the active site base, usually a carboxyl group. The hydroxide ion attack to the carbonyl carbon results to a "meta-stable" tetrahedral intermediate, which collapses after the C–N bond breaking. Both—and slight variants of these mechanisms—have received support, but significant differences are observed depending on the enzyme and substrate. Moreover, some enzymes, such as MshB, are differentially active with other metal ions, such as iron, cobalt, or manganese, making the role of the Zn cofactor difficult to rationalize. Among key, yet unsolved issues concerning its role are whether it coordinates or not to the carbonyl substrate and whether the active site water/hydroxide is bound to the ion (Hernick & Fierke, 2005).

In this work, we have applied our recently developed QM/MM HyDRA to study two key Mtb Zn hydrolases reaction mechanism. Our results not only highlight the potential of the algorithm to determine a protein reaction mechanism FEP but also shed light into the role of the Zn ion in both enzymes, providing key elements for future inhibitor design strategies.

2. COMPUTATIONAL METHODS

2.1 Theoretical Basis of HyDRA

The HyDRA is based on previous ideas which explored the use of multiple time steps to describe different system regions treated with different levels of theory (Woo, Margl, Blöchl, & Ziegler, 2002), applied in the context of MSMD simulations and JR. Recently, we have shown (Ramírez et al., 2014) that it is possible to obtain an accurate FEP in significant less computational time, when using JR with the HyDRA scheme. This provides an invaluable tool for QM/MM studies of enzymatic reactions especially those requiring significant protein reorganization along the chemical step.

As already mentioned, to obtain a FEP using JR, multiple steering nonequilibrium trajectories are performed and the resulting work profiles (performed by the driving force) for each independent trajectory are exponentially averaged (Eq. 1). Two important aspects should be highlighted here. First, that in order to obtain properly converged averages, a moderately large number of trajectories are required. Second, and more interesting, although there is no theoretical requirement about the speed at which the system must be driven along the reaction coordinate, too fast driving results in poor convergence. Usually, the external force is applied as a harmonic potential which minimum moves at constant velocity (ν), along the reaction coordinate (λ), according to Eq. (2)

$$\vec{F} = -k(\lambda - \lambda_0 - \nu \cdot \mathrm{d}t) \tag{2}$$

where λ_0 is the starting value of the reaction coordinate, which together with ν defines the center of the harmonic potential in each MD step. Therefore, the pulling speed determines the amount of steps and computational cost needed to perform each steered simulation. There is a key relation between the velocity and the convergence, and thus the accuracy of the obtained FEP. To understand it, let us consider first that for any value along the reaction coordinate, the obtained distribution of work values (from each pulling trajectory) is roughly Gaussian. Under near-equilibrium conditions (extremely slow pulling), the distribution is like a delta function and, thus, the external work equals the free energy. Pulling at higher speeds drives the system further from equilibrium and results in increased widths for the work distributions, and higher averages, but since obtaining accurate free energies requires a converged exponential average, the work values that need to be sampled are those in the lower tail of the distribution. Thus, faster speeds (large ν) result in the need of a larger number of trajectories. For any given pulling speed, the distribution is wider further from the starting conditions, since the system moves further from equilibrium with the reaction progress. As a consequence of this drift, results using MSMD-JR strategy systematically show that it is very difficult to get accurate FEPs if the pulling speed is too fast, no matter how many trajectories are performed (Xiong et al., 2006). Moreover, theoretical analysis (Pohorille et al., 2010) shows that if the standard deviation of the work values is around kT, about 1 in 6 trajectories samples values close to the free energy. The value decreases to 1 in 40 trajectories for $\sigma = 2k_{\mathrm{B}}T$ and becomes even larger for $\sigma = 5k_{\mathrm{B}}T$, strongly affecting FEP accuracy and convergence.

HyDRA solves most of these problems, allowing the MM system to relax (or at least come closer) to equilibrium during the pure MM steps. Even at relative fast pulling speeds, a reduction on the widths of the work values distribution (smaller SD) is observed, which increased accuracy and convergence of the FEP. This is possible since external force and thus work are only computed during the joint (QM/MM) step. In other words, HyDRA takes advantage of the fact that in the context of MSMD-JR, only the force on the reaction coordinate contributes to the free energy and the environment can relax freely at no cost, when it remains frozen (Ramírez et al., 2014).

The algorithm (shown schematically in Fig. 2) which can be easily implemented in any QM/MM code works as follows: first, both systems (MM and QM) move together in a joint (or QM–MM) step. In this step, the steering that drives the reaction and moves the system from equilibrium is performed. Subsequently, we allow the classical region to relax for a given number of pure MM steps, while the QM region remains fixed (or frozen). During these steps, the system moves closer to equilibrium (i.e., relaxes) but no force on the reaction coordinate, and thus, no work is performed. The key HyDRA parameter is the ratio between the MM and the joint steps, which is called the differential relaxation algorithm ratio (DRAr). A DRAr ratio of 1 is equivalent to conventional QM–MM MD simulation, while a DRAr of 4 implies that for each QM step, 4 pure MM relaxation

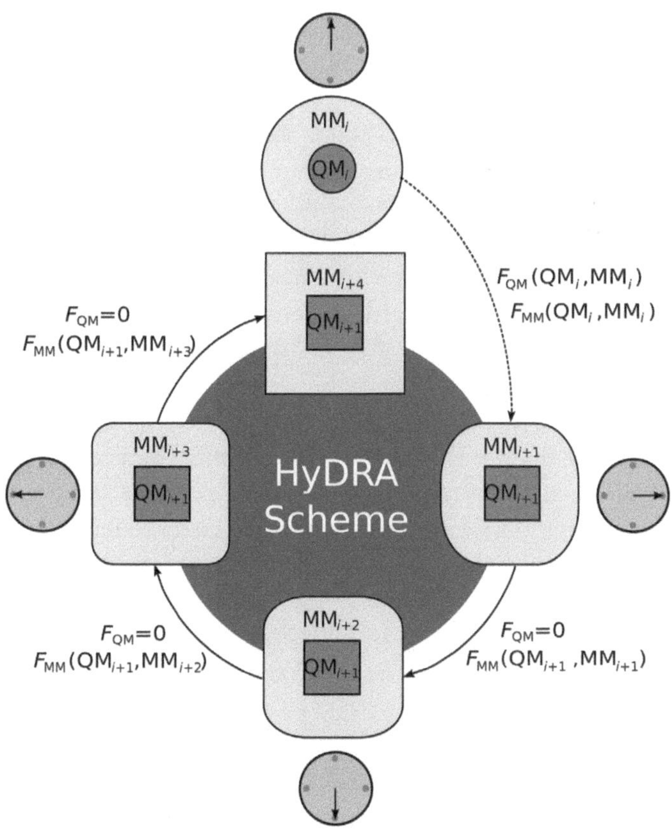

Figure 2 Hybrid differential relaxation algorithm scheme.

steps are performed. After the MM steps, both systems move together for another QM–MM step, in a new steering-relaxation cycle.

To implement HyDRA, we take advantage of the additive nature of the hybrid QM/MM hamiltonian, which allows to compute each force contribution separately. When simulation starts, both systems—the QM and the MM—are synchronized. In this initial conformation (subindex i on the scheme), all contribution to the forces is computed as usual. The forces acting on the QM atoms have contributions from the atoms in the QM subsystem in the initial conformation and also from the atoms in the MM region in this conformation. In our force notation, these are described as $F_{\mathrm{QM}}(\mathrm{QM}_i, \mathrm{MM}_i)$. The equivalent is true for the forces acting on the MM subsystem, $F_{\mathrm{MM}}(\mathrm{QM}_i, \mathrm{MM}_i)$. With this set of forces, the program performs a first dynamics step, leading to conformation $\mathrm{QM}_{i+1}, \mathrm{MM}_{i+1}$. Then the QM system is frozen and no forces acting on this region are allowed to act (forces acting on the QM subsystem are zeroed out, and thus, the QM subsystem remains fixed in the "$i+1$" conformation). Forces acting on the MM region are calculated as usual and atoms are moved. During the pure MM steps, the classical system moves in a fixed QM field, while going through conformations MM_{i+2}, MM_{i+3}, etc., until all DRAR steps have been performed and MM system arrives at $\mathrm{MM}_{i+\mathrm{DRAr}}$ ($\mathrm{DRAr}=4$ in Fig. 2). Now the system performs again a joint step and the cycle starts again. In HyDRA, each structure/conformation after a joint step is like any given QM–MM MD snapshot, while any structure after all MM steps have been performed represents a structure where the MM regions were allowed to relax to the frozen QM conformation.

2.2 Starting Structures

2.2.1 MshB

We started from the crystal structure of *wt*MshB (Maynes et al., 2003), PDBid 4EWL, which corresponds to the substrate-free protein. The Zn ion is coordinated by two water molecules: one closer to the active site base (Asp15) was retained, while the other was removed to allow substrate to coordinate to the ion. Substrate was built independently and docked by superimposition in place, using as a constraint that scissile bond carbonyl oxygen matches position of the removed water.

2.2.2 MA-Amidase

We started from the crystal structure of *wt*MA-amidase (Prigozhin et al., 2013), PDBid 4M6G, which is bound to the reaction product L-alanine-iso-D-glutamine and one additional water molecule near the active site.

The substrate was placed in the active site using as a template the structural alignment between MA-amidase Rv3717 and *E. coli* AmiD PDBid 3D2Y which was crystallized with the bound substrate anhydro-*N*-acetylmuramic acid-L-Ala-D-gamma-Glu-L-Lys, performing a biased docking.

2.3 Classical, DFT, and QM/MM Simulation Parameters

All classical parameters for both, the pure classical equilibration simulations and the MM system in the QM–MM dynamics, were taken from the AMBER force field ff99SB for the 20 amino acid residues (Case et al., 2014; Hornak et al., 2006) and TIP3P for the water molecules (Jorgensen, Chandrasekhar, Madura, Impey, & Klein, 1983). Classical force field parameters for the Zn ion and its coordination sphere were built using standard procedure as recommended by the AMBER package of programs. The QM–MM simulations were carried out at the density functional tight binding (SCC-DFTB) level of theory (Cui, Elstner, Kaxiras, Frauenheim, & Karplus, 2001) as implemented in the sander module of AMBER (de M. Seabra, Walker, Elstner, Case, & Roitberg, 2007).

Both complete systems were subjected to a relaxation protocol using classical MD. System was gently heated to 300 K and 50 ns of plain (NPT) MD were performed to allow the system to relax. Temperature and pressure were kept constant using Langevin thermostat and Berendsen barostat (Berendsen, Postma, van Gunsteren, DiNola, & Haak, 1984), using periodic boundary conditions. For all MD (and MD-QM/MM) simulations, we used an integration step of 1 fs. Starting from an equilibrated structure, 300-ps QM/MM-MD simulations were performed, and from these trajectory snapshots were taken for the MSMD simulations.

For both proteins, the QM subsystem consisted of the Zn ion, the side chains of the coordinated residues, the side chain of the active site base, the coordinated and/or reactive water molecules, and the amide bond of the substrate capped with methyl groups (i.e., the substrate corresponds to *N*-methyl-acetamide).

2.4 Free Energy Determination Simulation Strategy and Parameters

To determine the FEPs of each reaction step, we used combined MSMD-JR strategy outlined above. In all cases, several pulling speeds and DRAr were tested until reasonable convergence of the FEP was achieved. The whole reaction was divided into three steps, namely: hydroxide ion generation (activation step), nucleophilic attack (first reaction step), and C–N bond breaking (second reaction step).

2.4.1 Reaction Coordinate Definition

The reaction coordinate for the activation step corresponds to the difference between the water oxygen-transferring proton distance (dO_wH_t) and the distance between the transferring proton and the active base carboxylate oxygen acceptor (dH_tAcO):

$$RC_0 = dO_wH_t - dH_tAcO$$

The chosen reaction coordinate for the first reaction step was the distance between the hydroxide oxygen (O_{OH}) and the carbonyl carbon of the amide bond (C):

$$RC_1 = dO_{OH} - C$$

The chosen reaction coordinate for the second reaction step involved a combination of the amide bond distance (dCN), with a difference in the distances that account from the active site base (dH_tAcO) to amide N proton transfer (dNH_t)

$$RC_2 = dCN - (dNH_t - dH_tAcO).$$

Schematic representation of all three reaction coordinates is presented in Fig. 9.

2.4.2 MSMD Trajectories and Pulling Speed

For the hydroxide ion generation step, we performed 20 independent MSMD simulations, each of 3000 QM/MM steps, with a pulling speed of 0.4 Å/ps using a DRAr of 10. Simulations were performed in both forward and reverse directions and the reported free energy corresponds to optimal combination of both.

For the nucleophilic attack, we performed 20 independent MSMD simulations, each of 12,000 QM/MM steps, with a pulling speed of 0.1 Å/ps and using DRAr of 1, 4, and 10. For MshB, simulations were performed in both forward and reverse directions and the reported free energy corresponds to optimal combination of both. For MA-amidase, only forward reactions were considered, since back reactions went through a different path, showing a high hysteresis. For the C–N bond breaking, we performed 20 independent MSMD simulations, each of 5000 QM/MM steps and with a pulling speed of 1.34 Å/ps using DRAr of 10. For this step, only forward reactions were performed.

3. RESULTS

3.1 Mtb Zinc Hydrolases Display a Flexible Zinc Coordination Sphere

We begin our analysis with a brief structural analysis of the catalytic Zn active site in both Mtb Zn hydrolases. For both proteins, we first optimized several possible Zn coordination states using a QM/MM scheme. The states correspond to (A) Zn-$(His)_2$-Asp-Wat, (B) Zn-$(His)_2$-Asp-substrate CO, and (C) Zn-$(His)_2$-Asp-Wat-substrate CO as shown in Fig. 3. For comparison purposes, we optimized the same complexes isolated in vacuum. The key structural parameters for each complex are presented in Table 1.

If we look at the isolated complexes in vacuum, both tetracoordinated states (I and II) display tetrahedral coordination spheres with the acid as monodentate like ligand (dO1-Zn ~ 2.0 Å and dO2-Zn > 2.4 Å), and His-Zn-His/O angles in the 107–118° range. The fourth ligand, which is either water or the substrate carbonyl, is in each case very close to the Zn (dCO-Zn ~ 2.05 Å). Addition of a fifth ligand, as in states III, results in a trigonal bipyramidal coordination sphere, since the acid ligand remains monodentate. In both cases, one of the histidines becomes an axial ligand, while the other, the acid and either the water/substrate become equatorial ligands, remaining ligand either water/substrate acts as the other axial ligand (see Figure 3). This is evidenced mostly in the ligand coordination angles. For example, His(ax)-Zn-Xeq angles become closer to 90°, while His(ax)-Zn-Wat(ax) is 163°. Other point is that, addition of fifth ligand to the sphere results in a slight increase in His, water, and substrate coordination distances.

The QM/MM optimizations show that in both proteins, the tetracoordinated states are deviated from tetrahedral coordination structure adopted in vacuum, as evidenced mainly by the angles values, which deviate from expected 109°, especially so in MA-amidase, where His(ax)-Zn-Xeq angles are closer to 90°, the acid ligand remaining as monodentate. In MshB, the acid tends to become bidentate as evidenced by largest ZnO distance becoming significantly smaller (dZn-O 2.2.5–2.35 Å). As for the complexes in vacuum, incorporation of fifth ligand increases the trigonal/octahedral character of the coordination sphere and shows slightly elongated Zn-Wat/CO distances. Interestingly, Zn-Wat/CO distances seem to be exchanged between both proteins, which is reasonable, because their

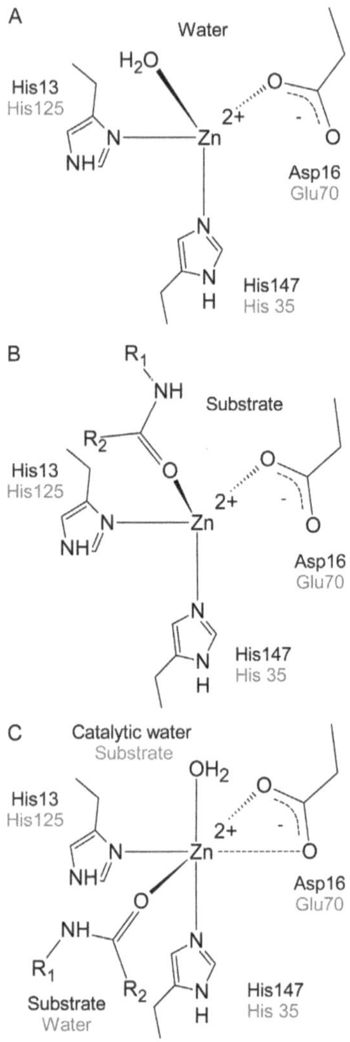

Figure 3 Scheme of Zn hydrolases active site in different coordination state. (A) Zn-(His)$_2$-Asp-Wat, (B) Zn-(His)$_2$-Asp-substrate, and (C) Zn-(His)$_2$-Asp-Wat-substrate. Residue labels for MshB and MA-amidase are labeled in black and grey respectively.

relative (axial–equatorial) positions are also exchanged (see Figure 3). Also noteworthy is the fact that incorporation of another sixth water ligand was unsuccessful, as the systems expel one of the waters.

In summary, Mtb Zn hydrolases are able to accommodate one or two external ligands, moving from a tetrahedral to trigonal/octahedral-like

Table 1 Geometrical Parameters of Catalytic Zn in Different Coordination States

	System	Zn-His	Zn-AcO[a]	Zn-Wat	Zn-CO	H-Zn-H	His(ax)[b]-Zn-Wat	His(ax)[b]-Zn-CO	His(ax)[b]-Zn-O1	Wat-Zn-CO
Vacuum	I	2.05–2.06	2.02–3.05	2.05	–	109	112	–	113	–
	II	2.08–2.07	2.04–2.45	–	2.04	115	–	107	118	–
	III	2.12–2.12	2.06–3.18	2.18	2.21	98	163	83	95	80
MshB	I	1.98–2.00	2.06–2.35	2.08	–	106	121	–	111	–
	II	2.05–1.96	2.13–2.25	–	2.12	112	–	113	103	–
	III	2.01–2.06	2.14–2.34	2.15	2.22	104	153	86	97	81
MA-Am	I	2.02–2.19	2.05–2.66	2.17	–	96	93	–	99	–
	II	2.09–2.05	2.03–2.81	–	2.11	96	–	112	104	–
	III	2.03–2.01	2.02–3.08	2.21	2.15	95	151	95	102	77

[a]The two values correspond to distances observed by each of the two oxygens of the acid ligand.
[b]Refers to the axial His ligand. Please refer to Figure 3 for further details.
Distances are in Å and angles in degrees.

coordination showing significant flexibility and differences in the mono–bi dentate coordination of the acid group. This is possible thanks to the restrictive influence of the protein interactions that maintain a *stressed* tetrahedral coordination, which can easily shift to a reactive penta/hexacoordinated state.

3.2 Hydroxide Ion Generation Step

Previous to the nucleophilic attack to the amide bond carbonyl, catalytic water must transfer a proton to the active site base to form the reactive hydroxide anion. Here, the first key difference between both Mtb hydrolases arises. In MshB, the reactive base is an aspartic acid (Asp15), which is not coordinated to the zinc ion but establishes a long-lasting hydrogen bond with the catalytic water, which in this case is coordinated to the Zn. This interaction is also responsible for the *stressed* tetrahedral and bipyramidal/octahedral coordination spheres, described above. On the other hand, in MA-amidase Rv3717 the general acid/base is a glutamic acid (Glu200). This residue is not coordinated to the zinc ion, although it is held in place through its interaction with Asn139 and neither is the catalytic water, held in place by the base.

For both cases, we computed the free energy for the proton transfer reaction (see Section 2). The results are presented in Table 2. The proton transfer reaction is almost barrierless but endergonic, which is not unexpected given the higher solution pK_a of water compared to an acid group (pK_a Wat of 7 vs. pK_a Asp/Glu of 4.5). The corresponding reaction free energy can thus be thought of as a pre-equilibrium step which determines the relative effective concentration of the "reactive" hydroxide-loaded enzyme with respect to water-bound "resting" state. In both proteins, there is a very tight hydrogen bond between catalytic water and the active site base, so that the proton can be considered to be partially shared, as evidenced by the short WatH-AcO distance of only 1.55–1.70 Å. The most interesting analysis, however, can be performed by comparing the two enzymes. In MshB, it is evident that water coordination to the Zn promotes hydroxide ion formation, DG is only ~6 kcal/mol. As expected, hydroxide interaction with the Zn ion is stronger than that of water, as evidenced by the shorter Zn-O distance and increased charge transfer. In MA-amidase, hydroxide ion formation requires ~11 kcal/mol. Interestingly, and possibly also as consequence of Zn coordination, hydroxide ion in MshB is significantly less negatively charged, when compared to MA-amidase.

Table 2 Structural and Energetic Parameters of the Hydroxide Ion Generation Step

Parameter	MshB	MA-Amidase
Initial dWatO-H	1.02	1.06
Final dWatO-H	1.52	1.64
Initial dAcO-H	1.71	1.55
Final dAcO-H	1.06	1.035
WatO-H-AcO angle[a]	154 → 163	166° → 164
Initial dWatO-Zn	2.13	–
Final dWatO-Zn	2.01	–
Initial qOH[b]	−0.15	−0.437
Final qOH[b]	−0.39	−0.911
$\Delta G^{\ddagger c}$	5.7	10.7

[a]Value is average ± SD along transfer reaction.
[b]Refers to the attacking OH^-.
[c]$\Delta G^{\ddagger}$ corresponds to the free activation energy.
Distances are in Å, angles in degrees, and energies in kcal/mol.

For comparison purposes, we also tested whether it is possible for either enzyme to produce hydroxide ion, transferring a proton from the coordinated water in state I type of complex to the co-coordinated acid, Glu70 in MA-amidase and Asp16 in MshB. However, in both cases proton transfer energy is huge (>50 kcal/mol), showing that Zn-coordinated acid cannot act as the active site base. Moreover, in MshB proton transfer forces the Asp16 to become monodentate, shifting coordination to tetrahedral and expelling the substrate from the coordination sphere.

3.3 Hydroxide Attack to Amide Carbonyl

Once formed, first reaction step corresponds to hydroxide ion attack to the carbonyl "carbon" of the amide bond that is going to be hydrolyzed. The resulting FEP is shown in Fig. 4A and B, for MshB and MA-amidase, respectively, while key structures are shown in Fig. 9. Results show that for both enzymes, hydroxide ion attack has a moderate high barrier, 14 kcal/mol in MshB and 21 kcal/mol in MA-amidase, which results in a "meta"-stable tetrahedral intermediate (that will be characterized in detail later), which lies 10 and 18 kcal/mol above the reactant in MshB and MA-amidase, respectively. Structures of the relevant configurations along

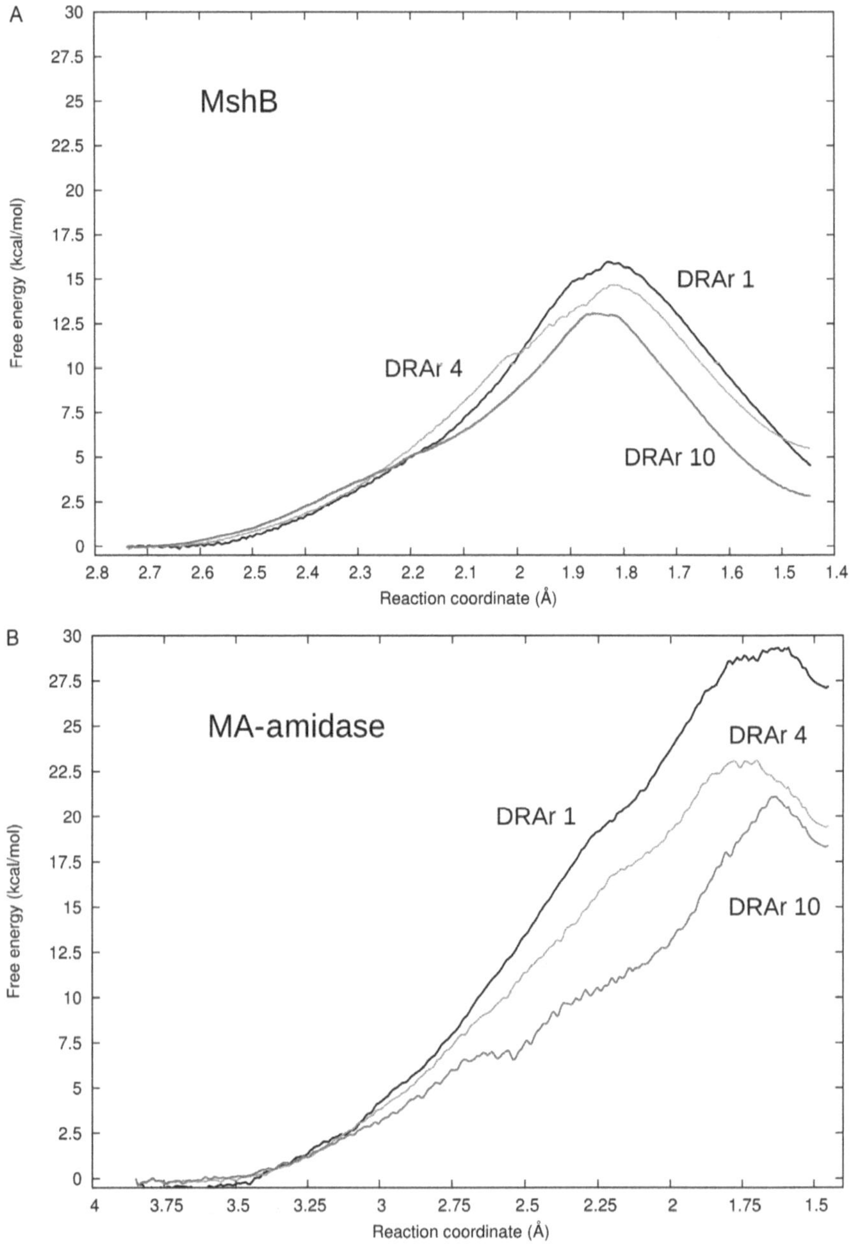

Figure 4 Free energy profiles for the hydroxide ion nucleophilic attack (first) reaction step for MshB (A) and MA-amidase (B). Profiles were obtained using 10 independent SMD simulations DRAr 1, 4, and 10 (shown in black, light grey and drak grey, respectively). Reaction coordinate is, in both cases, the distance between attacking OH^- and substrate C atom.

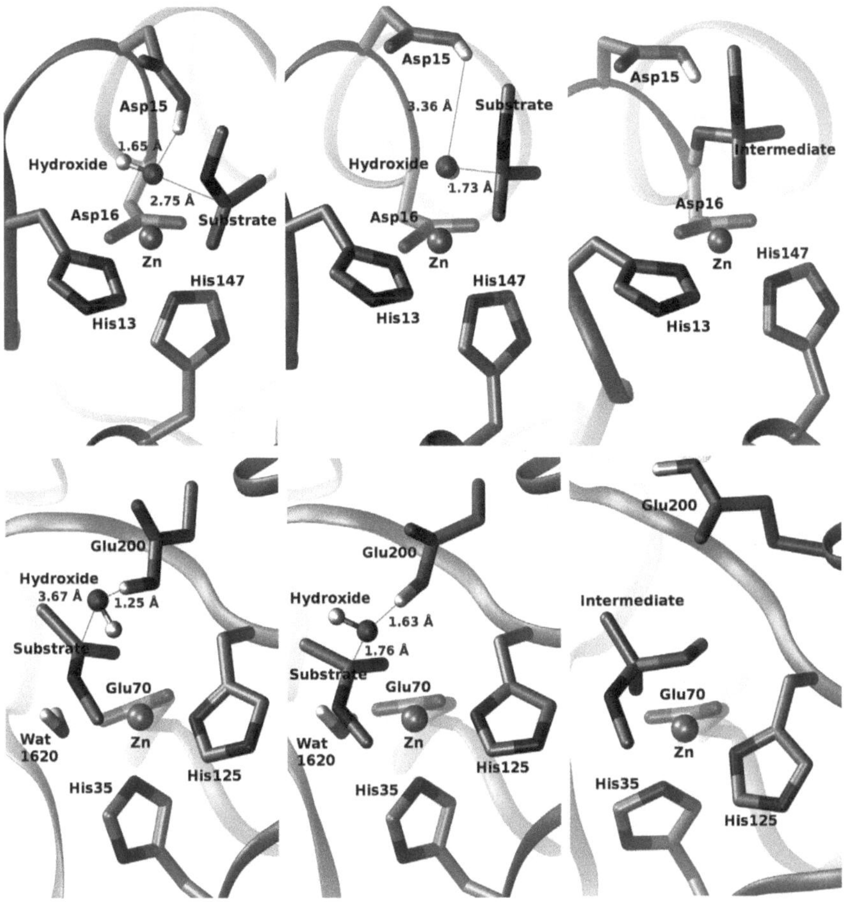

Figure 5 Structure of the reactant, transition state, and tetrahedral intermediate product for the nucleophilic attack step in MshB (upper panel) and MA-amidase (lower panel). Zn and hydroxide atoms are shown as balls and sticks, while the rest of the active site residues are shown as sticks. For the substrate, only scissile amide bond is shown.

the reaction coordinate (reactant, transition state, and tetrahedral intermediate product) are shown in Fig. 5.

3.3.1 Effect of DRAr

The effect of number of classical relaxation steps, for each QM steering speed, the so-called DRAr, is clearly evidenced in both enzymes and shows that for the selected pulling speed and number of trajectories, DRAr of 10 is needed to achieve best results. As shown in our previous work, effect of increasing the classical relaxation steps is to reduce both the barrier and the reaction free energy. Compared to standard SMD simulations, the effect of HyDRA is dramatic, reducing both the barrier heights and reaction free energies up to 10 kcal/mol. Its effect also highlights the relevance of the protein environment dynamical adjustment to the ongoing reaction. Interestingly, the effect is different in both enzymes, since while in MA-amidase most significant reduction is observed for the barrier, for MshB a largest change is observed for the reaction free energy, and thus the predicted stability of the intermediate. The origin of this difference is not clear but possibly arises from the role played by the Zn coordination to the attacking hydroxide (see below).

3.3.2 Detailed Mechanistic and Comparative Analysis Between MshB and MA-Amidase

To get insight into the origin of the lower barrier and higher stability of the intermediate in MshB compared to MA-amidase, it is interesting to analyze in detail how the electronic structure evolves along the reaction as shown in Table 3. Common to both enzymes are the following features that describe the nucleophilic attack, the first is that the "transition state" appears late, especially in MA-amidase once the new C–OH bond is already partially established (dC-OH ~ 1.7 Å). Crossing of the TS leads to significant changes in the electronic structure as evidenced by a change in the C hybridization from sp^2-like to sp^3-like (as evidenced by the change in the corresponding improper angle from ~180° to ~120°), a slight enlargement of the C–N bond (of ~0.1 Å) and most important an increase in the CO-Zn interaction (evidenced in the 0.15–0.36 Å shortening of the dZn-CO).

Although it is difficult to assign the energy cost to a principal variable and to determine what makes MshB more efficient, detailed analysis of the structural/electronic parameters shows some key aspects. First, it is clear that distance of the attacking OH to the scissile "C" and its change in hybridization possibly bears most of the energetic cost. In MA-amidase, the initial distance is larger since the nucleophile is not bound to the Zn ion, and this may partially account for higher barrier. Another interesting point is related to the fact that, due to hydroxide nucleophile coordination in MshB, for the reaction to happen the OH needs to slightly detach from Zn. The results clearly show that its release occurs mostly after the TS, and thus, this weakening of this interaction is not expected to contribute significantly to the reaction barrier. However, during formation of the "new" OH–C bond, hydroxide charge is significantly reduced and transferred to the carbonyl oxygen, the Zn, and the amide N. In MA-amidase, reduction in hydroxide charge is

Table 3 Structural Parameters, Energies, and Mulliken Charges Along the Hydroxide Attack Step

	MshB			**MA-Amidase**		
Parameter	**R**	**TS**	**I**	**R**	**TS**	**I**
dOH-C	2.75	1.83	1.49	3.67	1.68	1.46
dC-N	1.36	1.45	1.48	1.34	1.43	1.44
dZn-OH^{-a}	1.93	2.24	3.46	–	–	–
dZn-CO	2.36	2.09	1.93	2.12	2.04	1.98
dZn-OH2[b]	–	–	–	2.31	2.41	>3.5
"C" conf angle	178	134	120	177.28	130.53	118.24
qOH	−0.35	−0.24	−0.28	−0.665	−0.407	−0.217
qO(Sust)	−0.522	−0.643	−0.611	−0.464	−0.647	−0.755
qZn	0.390	0.43	0.37	0.446	0.416	0.341
qC	0.522	0.607	0.631	0.472	0.591	0.602
qN	−0.23	−0.30	−0.38	−0.16	−0.33	−0.38
$\Delta G^{\ddagger c}$	14.66			21.1		
ΔG	+9.75			+18.3		

[a]OH corresponds to the attacking nucleophile.
[b]OH2 corresponds to additional Zn-coordinated water which does not participate in the reaction.
[c]$\Delta G^{\ddagger}$ corresponds to the free activation energy.
Distances are in Å, angles in degrees, and energies in kcal/mol.

much larger than in MshB (since starts from a significantly smaller value), a fact that possibly also contributes to its higher barrier.

Another interesting point concerns the evolution of Zn coordination sphere since although, as already mentioned, in the reactive state both proteins display a trigonal/octahedral-like coordination, in the tetrahedral intermediate, after hydroxide ion attack, both proteins end with Zn in a tetrahedral coordination, which strengthens the Zn-CO interaction. However, while in MshB the attacking OH is finally released ($d \sim 3.46$ Å), in MA-amidase formation of the intermediate results in expulsion of the additional coordinated water (Wat1620) ($d > 3.5$ Å) which does not participate in the reaction.

3.3.3 Role of Substrate Carbonyl Coordination

To further analyze the role of substrate carbonyl to Zn coordination, we computed the FEP for the hydroxide nucleophilic attack step but starting from a conformation where substrate is not coordinated to the catalytic

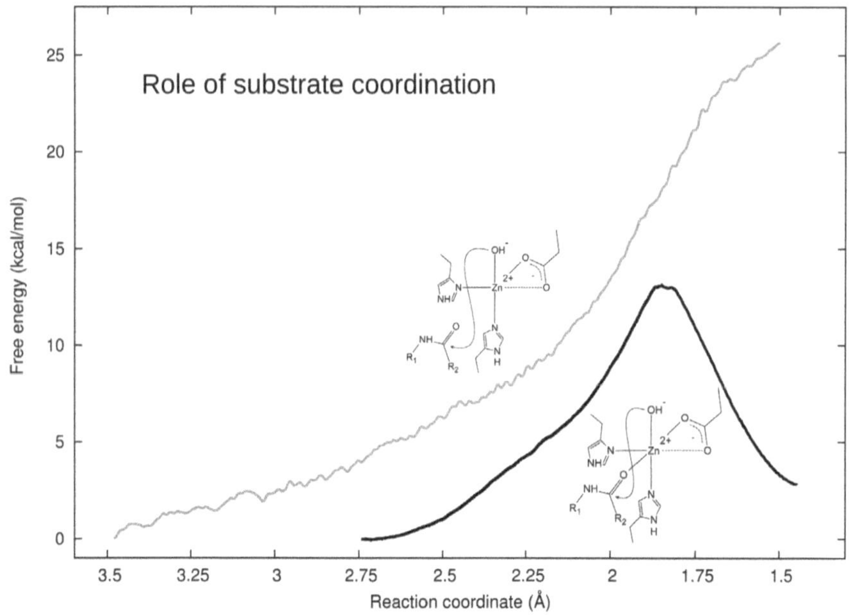

Figure 6 Free energy profiles for the hydroxide ion nucleophilic attack (first) reaction step for MshB, stating from substrate in the coordinated (black line) and non-coordinated (grey) state.

Zn. The results presented in Fig. 6 show that cofactor is critical for the establishment of "meta"-stable tetrahedral intermediate. In the absence of substrate-Zn coordination, for MshB, the reaction coordinate is longer, since the hydroxide and "C" are further apart. However, reaching the point where the Zn-bound coordinate starts requires only ~3 kcal/mol, the energy then raises in a similar manner, but no TS nor intermediate state is reached. Similar trend is observed for MA-amidase. These results point again for a primordial role of the Zn ion in stabilizing the negative charge developed in the tetrahedral intermediate (ex-carbonyl) oxygen upon nucleophilic attack.

3.4 C-N Amide Bond Breaking

3.4.1 Stability of Tetrahedral Intermediate

To proceed from the tetrahedral intermediate to the product, C–N bond breaking needs to be performed; key for this reaction is the reorganization of the hydrogen bond network to promote proton transfer to the amide NH. This strict requirement (barrier for C–N bond breaking is huge if no proton is transferred to the NH) results in a significantly stable intermediate which presumably lives long enough to be detected experimentally. In MshB, main stabilization of the substrate oxygen is due to its coordination to the zinc ion ($d \sim 2.00$) and a strong hydrogen bond with a water molecule ($d \sim 2.7$) that is held in its place during all the MD by hydrogen bonds with Arg68. Moreover, Arg68 also forms a hydrogen bond with the nonprotonated oxygen of Asp15 (the active site base). Similar to what is observed in MshB, in MA-amidase, the ex-carbonyl oxygen of the tetrahedral intermediate is stabilized by zinc coordination ($d \sim 1.97$) and also by a strong hydrogen bond with the additional Zn-coordinated water, held in place due to its interaction with Glu70. The ex-hydroxide, now a hydroxyl group bound to the amide bond carbon, is hydrogen bond to the Ala126 carbonyl backbone, which contributes to hold the intermediate in place for the C–N breaking step. In both proteins, the proton that will be transferred to the amide nitrogen is the same that was taken from the water to form the attacking nucleophile held in place by the active site base (Asp15 and Glu200 in MshB and MA-amidase).

3.4.2 FEPs of the C–N Bond Breaking Step

To analyze the C–N amide bond breaking step, we computed the corresponding FEPs using as the reaction coordinate a combination of C–N bond breaking and proton transfer step to the amide nitrogen, always with DRAr of 10. The results presented in Fig. 7 and Table 4 show that FEP is similar in both enzymes. A not unexpected fact, since for the residues involved in this step enzymes show no clear structural differences. The barrier is moderate, between 21 and 24 kcal/mol.

Analysis of the profiles in the context of the relevant structural parameters (presented in Fig. 7B, for MshB) shows that the reaction displays four distinct phases. In the first (zone 1), the free energy raises slowly, about 7 kcal/mol. This segment corresponds to the rotation of protonated base (Asp15 in MshB and Glu200 in MA-amidase) and reduction of the distances between each proton and its acceptor, generating the appropriate configuration for the double proton exchange. Subsequently, the energy rises steeply to reach the TS when the amide N takes the proton from the base, and the C–N bond breaks (zone 2). Short after the TS, there is a small decrease in the free energy due to relaxation of the newly formed NH_2, and proton transfer from intermediate OH to Asp15 (zone 3). Finally, the energy drops as a result of the relaxation of the entire system (zone 4). Interestingly, in MshB product is ~5 kcal/mol more stable than in MA-amidase, which may result from stronger stabilization of the product by Zn

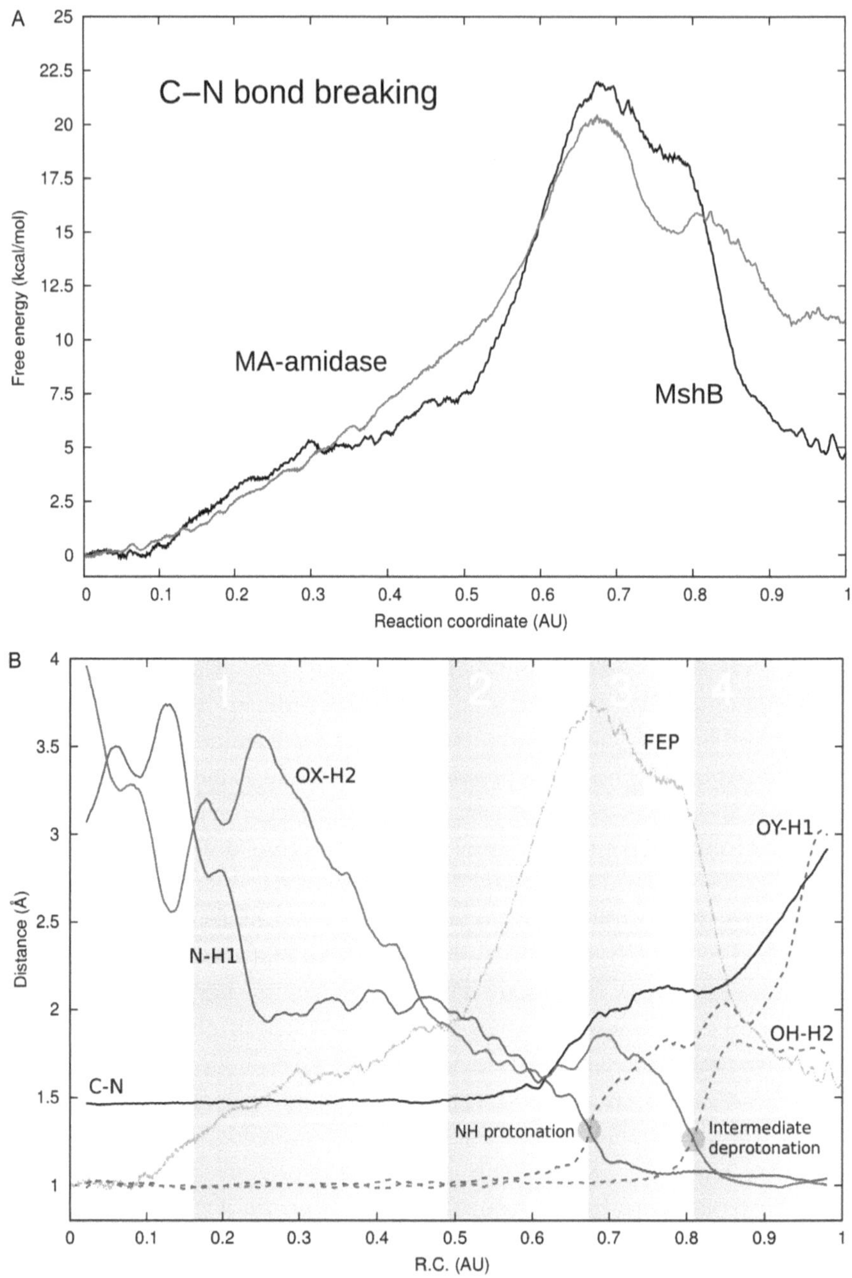

Figure 7 (A) Free energy profiles for the C–N bond breaking (second) reaction step for MshB (black line) and MA-amidase (grey line). Profiles were obtained using 10 independent SMD simulations and DRAr of 10. (B) Evolution of key structural parameters along the reaction coordinate for MshB, C–N bond (black, filled), N-H1 (dark grey, filled), OY-H1 (dotted, dark grey), OX-H2 (filled, light grey), and OH-H2 (dotted, light grey). FEP shown in grey at back.

Table 4 Structural Parameters, Energies, and Mulliken Charges Along the C–N Bond Breaking Step

	MshB			MA-Amidase		
Parameter	**R**	**TS**	**P**	**R**	**TS**	**P**
dC-N	1.48	1.97	2.99	1.44	1.47	2.97
dCO-Zn	1.93	2.04	1.99	1.98	1.97	2.14
"N" conf angle	146	121	106	143.31	127.38	106.11
"C" conf angle	120	138	175	118.24	120.69	167.58
dN-H1[a]	3.01	1.38	1.04	4.35	1.38	1.01
dOX-H2[b]	4.1	1.71	0.99	4.05	3.94	1.05
qN	−0.38	−0.31	−0.41	−0.356	−0.212	−0.427
qH1	0.35	0.37	0.36	0.351	0.369	0.395
qH2	0.35	0.30	0.18	0.368	0.30	0.166
qZn	0.37	0.39	0.37	0.371	0.391	0.364
$\Delta G^{\ddagger}$[c]	21.97			20.5		
ΔG	+4.5			+10.67		

[a]H1 is the proton transferred from the active site base to the amide N.
[b]H2 is the proton transferred from the tetrahedral intermediate to the active site base, while OX is the active site base proton acceptor (see Fig. 9 for details).
[c]$\Delta G^{\ddagger}$ corresponds to the free activation energy.
Distances are in Å, angles in degrees, and energies in kcal/mol.

coordination. It is interesting to note that in both cases, the recently formed carboxylate in the product remains coordinated to the Zn ion in a bidentate fashion. Relevant structures are shown in Fig. 8.

In summary, analysis of the parameters shows that "actual" C–N bond breaking and concomitant change in the C and N hybridization occur late, mostly after the TS. The key proton transfer to the amide N, on the contrary, occurs early and drives the reaction forward.

3.5 Alternative Mechanisms

For both enzymes, we also tested the anhydride mechanism as a possible alternative. In MA-amidase, carboxylate attack to the amide carbonyl "C" displays a barrier of more than 50 kcal/mol, regarding it as seemingly unlikely. In MshB, although the barrier is smaller (ca. 30 kcal/mol), it is still too high to compete against the hydroxide ion attack of only 14 kcal/mol. Also, in both cases the anhydride tetrahedral intermediate

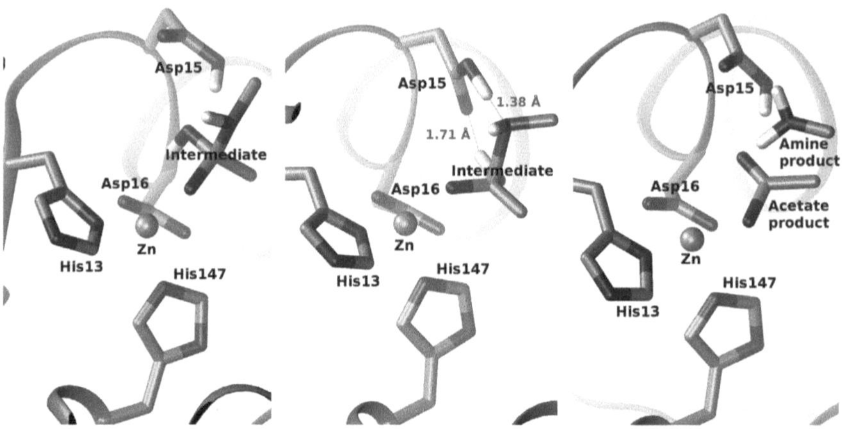

Figure 8 Structures of the tetrahedral intermediate, transition state, and products for C–N bond breaking.

does not seem to be stable, since it does not correspond to a clear local minimum in the corresponding FEP.

4. DISCUSSION

4.1 The Complete Mechanism of MshB and MA-Amidase Zn Hydrolases

The complete reaction mechanisms of mycobacterial Zn hydrolases, MshB and MA-amidase, are shown schematically in Fig. 9. Prereaction step requires formation of the hydroxide nucleophile, assisted by an active site base comprising the carboxylate side chain of either Asp15 or Glu200 residues. In the most effective (lower barrier) mechanism, as exemplified by MshB in this work, water/hydroxide ion nucleophile is tightly coordinated to the Zn ion. In MA-amidase in this work, we analyzed the reaction using a nonzinc-coordinated water/hydroxide as the nucleophile. It is important however to note that structure of the MA-amidase active complex was built *in silico*, by means of docking the substrate in place. Our results (data not shown) suggest that an alternative substrate-bound conformation is also possible where substrate carbonyl and water bound to the Zn ion are exchanged, leading to a conformation which resembles that of MshB (i.e., with Zn-coordinated water hydrogen bonded to the active site base Asp15). In light of our results, this more efficient conformation is probably the reactive one also in MA-amidase.

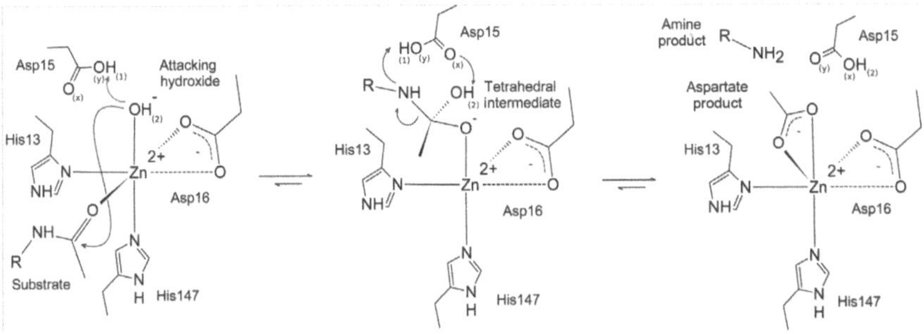

Figure 9 Proposed reaction mechanism for Zn hydrolases.

First reaction step corresponds to hydroxide anion nucleophilic attack to the carbonyl carbon of the scissile bond. This step has a moderately small barrier of 14 and 21 kcal/mol in MshB and MA-amidase, respectively, and leads to a "meta"-stable tetrahedral intermediate. Key to the hydroxide attack step is charge transfer from the attacking nucleophile to the substrate, which is mainly stabilized by the carbonyl oxygen Zn interaction. Hydroxide nucleophile coordination to the Zn lowers the barrier possibly by two means, first it makes the reaction coordinate shorter, since OH and substrate are both coordinated to the ion, and second lowering the initial negative charge in the nucleophile, which is already partially transferred to the ion.

The tetrahedral intermediate is stable in both enzymes, as judged both by the FEPs and by the fact that moderately long QM/MM MD simulations can be run, without observing it decaying back to the substrate or proceeding to the product. Its lifetime is key for allowing hydrogen bond network to rearrange, resulting in proper configuration for proton transfer to the amide N, which promotes C–N bond breaking (although about 7 kcal/mol corresponds to a local rearrangement, which suggests actual barrier could be smaller). The second step shows a moderately high barrier of 21 to 20 kcal/mol and is very similar in both enzymes. C–N bond breaking occurs late in the reaction, which is initiated by proton transfer.

In summary, both mycobacterial hydrolases follow the proposed water/hydroxide nucleophilic attack "two-step" mechanism, going through a "meta"-stable tetrahedral intermediate that interacts strongly with the Zn cofactor and which should be possible to detect experimentally.

4.2 Role of the Zn Ion in Catalysis

As seen during this work, Zn ion plays a key role throughout the whole reaction. To begin with, by coordinating substrate carbonyl Zn ion facilitates the appropriate location of the substrate for the nucleophilic attack to occur. In the hydroxide generation step, it polarizes the water molecule, sensibly lowering the activation energy of this step. This is evidenced in Table 2, where energy barrier for deprotonation of MshB-coordinated water is ~4 kcal lower than the one for deprotonation of MA-amidase non-coordinated water. Moreover, Zn ion produces the electrostatic stabilization of the tetrahedral intermediate, which has a negative charge. This is critical for the reaction to take place because, as shown in Fig. 2, nucleophilic attack never reaches a minimum when intermediate is not coordinated and, thus, not stabilized. Finally, the bidentate coordination of the aspartate product lowers the free energy of the products, increasing the equilibrium constant and driving the reaction forward.

4.3 Comparison with Other Zn Hydrolases

Having determined the mechanism and obtained the corresponding FEPs, it is interesting to compare our results with those obtained by other Zn hydrolases. Possibly, the best studied is carboxypeptidase-A (CPA). In a recent study using a DFTB-based QM/MM approach (Xu & Guo, 2009), authors reported that for peptide hydrolysis, Zn-coordinated water/hydroxide nucleophilic attack mechanism is the most likely, leading also to a tetrahedral intermediate with a barrier of 17.9 kcal/mol, which is similar to that reported here. Interestingly, the substrate carbonyl is not coordinated in the reactive structure and becomes coordinated during the reaction, strongly interacting in the tetrahedral intermediate. More interesting is that the same mechanism was also reported for termolysin (Blumberger, Lamoureux, & Klein, 2007), ACE (Zhang, Wu, & Xu, 2013), and the anthrax lethal factor peptidase (Smith, Smith, Yang, Xu, & Guo, 2010), all displaying barriers for the nucleophilic attack in the 13–18 kcal/mol range, and for the C–N bond breaking in the 5–14.7 kcal/mol range. In summary, the different studies of Zn hydrolases reactions point toward a conserved (or converged) mechanism among them.

4.4 Convergent Structural Evolution of Zn Hydrolases

To analyze the role of converged structural evolution of Zn hydrolases, we first analyzed their protein domain, family, or clan pertence according to PFAM. ACE (Masuyer, Schwager, Sturrock, Isaac, & Acharya, 2012), aminopeptidase (Yang, Liu, Lin, & Li, 2013), and termolysin (Gumpena, Kishor, Ganji, & Addlagatta, 2011) are part of the same clan (CL0126) but they are grouped into two different families (PF01401 and PF01433).

And although the last two share the same fold, they are clearly structurally different from ACE. Furthermore, histone deacetylase (Ciulli et al., 2008) belongs to Hist_deacetyl (PF00850) family, CPA (Tykvart et al., 2014) to Peptidase_M28 (PF04389), MshB to PIG-L (PF02585), and MA-amidase to Amidase_3 (PF01520). What is striking is that despite belonging to completely different families and displaying different structural folds, they share the basic conformation of the active site zinc ion conformation, and thus, we expect the underlying catalytic mechanism. Moreover, to further analyze the "essential" role of the Zn coordination sphere distortion, we analyzed the conformation of noncatalytic, i.e., structural Zn ions in protein structures. Although they display different coordination spheres, for example, His3-Asp (Wilder, Varney, Weiss, Gitti, & Weber, 2005; Zimmermann et al., 2009), His2-Cys2 (Foster et al., 1997), and Cys4 (Mossessova, Bickford, & Goldberg, 2003), they seem to be always in a "rigid" tetrahedral conformation which does not allow addition of a fifth ligand.

These observations strongly suggest that Zn hydrolases evolution has converged to a tetrahedrally distorted His2-Asp/Glu coordination sphere which allows for co-coordination of the water/hydroxide nucleophile and substrate carbonyl as required for optimal catalysis.

4.5 Final Remark on QM/MM Studies of Enzyme Reaction Mechanisms

The results presented here show the potential of the HyDRA method for obtaining enzymatic reaction FEPs. Although a detailed study of the 10–100 times reduction in computational cost was presented in our previous work (Ramírez, Zeida, Jara, Roitberg, & Martí, 2014), it is interesting to analyze the performance of the HyDRA method in the context of similar works on previously mentioned enzymes. Despite being difficult to make a detailed comparison when different codes, levels of theory, and systems were employed, the total number of QM/MM steps required to obtain the profile can be used as a code-system-independent parameter. The number of steps required to obtain the FEP for carboxypeptidase A, thermolysin, and ACE was roughly 2, 1.5, and 8 million, respectively. Employing a similar number of classical steps-but fewer quantum steps-, we obtained the FEP for each reaction. The HyDRA method, using DRAr of 10, and only forward reactions allowed us to obtain the profiles in 400,000 steps, which is about five times faster than the CPA and thermolysin cases and 20 times faster compared to ACE. These results highlight the tremendous power and advantage of the HyDRA strategy.

5. CONCLUSIONS

Our results show that both mycobacterial Zn hydrolases, MshB and MA-Amidase, work through a general water/hydroxide nucleophilic attack two-step reaction mechanism, whose key features are: Zn-coordinated hydroxide formation through a conserved active site base; nucleophilic attack to the scissile carbonyl to yield a meta-stable tetrahedral intermediate (step 1), whose stability is promoted by the Zn ion which stabilizes (and accepts) negative charge developed in the ex-carbonyl oxygen; and breaking of the C–N bond promoted by active site base to amide nitrogen proton transfer. This mechanism, which is also operative in several other Zn hydrolases, points to a convergent evolution in active sites for these types of enzymes.

ACKNOWLEDGMENTS

Computer power was provided by High Performance Computer (HPC) Center at the University of Florida. Research was funded by grants PICTO-2012-0057 and PIP 2012-2014 #112201101 00850 to M.A.M.

REFERENCES

Berendsen, H. J. C., Postma, J. P. M., van Gunsteren, W. F., DiNola, A., & Haak, J. R. (1984). Molecular dynamics with coupling to an external bath. *The Journal of Chemical Physics*, *81*(8), 3684. http://dx.doi.org/10.1063/1.448118.

Blumberger, J., Lamoureux, G., & Klein, M. L. (2007). Peptide hydrolysis in thermolysin: Ab initio QM/MM investigation of the Glu143-assisted water addition mechanism. *Journal of Chemical Theory and Computation*, *3*(5), 1837–1850. http://dx.doi.org/10.1021/ct7000792.

Case, D. A., Babin, V., Berryman, J. T., Betz, R. M., Cai, Q., Cerutti, D. S., et al. (2014). *The amber molecular dynamics package*. AMBER 14, San Francisco: University of California. Retrieved from, http://ambermd.org/.

Ciulli, A., Scott, D. E., Ando, M., Reyes, F., Saldanha, S. A., Tuck, K. L., et al. (2008). Inhibition of Mycobacterium tuberculosis pantothenate synthetase by analogues of the reaction intermediate. *Chembiochem: A European Journal of Chemical Biology*, *9*(16), 2606–2611. http://dx.doi.org/10.1002/cbic.200800437.

Cui, Q., Elstner, M., Kaxiras, E., Frauenheim, T., & Karplus, M. (2001). A QM/MM implementation of the self-consistent charge density functional tight binding (SCC-DFTB) method. *The Journal of Physical Chemistry. B*, *105*(2), 569–585. http://dx.doi.org/10.1021/jp0029109.

de M. Seabra, G., Walker, R. C., Elstner, M., Case, D. A., & Roitberg, A. E. (2007). Implementation of the SCC-DFTB method for hybrid QM/MM simulations within the amber molecular dynamics package. *The Journal of Physical Chemistry. A*, *111*(26), 5655–5664. http://dx.doi.org/10.1021/jp070071l.

Ferrer, S., Ruiz-Pernía, J., Martí, S., Moliner, V., Tuñón, I., Bertrán, J., et al. (2011). Hybrid schemes based on quantum mechanics/molecular mechanics simulations goals to success, problems, and perspectives. *Advances in Protein Chemistry and Structural Biology, 85*, 81–142. http://dx.doi.org/10.1016/B978-0-12-386485-7.00003-X.

Foster, M. P., Wuttke, D. S., Radhakrishnan, I., Case, D. A., Gottesfeld, J. M., & Wright, P. E. (1997). Domain packing and dynamics in the DNA complex of the N-terminal zinc fingers of TFIIIA. *Nature Structural Biology*, *4*(8), 605–608. Retrieved from, http://www.ncbi.nlm.nih.gov/pubmed/9253405.

Gumpena, R., Kishor, C., Ganji, R. J., & Addlagatta, A. (2011). Discovery of α, β- and α, γ-diamino acid scaffolds for the inhibition of M1 family aminopeptidases. *ChemMedChem*, *6*(11), 1971–1976. http://dx.doi.org/10.1002/cmdc.201100298.

Hénin, J., & Chipot, C. (2004). Overcoming free energy barriers using unconstrained molecular dynamics simulations. *The Journal of Chemical Physics*, *121*(7), 2904–2914. http://dx.doi.org/10.1063/1.1773132.

Hernick, M., & Fierke, C. A. (2005). Zinc hydrolases: The mechanisms of zinc-dependent deacetylases. *Archives of Biochemistry and Biophysics*, *433*(1), 71–84. http://dx.doi.org/10.1016/j.abb.2004.08.006.

Hornak, V., Abel, R., Okur, A., Strockbine, B., Roitberg, A., & Simmerling, C. (2006). Comparison of multiple Amber force fields and development of improved protein backbone parameters. *Proteins*, *65*(3), 712–725. http://dx.doi.org/10.1002/prot.21123.

Huang, X., & Hernick, M. (2012). Examination of mechanism of N-acetyl-1-D-myo-inosityl-2-amino-2-deoxy-α-D-glucopyranoside deacetylase (MshB) reveals unexpected role for dynamic tyrosine. *The Journal of Biological Chemistry*, *287*(13), 10424–10434. http://dx.doi.org/10.1074/jbc.M111.320184.

Huang, X., Kocabas, E., & Hernick, M. (2011). The activity and cofactor preferences of N-acetyl-1-D-myo-inosityl-2-amino-2-deoxy-alpha-D-glucopyranoside deacetylase (MshB) change depending on environmental conditions. *The Journal of Biological Chemistry*, *286*(23), 20275–20282. http://dx.doi.org/10.1074/jbc.M111.234229.

Jarzynski, C. (1997). Nonequilibrium equality for free energy differences. *Physical Review Letters*, *78*(14), 2690–2693. http://dx.doi.org/10.1103/PhysRevLett.78.2690.

Jorgensen, W. L., Chandrasekhar, J., Madura, J. D., Impey, R. W., & Klein, M. L. (1983). Comparison of simple potential functions for simulating liquid water. *The Journal of Chemical Physics*, *79*(2), 926. http://dx.doi.org/10.1063/1.445869.

Kamerlin, S. C. L., & Warshel, A. (2010). At the dawn of the 21st century: Is dynamics the missing link for understanding enzyme catalysis? *Proteins*, *78*(6), 1339–1375. http://dx.doi.org/10.1002/prot.22654.

Koul, A., Arnoult, E., Lounis, N., Guillemont, J., & Andries, K. (2011). The challenge of new drug discovery for tuberculosis. *Nature*, *469*(7331), 483–490. http://dx.doi.org/10.1038/nature09657.

Kumar, A., Kumar, S., Kumar, D., Mishra, A., Dewangan, R. P., Shrivastava, P., et al. (2013). The structure of Rv3717 reveals a novel amidase from Mycobacterium tuberculosis. *Acta Crystallographica. Section D, Biological Crystallography*, *69*(Pt. 12), 2543–2554. http://dx.doi.org/10.1107/S0907444913026371.

Laio, A., & Parrinello, M. (2002). Escaping free-energy minima. *Proceedings of the National Academy of Sciences of the United States of America*, *99*(20), 12562–12566. http://dx.doi.org/10.1073/pnas.202427399.

Leach, A. R. (2001). *Molecular modelling: Principles and applications* (2nd ed.). Dorset Press, Dorchester, England: Pearson.

Liphardt, J., Dumont, S., Smith, S. B., Tinoco, I., & Bustamante, C. (2002). Equilibrium information from nonequilibrium measurements in an experimental test of Jarzynski's equality. *Science (New York, N.Y.)*, *296*(5574), 1832–1835. http://dx.doi.org/10.1126/science.1071152.

Masuyer, G., Schwager, S. L. U., Sturrock, E. D., Isaac, R. E., & Acharya, K. R. (2012). Molecular recognition and regulation of human angiotensin-I converting enzyme

(ACE) activity by natural inhibitory peptides. *Scientific Reports*, *2*:717. http://dx.doi.org/10.1038/srep00717.

Maynes, J. T., Garen, C., Cherney, M. M., Newton, G., Arad, D., Av-Gay, Y., et al. (2003). The crystal structure of 1-D-myo-inosityl 2-acetamido-2-deoxy-alpha-D-glucopyranoside deacetylase (MshB) from Mycobacterium tuberculosis reveals a zinc hydrolase with a lactate dehydrogenase fold. *The Journal of Biological Chemistry*, *278*(47), 47166–47170. http://dx.doi.org/10.1074/jbc.M308914200.

Mossessova, E., Bickford, L. C., & Goldberg, J. (2003). SNARE selectivity of the COPII coat. *Cell*, *114*(4), 483–495. Retrieved from, http://www.ncbi.nlm.nih.gov/pubmed/12941276.

Newton, G. L., Av-Gay, Y., & Fahey, R. C. (2000). N-Acetyl-1-D-myo-inosityl-2-amino-2-deoxy-alpha-D-glucopyranoside deacetylase (MshB) is a key enzyme in mycothiol biosynthesis. *Journal of Bacteriology*, *182*(24), 6958–6963. Retrieved from, http://www.pubmedcentral.nih.gov/articlerender.fcgi?artid=94821&tool=pmcentrez&rendertype=abstract.

Newton, G. L., Buchmeier, N., & Fahey, R. C. (2008). Biosynthesis and functions of mycothiol, the unique protective thiol of Actinobacteria. *Microbiology and Molecular Biology Reviews: MMBR*, *72*(3), 471–494. http://dx.doi.org/10.1128/MMBR.00008-08.

Park, S., Khalili-Araghi, F., Tajkhorshid, E., & Schulten, K. (2003). Free energy calculation from steered molecular dynamics simulations using Jarzynski's equality. *The Journal of Chemical Physics*, *119*(6), 3559. http://dx.doi.org/10.1063/1.1590311.

Pohorille, A., Jarzynski, C., & Chipot, C. (2010). Good practices in free-energy calculations. *The Journal of Physical Chemistry. B*, *114*(32), 10235–10253. http://dx.doi.org/10.1021/jp102971x.

Prigozhin, D. M., Mavrici, D., Huizar, J. P., Vansell, H. J., & Alber, T. (2013). Structural and biochemical analyses of Mycobacterium tuberculosis N-acetylmuramyl-L-alanine amidase Rv3717 point to a role in peptidoglycan fragment recycling. *The Journal of Biological Chemistry*, *288*(44), 31549–31555. http://dx.doi.org/10.1074/jbc.M113.510792.

Ramírez, C. L., Zeida, A., Jara, G. E., Roitberg, A. E., & Martí, M. A. (2014). Improving efficiency in SMD simulations through a hybrid differential relaxation algorithm. *Journal of Chemical Theory and Computation*, *10*(10), 4609–4617. http://dx.doi.org/10.1021/ct500672d.

Smith, C. R., Smith, G. K., Yang, Z., Xu, D., & Guo, H. (2010). Quantum mechanical/molecular mechanical study of anthrax lethal factor catalysis. *Theoretical Chemistry Accounts*, *128*(1), 83–90. http://dx.doi.org/10.1007/s00214-010-0765-z.

Tykvart, J., Schimer, J., Bařinková, J., Pachl, P., Poštová-Slavětínská, L., Majer, P., et al. (2014). Rational design of urea-based glutamate carboxypeptidase II (GCPII) inhibitors as versatile tools for specific drug targeting and delivery. *Bioorganic & Medicinal Chemistry*, *22*(15), 4099–4108. http://dx.doi.org/10.1016/j.bmc.2014.05.061.

Van der Kamp, M. W., & Mulholland, A. J. (2013). Combined quantum mechanics/molecular mechanics (QM/MM) methods in computational enzymology. *Biochemistry*, *52*(16), 2708–2728. http://dx.doi.org/10.1021/bi400215w.

Warshel, A., & Levitt, M. (1976). Theoretical studies of enzymic reactions: Dielectric, electrostatic and steric stabilization of the carbonium ion in the reaction of lysozyme. *Journal of Molecular Biology*, *103*(2), 227–249. http://dx.doi.org/10.1016/0022-2836(76)90311-9.

Wilder, P. T., Varney, K. M., Weiss, M. B., Gitti, R. K., & Weber, D. J. (2005). Solution structure of zinc- and calcium-bound rat S100B as determined by nuclear magnetic resonance spectroscopy. *Biochemistry*, *44*(15), 5690–5702. http://dx.doi.org/10.1021/bi0475830.

Woo, T. K., Margl, P., Blöchl, P. E., & Ziegler, T. (2002). Sampling phase space by a combined QM/MM ab initio Car–Parrinello molecular dynamics method with different

(multiple) time steps in the quantum mechanical (QM) and molecular mechanical (MM) domains. *The Journal of Physical Chemistry. A, 106*(7), 1173–1182. http://dx.doi.org/10.1021/jp0135860.
Wyckoff, T. J., Taylor, J. A., & Salama, N. R. (2012). Beyond growth: Novel functions for bacterial cell wall hydrolases. *Trends in Microbiology, 20*(11), 540–547. http://dx.doi.org/10.1016/j.tim.2012.08.003.
Xiong, H., Crespo, A., Marti, M., Estrin, D., & Roitberg, A. E. (2006). Free energy calculations with non-equilibrium methods: Applications of the Jarzynski relationship. *Theoretical Chemistry Accounts, 116*(1–3), 338–346. http://dx.doi.org/10.1007/s00214-005-0072-2.
Xu, D., & Guo, H. (2009). Quantum mechanical/molecular mechanical and density functional theory studies of a prototypical zinc peptidase (carboxypeptidase A) suggest a general acid-general base mechanism. *Journal of the American Chemical Society, 131*(28), 9780–9788. http://dx.doi.org/10.1021/ja9027988.
Yang, Y., Liu, C., Lin, Y.-L., & Li, F. (2013). Structural insights into central hypertension regulation by human aminopeptidase A. *The Journal of Biological Chemistry, 288*(35), 25638–25645. http://dx.doi.org/10.1074/jbc.M113.494955.
Zhang, C., Wu, S., & Xu, D. (2013). Catalytic mechanism of angiotensin-converting enzyme and effects of the chloride ion. *The Journal of Physical Chemistry. B, 117*(22), 6635–6645. http://dx.doi.org/10.1021/jp400974n.
Zheng, L., Chen, M., & Yang, W. (2008). Random walk in orthogonal space to achieve efficient free-energy simulation of complex systems. *Proceedings of the National Academy of Sciences of the United States of America, 105*(51), 20227–20232. http://dx.doi.org/10.1073/pnas.0810631106.
Zimmermann, M., Clarke, O., Gulbis, J. M., Keizer, D. W., Jarvis, R. S., Cobbett, C. S., et al. (2009). Metal binding affinities of Arabidopsis zinc and copper transporters: Selectivities match the relative, but not the absolute, affinities of their amino-terminal domains. *Biochemistry, 48*(49), 11640–11654. http://dx.doi.org/10.1021/bi901573b.

CHAPTER THREE

A Practical Quantum Mechanics Molecular Mechanics Method for the Dynamical Study of Reactions in Biomolecules

Jesús I. Mendieta-Moreno*,†, Iñigo Marcos-Alcalde†, Daniel G. Trabada*, Paulino Gómez-Puertas†, José Ortega*,[1], Jesús Mendieta†,‡

*Departamento de Física Teórica de la Materia Condensada and Condensed Matter Physics Center (IFIMAC), Universidad Autónoma de Madrid, Madrid, Spain
†Molecular Modelling Group, Center of Molecular Biology "Severo Ochoa" (CSIC-UAM), Madrid, Spain
‡Biomol-Informatics SL, Campus UAM, Madrid, Spain
[1]Corresponding author: e-mail address: jose.ortega@uam.es

Contents

Abstract

Quantum mechanics/molecular mechanics (QM/MM) methods are excellent tools for the modeling of biomolecular reactions. Recently, we have implemented a new QM/MM method (Fireball/Amber), which combines an efficient density functional theory method (Fireball) and a well-recognized molecular dynamics package (Amber), offering an excellent balance between accuracy and sampling capabilities. Here, we present a detailed explanation of the Fireball method and Fireball/Amber implementation. We also discuss how this tool can be used to analyze reactions in biomolecules using steered molecular dynamics simulations. The potential of this approach is shown by the analysis of a reaction catalyzed by the enzyme triose-phosphate isomerase (TIM). The

Advances in Protein Chemistry and Structural Biology, Volume 100
ISSN 1876-1623
http://dx.doi.org/10.1016/bs.apcsb.2015.06.003

conformational space and energetic landscape for this reaction are analyzed without *a priori* assumptions about the protonation states of the different residues during the reaction. The results offer a detailed description of the reaction and reveal some new features of the catalytic mechanism. In particular, we find a new reaction mechanism that is characterized by the intramolecular proton transfer from O1 to O2 and the simultaneous proton transfer from Glu 165 to C2.

1. INTRODUCTION

One of the more interesting challenges at present in Biophysics and Molecular Biology is to understand how the proteins work at the atomic level. Processes such as conformational changes (Karplus & McCammon, 2002) or ligand binding (Gilson & Zhou, 2007) can be studied by molecular mechanics (MM) simulations using empirical force fields. Biomolecular reactions, however, require chemical bonds to be broken and formed, processes that are not properly described using MM force fields. Chemical bonds are intrinsically related to the quantum electronic structure of the material and therefore quantum mechanics (QM) techniques are required in these cases (Martin, 2004; Marx & Hutter, 2009). Unfortunately, the computational cost of QM calculations for biomolecular systems is prohibitive (Adcock & Mccammon, 2006). By describing the active region with a QM method and the surroundings with a MM method, QM/MM approaches offer the required accuracy to model biomolecular reactions with reasonable computational costs (Field, Bash, & Karplus, 1990; Senn & Thiel, 2009; Warshel & Levitt, 1976).

The accuracy and computational efficiency of QM/MM calculations critically depend on the QM method used. In most of the biomolecular applications, a detailed study of the conformational space for the reaction is essential to understand, e.g., the catalytic mechanisms. Computationally efficient QM methods, such as semi-empirical approaches, enable wider conformational sampling (Martín-García, Mendieta-Moreno, López-Viñas, Gómez-Puertas, & Mendieta, 2012), but usually do not offer enough accuracy. Conversely, high-accuracy quantum chemistry methods greatly increase the computational costs, significantly reducing the conformational sampling capabilities (Zhang, 2005). Recently, we have implemented a new QM/MM technique (Mendieta-Moreno et al., 2014) based on the combination of Fireball (QM) (Demkov, Ortega, Sankey, & Grumbach, 1995; Jelínek, Wang, Lewis, Sankey, & Ortega, 2005; Lewis et al., 2001, 2011;

Sankey & Niklewski, 1989) and AMBER (MM) (Case et al., 2012; Salomon-Ferrer, Case, & Walker, 2012). FIREBALL is a density functional theory (DFT) molecular dynamics (MD) technique that is very well suited for QM/MM biomolecular applications. This method allows molecular dynamics (MD) simulations involving relatively large QM regions ($\sim 10^2$ atoms) with good accuracy (Mendieta-Moreno et al., 2014). AMBER is a well-recognized MD package for the simulation of biomolecules which offers a wide range of MD tools. In particular, the *sander* program included in AMBER offers powerful techniques for exhaustive sampling of the conformational space using biased MD approaches (umbrella sampling, replica exchange, nudged elastic band, targeted MD, steered MD, ...), which can be used directly in the QM/MM applications (Case et al., 2012).

This chapter describes the FIREBALL/AMBER implementation and highlights its most interesting features. In Section 2, detailed descriptions of FIREBALL and the FIREBALL/AMBER implementation are provided. This method can be used to analyze reactions in biomolecules sampling the conformational space by means of long QM/MM MD simulations. As explained in Section 3, we use steered MD to generate of the order of $\sim 10^6$ individual structures that allow us to calculate free energy maps for the reaction. In Section 4, we analyze the reversible interconversion between dihydroxyacetone phosphate (DHAP) and glyceraldehyde 3-phosphate (GAP) catalyzed by the triose-phosphate isomerase (TIM) using these techniques. Finally, in Section 5 we present our conclusions.

2. DESCRIPTION OF THE METHOD

2.1 QM Method: FIREBALL

In QM/MM MD methods, we need to obtain the quantum electronic structure of the system for each time step ($t_i = m_i \Delta t$; $\Delta t \sim 10^{-15}$ s) in the simulation. This is the reason why these methods are so computationally demanding. The most practical QM/MD methods are based on DFT that offers a considerable simplification of the quantum many-body problem (Hohenberg & Kohn, 1964; Kohn & Sham, 1965; Martin, 2004). In the following, we will use the acronym DFT/MM to specify a QM/MM technique in which a DFT method is used for the QM calculation. FIREBALL is a DFT method that is specifically designed for the study of complex systems using MD simulations (Demkov et al., 1995; Jelínek et al., 2005; Lewis et al., 2001, 2011; Sankey & Niklewski, 1989). In this method, the electronic structure is obtained solving the DFT equations (see below) using a basis set of atomic-like orbitals localized in the atoms of the system. These orbitals are short-ranged, numerical, and are optimized so that reasonably good accuracy can be obtained with a low number of orbitals per atom (Basanta, Dappe, Jelínek, & Ortega, 2007; Sankey & Niklewski, 1989). Recently, we have developed basis sets of optimized atomic-like orbitals for biomolecular systems containing H, C, N, O, P, S, and Mg atoms. In particular, our minimal basis set for biomolecules includes one s orbital for H; sp^3 orbitals for C, N, and O; and sp^3d^5 orbitals for P, S, and Mg. In order to optimize the basis set, we have analyzed different sets of organic molecules relevant for biological systems and have considered intermolecular energies and distances as well as intramolecular distances. In particular, we have used the S66 reference dataset (Řezáč, Riley, & Hobza, 2011) for intermolecular energies and distances as well as the ionic hydrogen bond dataset (Řezáč & Hobza, 2012) for complexes featuring ionic hydrogen bonds (hydrogen bonds between ionic and neutral groups). Details will be published elsewhere.

Most DFT methods use periodic boundary conditions in some way (plane-waves basis sets, representation of the potential and electron density on a periodic grid, use of fast Fourier transforms, etc.), since these techniques were first developed in Solid State Physics to analyze the paradigmatic case of the crystalline solid with atoms arranged in a periodic structure. In biomolecular systems, however, the atoms are not arranged in any periodic way and there is no advantage to use any artificial periodicity in the calculations. FIREBALL is a DFT technique completely formulated in *real-space*, i.e., it does not require the use of supercells (periodic boundary conditions). Due to this property, FIREBALL is well suited for biomolecular simulations in combination with MM calculations (that are also real-space methods).

In the FIREBALL method, the electronic structure is obtained in each time step of the simulation via solution of the DFT Schrödinger-like equation (Martin, 2004):

$$\hat{H}\psi_i(\mathbf{r}) = \left(-\frac{\hbar^2}{2m_e}\nabla^2 + V_{\text{ion}}(\mathbf{r}) + V_{\text{ee}}(\mathbf{r}) + V_{\text{xc}}(\mathbf{r})\right)\psi_i(\mathbf{r}) = \epsilon_i\psi_i(\mathbf{r}). \quad (1)$$

The first term represents the kinetic energy of the electrons, V_{ion}, is the ionic potential:

$$V_{\text{ion}}(\mathbf{r}) = \sum_\alpha \frac{-eZ_\alpha}{|\mathbf{r} - \mathbf{R}_\alpha|}, \quad (2)$$

(eZ_α is the nuclear or pseudopotential charge on atom α at position $\mathbf{R}_\alpha$; $-e$ is the charge of the electron),

$$V_{ee}(\mathbf{r}) = e\int \frac{\rho(\mathbf{r}')}{|\mathbf{r}-\mathbf{r}'|}d^3r' \tag{3}$$

is the average electron potential (i.e., the Hartree potential). Finally, the exchange-correlation potential V_{xc} is the functional derivative of the DFT exchange-correlation energy $E_{xc}[\rho(\mathbf{r})]$, $V_{xc}=\delta E_{xc}/\delta\rho(\mathbf{r})$ (Martin, 2004). In these equations, $\rho(\mathbf{r})$ is the electron density (Foulkes & Haydock, 1989; Harris, 1985), which is determined in a self-consistent way in terms of the orbital occupation numbers (Demkov et al., 1995).

In the FIREBALL method, a basis set of numerical atomic-like orbitals, ϕ_μ, is used to solve Eq. (1) (Basanta et al., 2007; Sankey & Niklewski, 1989):

$$\psi_i(\mathbf{r}) = \sum_\mu c_{i\mu}\phi_\mu(\mathbf{r}-\mathbf{R}_\alpha); \tag{4}$$

μ is a combined index, $\mu \equiv (\alpha,l,m)$ that refers to the particular atom, α (at $\mathbf{R}_\alpha$), atomic subshell l (e.g., 3s,4s,3p,3d,), and angular component m (e.g., p_x,p_y,p_z). The electron density $\rho(\mathbf{r})$ is also written in terms of the orbitals ϕ_μ as a sum of atomic-like densities, $\rho_\alpha(\mathbf{r})$, for the different atoms in the system (Demkov et al., 1995):

$$\rho(\mathbf{r}) = \sum_\alpha \rho_\alpha(\mathbf{r}) = \sum_\mu n_\mu|\phi_\mu(\mathbf{r}-\mathbf{R}_\alpha)|^2. \tag{5}$$

In practice, the atomic densities ρ_α are approximated to be spherically symmetric around each atomic site $\mathbf{R}_\alpha$. The electron density $\rho(\mathbf{r})$ in Eq. (5) is determined in a self-consistent way in terms of the orbital occupation numbers n_μ (Demkov et al., 1995; García-Vidal et al., 1994; Schönhammer, Gunnarsson, & Noack, 1995). For this purpose, output orbital occupation numbers n_μ^{out} are obtained from the output electron density $\rho^{\text{out}}(\mathbf{r})$ as obtained from the occupied eigenvectors ψ_i in Eq. (1):

$$\rho^{\text{out}}(\mathbf{r}) = \sum_{i\in\text{occ}} |\psi_i(\mathbf{r})|^2 \to \sum_\mu n_\mu^{\text{out}}|\phi_\mu(\mathbf{r}-\mathbf{R}_\alpha)|^2. \tag{6}$$

Different methods can be used to project [arrow in Eq. (6)] the output electron density from Eq. (1) into the form given by Eq. (5) using, for example, Löwdin Orbitals (Carlson & Keller, 1957; Demkov et al., 1995; Löwdin, 1950) or Natural Atomic Orbitals (Reed, Weinstock, & Weinhold,

1985). Equations (1–6) are iteratively solved until self-consistency: in the self-consistent solution $n_\mu^{\rm out}$ and the input orbital occupation numbers n_μ must coincide (Demkov et al., 1995). In each of these iterations, the eigenvectors ψ_i are obtained from the diagonalization of the $N \times N$ Hamiltonian matrix in Eq. (1), $H_{\mu,\nu} = \langle \phi_\mu | \hat{H} | \phi_\mu \rangle$, where N is the total number of orbitals ϕ_μ in the basis set.

Once these equations have been solved, the total energy and forces are calculated as

$$E_{\rm tot}[\rho(\mathbf{r})] = E_{\rm sum} - E_{\rm ee}[\rho(\mathbf{r})] + E_{\rm xc}[\rho(\mathbf{r})] - \int \rho(\mathbf{r}) V_{\rm xc}[\rho(\mathbf{r})] d^3r + E_{\rm ion-ion}; \tag{7}$$

$$\mathbf{F}_\alpha = -\frac{\partial E_{\rm tot}}{\partial \mathbf{R}_\alpha}. \tag{8}$$

In Eq. (7), $E_{\rm sum}$ is a sum over the occupied eigenvalues, $E_{\rm sum} = \sum_{i \in {\rm occ}} \epsilon_i$, of the self-consistent solution of Eq. (1), $E_{\rm ee}$ is the average electron–electron interaction energy:

$$E_{\rm ee}[\rho(\mathbf{r})] = \frac{e^2}{2} \int \frac{\rho(\mathbf{r})\rho(\mathbf{r}')}{|\mathbf{r} - \mathbf{r}'|} d^3r d^3r', \tag{9}$$

$E_{\rm ion-ion}$ is the ion–ion interaction energy:

$$E_{\rm ion-ion} = \frac{e^2}{2} \sum_{i,j} \frac{Z_\alpha Z_\beta}{|\mathbf{R}_\alpha - \mathbf{R}_\beta|}, \tag{10}$$

and $E_{\rm xc}[\rho(\mathbf{r})]$ is the exchange-correlation energy (Martin, 2004).

The computational efficiency of FIREBALL is related to both the use of optimized basis sets of atomic-like orbitals and the representation of $\rho(\mathbf{r})$ as a sum of atomic-like densities[1] (Eq. 5). In this way, four-center integrals are not required for the solution of Eq. (1) or the calculation of $E_{\rm tot}$ and the forces $\mathbf{F}_\alpha$, and all the two- and three-center interactions are tabulated beforehand and placed in interpolation data tables which are no larger than two-dimensional (Lewis et al., 2011; Sankey & Niklewski, 1989). A practical tabulation-interpolation scheme is used, and all the matrix elements required during the MD simulation are evaluated by looking up the necessary information from the data tables, which are read at the beginning

[1] This approximation is related to the use of a self-consistent version of the Harris–Foulkes functional (Demkov et al., 1995; Foulkes & Haydock, 1989; Harris, 1985).

of the calculation and stored in memory throughout the MD simulation. In order to also use this tabulation-interpolation scheme for the calculation of the exchange-correlation contributions, which are highly nonlinear, different approximations have been developed (Horsfield, 1997; Jelínek et al., 2005; Lewis et al., 2001; Sankey & Niklewski, 1989).

Regarding the calculation of the forces $\mathbf{F}_\alpha$, Eq. (8), the most difficult term is the contribution from E_{sum}. Using a variant of the Hellmann–Feynman theorem (Sankey & Niklewski, 1989), this contribution is calculated in terms of the derivatives of the Hamiltonian matrix elements, $H_{\mu,\nu}$. These matrix elements are precalculated and stored in data tables and thus the required derivatives $\partial H_{\mu,\nu}/\partial \mathbf{R}_\alpha$ are readily available (see Sankey & Niklewski, 1989 for details).

2.2 FIREBALL/AMBER

In QM/MM methods, the system is divided into a chemically active region (e.g., the reactant molecules and catalytic site residues) and the *environment* (i.e., rest of the protein, water molecules, ions). The active region (or QM region) typically contains about 10^2 atoms and is described using a QM approach. For practical reasons, the environment ($\sim 10^5$–10^6 atoms), or MM region, is described using a MM method. In the study reactions in biomolecules, it is particularly important to properly take into account the effect of the electrostatic potential of the environment on the electronic structure of the active region, so we pay special attention to this contribution in the following discussion. In our FIREBALL/AMBER method (Mendieta-Moreno et al., 2014), we consider the electrostatic interaction between the active region and the environment at the QM level, including in the QM Hamiltonian $H_{\mu,\nu}$ the electrostatic potential due to the force-field partial charges, eq_k, on the MM atoms. This yields:

$$H_{\mu,\nu} = H^{\text{A}}_{\mu,\nu} + \delta H^{\text{E}}_{\mu,\nu} \tag{11}$$

$H^{\text{A}}_{\mu,\nu}$ is the contribution of the atoms in the active region to the Hamiltonian matrix element, and $\delta H^{\text{E}}_{\mu,\nu}$ the electrostatic potential contribution due to the MM partial charges eq_k in the environment:

$$\delta H^{\text{E}}_{\mu,\nu} = \sum_k h^k_{\mu,\nu} = \sum_k -eq_k \int \frac{\phi_\mu(\mathbf{r}-\mathbf{R}_\alpha)\phi_\nu(\mathbf{r}-\mathbf{R}_\beta)}{|\mathbf{r}-\mathbf{R}_k|} d^3\mathbf{r} \tag{12}$$

$\phi_\mu(\mathbf{r} - \mathbf{R}_\alpha)$ and $\phi_\nu(\mathbf{r} - \mathbf{R}_\beta)$ are atomic-like basis set orbitals centered on the atoms placed at $\mathbf{R}_\alpha$ and $\mathbf{R}_\beta$, and $h^k_{\mu,\nu}$ is the contribution to the Hamiltonian

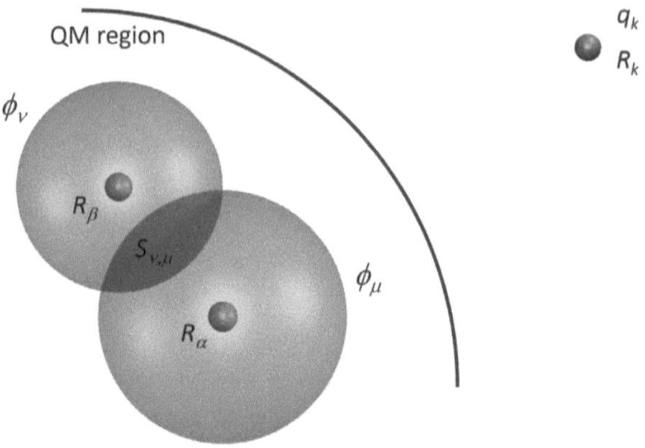

Figure 1 Scheme representing the electrostatic interaction between the MM charge, eq_k, at $\mathbf{R}_k$ and the overlap charge, $-eS_{\mu,\nu}$, yielding the contribution $h^k_{\mu,\nu}$ to the QM Hamiltonian matrix element $H_{\mu,\nu}$ between orbitals $\phi_\mu(\mathbf{r}-\mathbf{R}_\alpha)$ and $\phi_\nu(\mathbf{r}-\mathbf{R}_\beta)$ (see Eqs. 12 and 13).

matrix element $H_{\mu,\nu}$ due to the electrostatic potential of a point charge eq_k at $\mathbf{R}_k$ (see Fig. 1). For practical reasons (Demkov et al., 1995), $h^k_{\mu,\nu}$ is approximated as:

$$h^k_{\mu,\nu} = -eq_k\left(\frac{g_\mu}{|\mathbf{R}_k - \mathbf{R}_\alpha|} + \frac{g_\nu}{|\mathbf{R}_k - \mathbf{R}_\beta|}\right), \tag{13}$$

with *effective overlap charges* g_μ, g_ν

$$g_\mu = \frac{1}{2}S_{\mu,\nu} - \frac{p_{\mu,\nu}}{|\mathbf{R}_\alpha - \mathbf{R}_\beta|}, \tag{14}$$

$$g_\nu = \frac{1}{2}S_{\mu,\nu} + \frac{p_{\mu,\nu}}{|\mathbf{R}_\alpha - \mathbf{R}_\beta|}; \tag{15}$$

$S_{\mu,\nu}$ is the overlap:

$$S_{\mu,\nu} = \int \phi_\mu(\mathbf{r} - \mathbf{R}_\alpha)\phi_\nu(\mathbf{r} - \mathbf{R}_\beta)d^3\mathbf{r}, \tag{16}$$

and $p_{\mu,\nu}$ is the component along the $(\mathbf{R}_\beta - \mathbf{R}_\alpha)$ direction of the dipole moment $\mathbf{P}_{\mu,\nu}$ with respect to the midpoint, $\mathbf{R}_\mathrm{m}$, between atoms α and β:

$$\mathbf{P}_{\mu,\nu} = \int [\mathbf{r} - \mathbf{R}_\mathrm{m}]\phi_\mu(\mathbf{r} - \mathbf{R}_\alpha)\phi_\nu(\mathbf{r} - \mathbf{R}_\beta)d^3\mathbf{r} \tag{17}$$

$(\mathbf{R}_\mathrm{m} = (\mathbf{R}_\alpha + \mathbf{R}_\beta)/2)$. Physically, the approximation of Eq. (13) includes monopole and dipole far-field effects (Demkov et al., 1995).

The electrostatic potential due to the environment also interacts with the nuclear (or pseudopotential) charges of the QM atoms, eZ_α, yielding the following contribution to the total energy (Mendieta-Moreno et al., 2014):

$$E^{\text{nuc}}_{\text{QM-MM}} = \sum_{\alpha,k} \frac{e^2 q_k Z_\alpha}{|\mathbf{R}_k - \mathbf{R}_\alpha|}. \tag{18}$$

In addition to the electrostatic term, the interaction between the QM and MM regions also presents bonded and nonbonded contributions, in similarity with the MM force field (Senn & Thiel, 2009). In most of the applications, the frontier between the QM and MM regions goes through covalent bonds, resulting in bonds that need to be properly saturated in the QM calculation. In the calculations presented in Section 4, we have used the link atom method (Field, Albe, Bret, Proust-De Martin, & Thomas, 2000; Walker, Crowley, & Case, 2008), adding H atoms to saturate the broken bonds at the frontier between the QM and MM regions. The nonbonded van der Waals interactions between QM and MM atoms are described using the Lennard-Jones parameters from the MM force field, and the long-range electrostatic interactions are calculated using the Particle Mesh Ewald method (Walker et al., 2008). We have used Löwdin Orbitals to determine the electron charges in the QM region, n_μ. Finally, in the simulations presented in Section 4, we have used the BLYP exchange-correlation functional (Becke exchange (Becke, 1988) with Lee–Yang–Parr correlation (Lee, Yang, & Parr, 1988)); these exchange-correlation contributions are calculated with the help of the multicenter weighted exchange-correlation density approximation (McWEDA) (Jelínek et al., 2005).

3. DYNAMICAL ANALYSIS OF REACTIONS IN BIOMOLECULES

Biomolecular systems have a large number of atoms and work at physiological temperature. Therefore, they present a high number of degrees of freedom, and for a given macroscopic condition, there is a huge number of different accessible structures. In principle, MD simulations allow us to take into account the high number of degrees of freedom in these systems. However, long DFT/MM MD simulations for biomolecular systems are still rare due to their computational cost; DFT/MM studies of reactions in biomolecules are usually performed using minimization techniques. Note that this approach assumes that a single energy-minimized structure represents each state, not taking into account the plethora of different accessible structures.

In order to analyze a given biomolecular reaction, we must first determine the geometrical coordinates that are related to the key structural changes taking place in the reaction. The values of these coordinates define different states of the system, each of which corresponding to an ensemble of different *microscopic* structures. Therefore, instead of a single atomic structure and energy value, an ensemble of different structures and energy values is associated with each state. Thus, we have to analyze the variation of the free energy as a function of the chosen reaction coordinates. Alternatively, the corresponding potential of mean force can also be used. For this purpose, several methodologies, such as WHAM, thermodynamic integration, metadynamics, free energy perturbation, are available (Jorge, Garrido, Queimada, Economou, & MacEdo, 2010; Kumar, Rosenberg, Bouzida, Swendsen, & Kollman, 1992; Laio & Parrinello, 2002; Park, Khalili-Araghi, Tajkhorshid, & Schulten, 2003; Zwanzig, 1954).

In DFT/MM studies of enzymatic reactions, these statistical considerations can play an important role but are infrequently taken into account (principally when a large portion of the conformational space is to be studied) due to the high computational cost of DFT methods. The use of semiempirical QM methods such as AM1 and PM3 (Stewart, 2004) can reduce this cost and make a wider conformational sampling possible, but they have a low accuracy, especially for the calculation of barriers involved in reactions. The Fireball/Amber implementation allows us to sample large portions of the conformational space with a higher accuracy and a similar computational cost, using DFT/MM MD simulations.

In our approach, we typically define two director geometrical parameters which adequately represent the different states of the process under study. For example, if an enzymatic reaction is essentially characterized by the formation of one bond and the disruption of another, the bond distances for these two bonds could be used as reaction coordinates (see, e.g., Mendieta-Moreno et al., 2014). Once the appropriate reaction coordinates have been defined, we explore the conformational space using an adaptively biased MD approach that presents some characteristics of steered MD as well as umbrella sampling. In this approach, we generate several MD trajectories where one of the reaction coordinates is constrained with an appropriate harmonic potential while the other reaction coordinate is adiabatically changed by means of another harmonic potential whose position is slowly moving with constant speed during the simulation time, as in steered MD. Combining a set of equidistant parallel MD trajectories, we can perform a highly dense sampling of the conformational space of interest. In a typical calculation, we obtain around 2×10^6 individual structures, each with their own values for the energy and set of reaction coordinates.

In the next step, we create a uniform grid of points in the plane defined by the two reaction coordinates and assign each of the individual structures in our sampling to the closest grid point. In this way, each grid point is associated to a group of atomic structures; typically, each of these groups contains $\sim 10^3$ different atomic structures, on average. With all this information, the free energy landscape for the reaction is calculated as follows. First, we calculate the value of the function Q for each point (x,y) on the grid

$$Q(x,y) = \frac{1}{N_i(x,y)} \sum_i \exp\left(\frac{-E_i(x,y)}{k_B T}\right), \tag{19}$$

x and y are the values of the reaction coordinates, $E_i(x,y)$ are the potential energy values for all the atomic structures corresponding to the grid point (x,y), and $N_i(x,y)$ is the number of atomic structures in the sum. The free energy landscape is then obtained as

$$F(x,y) = -k_B T \ \ln[Q(x,y)]. \tag{20}$$

Alternatively, the energetic landscape can also be obtained from the following potential of mean force:

$$P(x,y) = \frac{\sum_i E_i(x,y) \exp\left(\frac{-E_i(x,y)}{k_B T}\right)}{\sum_i \exp\left(\frac{-E_i(x,y)}{k_B T}\right)}. \tag{21}$$

The final map (see, e.g., Fig. 5) is obtained applying a 3D local regression (LOESS) method to the data to remove the irregularities inherent in the statistical sampling data (Cleveland & Devlin, 1986), since the variations in this energetic landscape should be smooth as a function of the reaction coordinates.

4. CATALYTIC MECHANISM OF TIM

4.1 Introduction

As an example of our approach to analyze enzymatic reactions, we study the reversible interconversion between DHAP and GAP catalyzed by the TIM. This enzyme has been extensively studied by many different techniques including theoretical calculations. However, some details of its catalytic mechanisms are still not fully elucidated. It is generally accepted that the transfer of a proton from C1 atom of DHAP to the glutamic 165 yields an enediolate intermediate which can be detected also in the nonenzymatic reaction (Hall & Knowles, 1975). However, there is no such general agreement about the catalytic mechanism by which the enediolate intermediate yields glyceraldehyde, the final product (see Fig. 2).

Several hypotheses have been proposed for the proton transfer paths during the second step of the reaction. In one of the proposed paths (Bash et al., 1991), His 95 could act as a catalytic acid donating a proton to the carboxyl oxygen in C2. After this step, the proton of the alcohol group in O1 is transferred to His 95 and C2 accepts the proton from Glu 165 recovering the enzyme the original configuration. RMN data support, however, that the imidazole ring of His 95 is not charged over the entire pH range of the enzymatic activity (Lodi & Knowles, 1991); thus, the proposed path would imply the formation of an imidazolate anion, in spite of the very high pK_a value ($\approx$14) associated with this process (Walba & Isensee, 1961). Other paths have been proposed in order to avoid the formation of the energetically unfavorable imidazolate anion, including the transfer of the proton from Glu 165 to the carboxyl oxygen in C2 and the subsequent protonation of C2 from the C1 alcohol group through Glu 165 (Cui & Karplus, 2002). Other residues such as Lys 12 (Go, Amyes, & Richard, 2010) and Glu 97

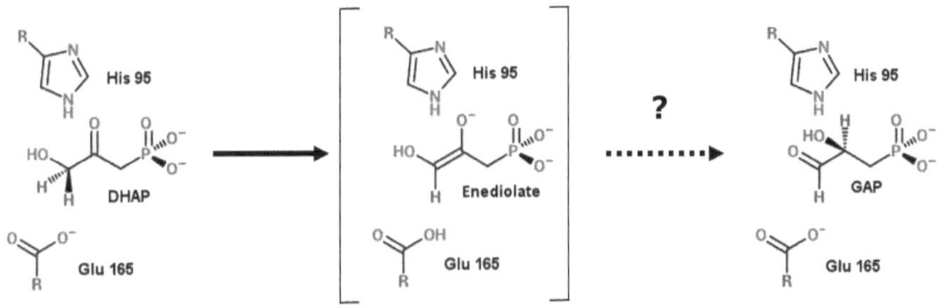

Figure 2 Scheme for the conversion of DHAP to GAP catalyzed by TIM.

(Samanta, Murthy, Balaram, & Balaram, 2011) have also been proposed to be involved in the catalytic mechanism to avoid the formation of the imidazole anion. The internal proton transfer from O1 to O2 without involving other groups has also been considered, but it seems to be energetically unfavorable in relation with other paths, at least while Glu 165 stays protonated (Cui & Karplus, 2002).

The uncertainties about the proton transfer mechanisms could be due to the fact that the energetic profiles obtained for different proton transfer paths were calculated by means of energy-minimization techniques in which the protonation state was defined *a priori*. The use of MD techniques to calculate the potential of mean force for a conformational space defined by reaction coordinates that do not presuppose any particular protonation state for the different residues can help to clarify the proton transfer mechanisms. The use of MD simulations (e.g., at $T=300$ K) can also help to take into account important dynamical effects, such as the formation of low-barrier hydrogen bonds where the hydrogen can freely move in the space between the two heavy atoms (Cleland & Kreevoy, 1994; Ishikita & Saito, 2014).

During the reaction catalyzed by TIM, the orbital hybridization of the carbon atom C1 changes from sp^3 in DHAP to sp^2 in GAP, while the orbital hybridization of the carbon atom C2 changes from sp^3 to sp^2 (see Fig. 3). The molecular orbital rearrangement modifies the geometry of C1 and C2 carbons from planar (sp^2) to tetrahedral (sp^3), and vice versa (Fig. 3). Therefore, the dihedral angles for the C1 and C2 atoms are a good choice of reaction coordinates to monitor the reaction. Using these parameters to define the conformational space, no assumption is made about the protonation state of the different residues involved in the reaction.

4.2 Results

After 10 ns of classical MD stabilization ($T=300$ K) of the system based on the X-ray structure of TIM (pdb: 1NEY) (Jogl, Rozovsky, McDermott, & Tong, 2003), we define the atoms to be included in the QM region (see Fig. 4). This region includes the substrate (DHAP); the side chain of catalytic residues Glu 165, His 95; the main chain amino group of residue 97 that stabilizes His 95; the main chain amino group of residues 171, 211, and 233 which stabilize the phosphate group of DHAP; and the side chain of Lys 12. This last side chain has been included in the QM region because the mutation of this residue has a dramatic effect on the catalysis

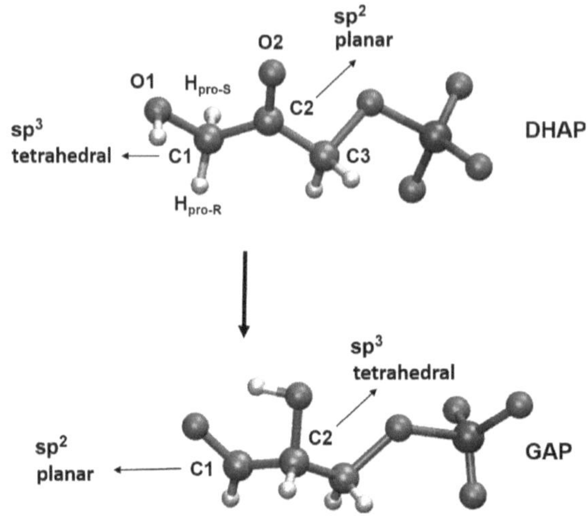

Figure 3 Orbital hybridization and geometry of dihydroxyacetone phosphate (DHAP) and glyceraldehyde 3-phosphate (GAP). The atoms involved in the two dihedral angles used as reaction coordinates are O1–C2–$H_{pro\text{-}S}$–C1 for the C1 dihedral angle and C1–C3–O2–C2 for the C2 dihedral angle.

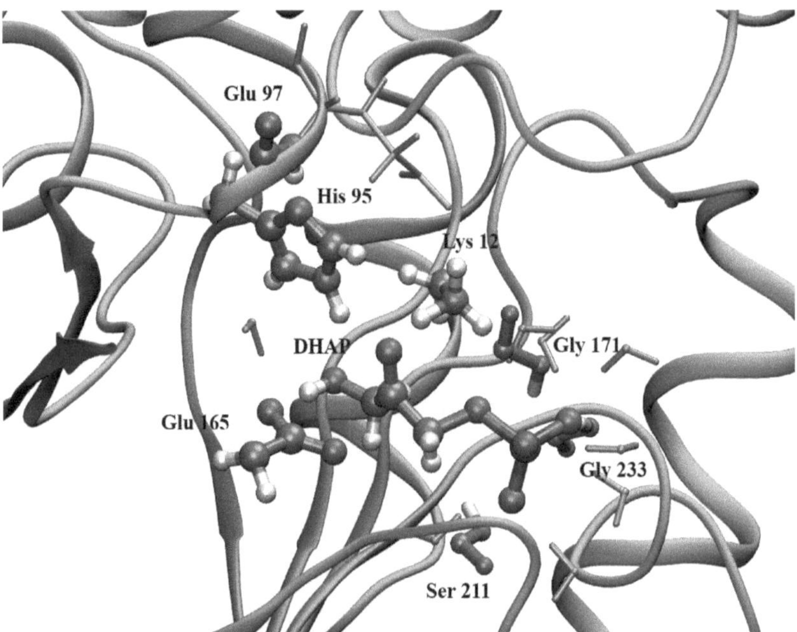

Figure 4 Active center for the reaction catalyzed by triose-phosphate isomerase. Atoms in QM region are represented in ball and sticks, the rest of the protein is presented in gray, and some solvent molecules present in the active site in blue (gray in the print version).

modifying the ratio K_{cat}/K_m (Go, Koudelka, Amyes, & Richard, 2010). Figure 4 shows a global vision of the QM region inside the protein structure.

In order to adapt the system from classical MD to DFT/MM MD, a 100-ps nonrestraint DFT/MM MD simulation was performed. Once the system is stable, we can initiate the sampling of the conformational space defined by the C1 and C2 dihedral angles (α_1 and α_2, respectively) previously established. Figure 5 shows the energetic landscape for the reaction catalyzed by TIM obtained using the techniques described in Sections 2 and 3. This landscape presents three energy minima, which can be associated with the substrate (DHAP), the intermediate state, and the product of the reaction (GAP), respectively.

The lowest energy basin corresponds to DHAP, the initial state of the reaction. The energy value of this minimum was taken as energy reference. This minimum is centered around $\alpha_1 = 30°$, which corresponds to sp^3 hybridization for the C1 atom. However, as a result of the interaction

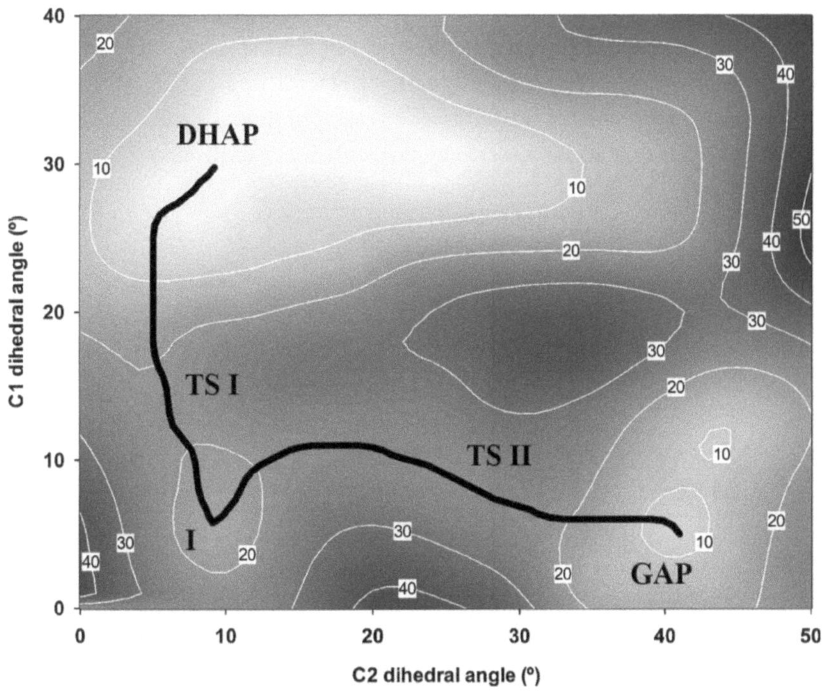

Figure 5 Free energy landscape in kcal/mol of TIM-catalyzed reaction between DHAP and GAP. The *x* and *y* axes are the two reaction coordinates, i.e., the C2 and C1 dihedral angles, α_2 and α_1, respectively. The most probable pathway is represented as a cyan (black in the print version) line.

between the negatively charged O2 atom of DHAP and some residues in the active site (Lys 12 and His 95), the range of values covered by the C2 dihedral angle α_2 in this basin is quite extensive. The second minimum presents C1 and C2 dihedral angle values corresponding to an sp^2 hybridization, $\alpha_2,\alpha_1 \sim 5°–10°$. This corresponds to the generally accepted enediolate intermediate of the reaction (see Fig. 2). The third minimum is centered around $\alpha_1 = 5°$ and $\alpha_2 = 40°$. These values are compatible with sp^2 hybridization for the C1 atom and sp^3 hybridization for the C2 atom, as corresponds to GAP, the final state of the reaction. Figure 5 also shows the lower energy-barrier path that corresponds with the most probable reaction pathway. Figure 6 shows the energy profile corresponding to this path.

The reaction path can be divided into two different steps. The first step goes from the initial state to the enediolate intermediate state through a 21-kcal/mol energy barrier. A similar barrier for this first step has been obtained in a free energy computational study (Hu, Lu, & Yang, 2007). The second step goes from the intermediate state to GAP, the reaction product. The energy barrier for this second step presents a value of 23 kcal/mol. We have also performed a DFT/MM MD simulation along our reaction path (using steered MD) in order to have a further insight into the reorganization of atoms during the catalytic mechanism (see Fig. 7).

The first step is mainly related to the rearrangement of the orbitals of the C1 carbon, which can be described through the variation of α_1 from 30°

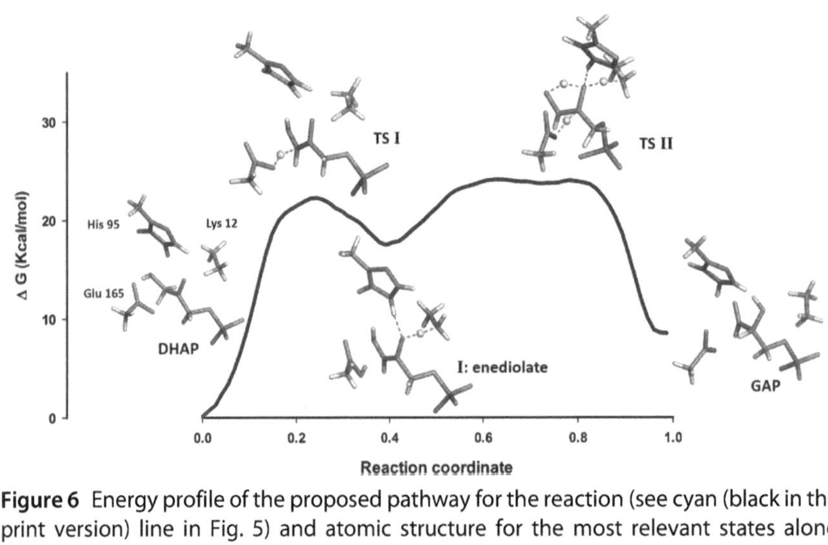

Figure 6 Energy profile of the proposed pathway for the reaction (see cyan (black in the print version) line in Fig. 5) and atomic structure for the most relevant states along the path.

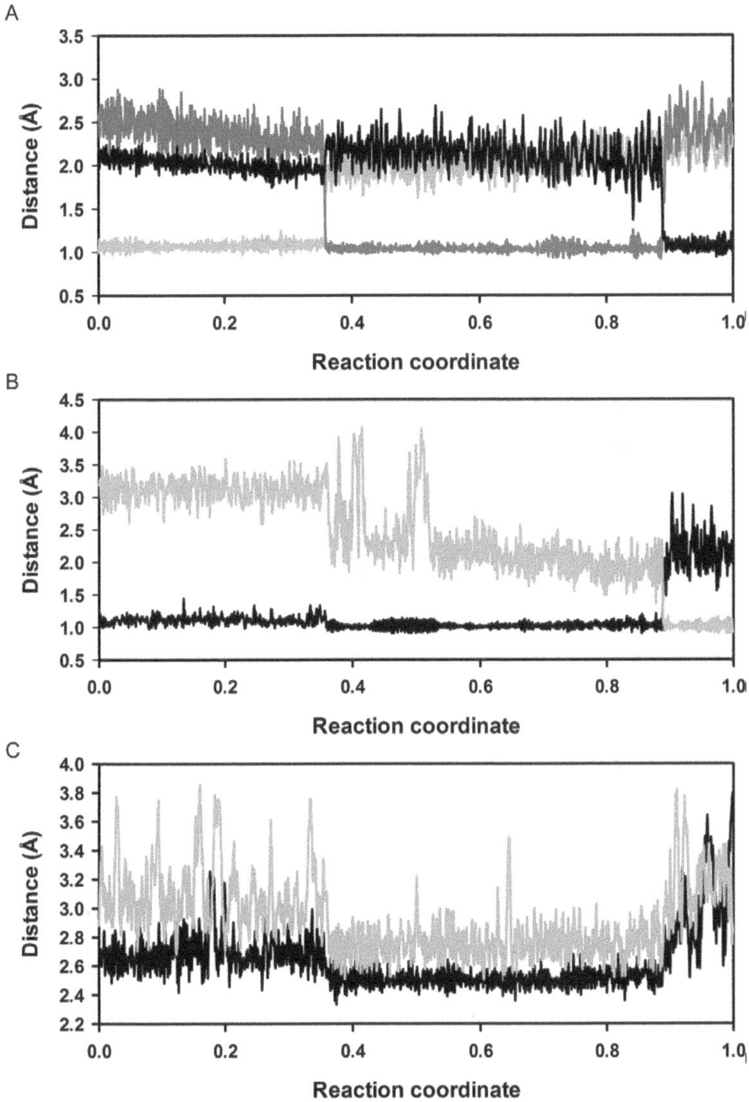

Figure 7 Distances measured during a DFT/MM MD simulation along the proposed pathway (Fig. 6). During the reaction, $H_{pro\text{-}R}$ is jumping from C1 to C2, while another H jumps from O1 to O2. (A) Distances between $H_{pro\text{-}R}$ and C1 (cyan (light gray in the print version)), the O in Glu 165 (green (dark gray in the print version)) or C2 (black). (B) Distances between H and O1 (blue (black in the print version)) or O2 (red (gray in the print version)); this H is initially bonded to O1. (C) Distances for the hydrogen bonds between O2 and nearby residues that stabilize the intermediate and transition states: distance between O2 and N of His 95 (red (gray in the print version)) and distance between O2 and N of Lys 12 (black).

(sp^3) to 5° (sp^2). During this process, the Glu 165 acts as catalytic base accepting the $H_{pro\text{-}R}$ from C1 (see Figs. 2 and 6). The enediolate forms a H-bond between DHAP O2 and the ϵN of His 95. Also, we find that the H-bond formed between the keto group of DHAP and Lys 12 adopts the character of a low-barrier H-bond where a hydrogen is shared between the ζN of Lys and the DHAP O2. These two H-bonds are the main interactions stabilizing the intermediate state.

The second step of the path undergoes through the rearrangement of the orbitals of the C2 carbon, which can be described through the variation of α_2 from 10° (sp^2) to 40° (sp^3). Along this process, the low-barrier H-bond of Lys 12 and the H-bond of His 95 stabilize the increasingly negative charge of the O2, which reaches a maximum when the reaction passes over the saddle point. In this transition state, the H is donated from O1 to O2 through a direct intramolecular transfer. *Simultaneously*, the Glu 165 acts as a catalytic acid, transferring the former $H_{pro\text{-}R}$ to the enediolate C2 atom, yielding GAP. Once the proton rearrangement has finished, the interaction between His 95 and O2 is lost. In addition, the low-barrier H-bond formed by Lys 12 becomes a typical H-bond (Fig. 7).

4.3 Discussion

The reaction mechanism described here presents some important differences with other previously proposed mechanisms. The proton from the alcohol group in C1 is directly transferred to the carboxyl oxygen in C2. This proton transfer takes place simultaneously with the proton transfer from Glu 165 to C2 (see Fig. 7). Notice also that during this process, the formation of the energetically unfavorable imidazolate anion is not required.

The intramolecular proton transfer from O1 to O2 has been considered previously and was found energetically unfavorable in relation to the transfer through the imidazole anion (Cui & Karplus, 2002). This analysis was performed using an energy-minimization approach; interestingly, Glu 165 stays protonated during these calculations. The different results obtained in our MD simulations are most likely related to the different computational approaches (energy minimization vs. dynamical simulation). In our dynamical approach, we do not make any assumptions about the protonation states during the process, allowing the protons to move freely. Thus, the simultaneous transfer of the two protons in the second step of the reaction is an unexpected outcome of our dynamical exploration. Since this is a dynamical process, it is hard to observe with minimization studies.

Another important dynamical effect observed in our simulations is the formation of the low-barrier H-bond between O2 and Lys 12 (see Fig. 7) that stabilizes the second transition state, facilitating the simultaneous proton transfer.

5. CONCLUSIONS

We have presented a detailed description of a practical DFT/MM method for the analysis of reactions in biomolecules using MD simulations. This technique combines the local-orbital DFT method Fireball and the Amber suite of programs for the simulation of biomolecular systems (Mendieta-Moreno et al., 2014). Due to the good balance between computational efficiency and accuracy, this method can be used to analyze reactions in biomolecules, sampling an appropriate conformational space by means of long DFT/MM MD simulations with relatively large QM regions ($\sim 10^2$ atoms). In our dynamical analysis of these reactions, the conformational space is explored using an adaptively biased MD approach that presents characteristics of steered MD and umbrella sampling. In a typical calculation, we obtain with this procedure around 2×10^6 individual structures. This information allows us to generate a free energy landscape for the reaction and get new insights into the possible reaction mechanisms.

As an example of this approach, we have studied the conversion between DHAP and GAP catalyzed by the TIM. In our analysis, we have chosen as reaction coordinates the dihedral angles for the DHAP C1 and C2 atoms. In this way, no assumption is made about the protonation state of the different residues present in the QM region along the MD simulations. The free energy landscape presents three energy minima, which can be associated with the substrate (DHAP), the intermediate state (enediolate), and the product of the reaction (GAP). The first step of the reaction (DHAP → enediolate) presents an energy barrier of 21 kcal/mol, in agreement with previous calculations (Hu et al., 2007). Regarding the second step (enediolate → GAP), we find a new reaction mechanism that is characterized by the intramolecular proton transfer from O1 to O2 and the simultaneous proton transfer from Glu 165 to C2. This is a dynamical mechanism not found in previous studies based on energy-minimization techniques. We also find the formation of a low-barrier H-bond between O2 and Lys 12 that stabilizes the transition state from the intermediate state to the final state.

This work highlights the importance of the appropriate exploration of the conformational space for the analysis of reactions in biomolecules, using DFT/MM MD simulations to take into account the dynamically accessible structures.

ACKNOWLEDGMENTS

This work is supported by the Spanish Ministerio de Economía y Competitividad (MINECO), projects MAT2014-59966-R (J.O.) and IPT2011- 0964-900000 (P.G-P.). The computational support of the Centro de Computacion Cientifica-CCC-UAM is acknowledged. Work at Biomol-Informatics was partially financed by the European Social Fund.

REFERENCES

Adcock, S. A., & Mccammon, J. A. (2006). Molecular dynamics: Survey of methods for simulating the activity of proteins. *Chemical Reviews*, *106*(5), 1589–1615.

Basanta, M. A., Dappe, Y. J., Jelínek, P., & Ortega, J. (2007). Optimized atomic-like orbitals for first-principles tight-binding molecular dynamics. *Computational Materials Science*, *39*, 759–766.

Bash, P. A., Field, M. J., Davenport, R. C., Petsko, G. A., Ringe, D., & Karplus, M. (1991). Computer simulation and analysis of the reaction pathway of triosephosphate isomerase. *Biochemistry*, *30*(24), 5826–5832.

Becke, A. D. (1988). Density-functional exchange-energy approximation with correct asymptotic behavior. *Physical Review A*, *38*, 3098.

Carlson, B. C., & Keller, J. M. (1957). Orthogonalization procedures and the localization of wannier functions. *Physical Review*, *105*, 102–103.

Case, D. A., Darden, T. A., Cheatham, T. E., III, Simmerling, C. L., Wang, J., Duke, R. E., et al. (2012). *AMBER 12*. San Francisco: University of California.

Cleland, W. W., & Kreevoy, M. M. (1994). Low-barrier hydrogen bonds. *Science*, *264*, 1887–1890.

Cleveland, W. S., & Devlin, S. J. (1986). Locally weighted regression: An approach to regression analysis by local fittings. *Journal of the American Statistical Association*, *83*(403), 596–610.

Cui, Q., & Karplus, M. (2002). Quantum mechanical/molecular mechanical studies of the triosephosphate isomerase-catalyzed reaction: Verification of methodology and analysis of reaction mechanisms. *Journal of Physical Chemistry B*, *106*(7), 1768–1798.

Demkov, A. A., Ortega, J., Sankey, O. F., & Grumbach, M. P. (1995). Electronic structure approach for complex silicas. *Physical Review B*, *52*, 1618–1630.

Field, M. J., Albe, M., Bret, C., Proust-De Martin, F., & Thomas, A. (2000). The dynamo library for molecular simulations using hybrid quantum mechanical and molecular mechanical potentials. *Journal of Computational Chemistry*, *21*(12), 1088–1100.

Field, M. J., Bash, P. A., & Karplus, M. (1990). A combined quantum-mechanical and molecular mechanical potential for molecular-dynamics simulations. *Journal of Computational Chemistry*, *11*, 700–733.

Foulkes, W. M. C., & Haydock, R. (1989). Tight-binding models and density-functional theory. *Physical Review B*, *39*, 12520–12536.

García-Vidal, F. J., Merino, J., Pérez, R., Rincón, R., Ortega, J., & Flores, F. (1994). Density-functional approach to lcao methods. *Physical Review B*, *50*, 10537–10547.

Gilson, M. K., & Zhou, H. X. (2007). Calculation of protein-ligand binding affinities. *Annual Review of Biophysics and Biomolecular Structure*, *36*, 21–42.

Go, M. K., Amyes, T. L., & Richard, J. P. (2010). Rescue of K12G triosephosphate isomerase by ammonium cations: The reaction of an enzyme in pieces. *Journal of the American Chemical Society*, *132*(38), 13525–13532.

Go, M., Koudelka, A., Amyes, T., & Richard, J. (2010). Role of Lys-12 in catalysis by triosephosphate isomerase: A two-part substrate approach. *Biochemistry*, *49*(25), 5377–5389.

Hall, A., & Knowles, J. R. (1975). The uncatalyzed rates of enolization of dihydroxyacetone phosphate and of glyceraldehyde 3-phosphate in neutral aqueous solution. The quantitative assessment of the effectiveness of an enzyme catalyst. *Biochemistry*, *14*(19), 4348–4353.

Harris, J. (1985). Simplified method for calculating the energy of weakly interacting fragments. *Physical Review B*, *31*, 1770.

Hohenberg, P., & Kohn, W. (1964). Inhomogeneous electron gas. *Physical Review*, *136*, B864–B871.

Horsfield, A. P. (1997). Efficient ab initio tight binding. *Physical Review B*, *56*, 6594–6602.

Hu, H., Lu, Z., & Yang, W. (2007). QM/MM minimum free energy path: Methodology and application to triosephosphate isomerase. *Journal of Chemical Theory and Computation*, *3*(2), 390–406.

Ishikita, H., & Saito, K. (2014). Proton transfer reactions and hydrogen-bond networks in proton transfer reactions and hydrogen-bond networks in protein environments. *Journal of the Royal Society Interface*, *11*(91), 20130518.

Jelínek, P., Wang, H., Lewis, J., Sankey, O., & Ortega, J. (2005). Multicenter approach to the exchange-correlation interactions in ab initio tight-binding methods. *Physical Review B*, *71*, 235101.

Jogl, G., Rozovsky, S., McDermott, A. E., & Tong, L. (2003). Optimal alignment for enzymatic proton transfer: Structure of the Michaelis complex of triosephosphate isomerase at 1.2-A resolution. *Proceedings of the National Academy of Sciences of the United States of America*, *100*(1), 50–55.

Jorge, M., Garrido, N. M., Queimada, A. J., Economou, I. G., & MacEdo, E. A. (2010). Effect of the integration method on the accuracy and computational efficiency of free energy calculations using thermodynamic integration. *Journal of Chemical Theory and Computation*, *6*(4), 1018–1027.

Karplus, M., & McCammon, J. A. (2002). Molecular dynamics simulations of biomolecules. *Nature Structural Biology*, *9*, 646–652.

Kohn, W., & Sham, L. J. (1965). Self-consistent equations including exchange and correlation effects. *Physical Review*, *140*, A1133.

Kumar, S., Rosenberg, J. M., Bouzida, D., Swendsen, R. H., & Kollman, P. A. (1992). The weighted histogram analysis method for free-energy calculations on biomolecules. I. The method. *Journal of Computational Chemistry*, *13*(8), 1011–1021 (John Wiley & Sons, Inc.).

Laio, A., & Parrinello, M. (2002). Escaping free-energy minima. *Proceedings of the National Academy of Sciences of the United States of America*, *99*(20), 12562–12566.

Lee, C., Yang, W., & Parr, R. G. (1988). Development of the Colle-Salvetti correlation-energy formula into a functional of the electron density. *Physical Review B*, *37*, 785.

Lewis, J. P., Glaesemann, K. R., Voth, G. A., Fritsch, J., Demkov, A. A., Ortega, J., et al. (2001). Further developments in the local-orbital density-functional-theory tight-binding method. *Physical Review B*, *64*, 195103.

Lewis, J. P., Jelinek, P., Ortega, J., Demkov, A. A., Trabada, D. G., Haycock, B., et al. (2011). Advances and applications in the FIREBALL ab initio tight-binding molecular-dynamics formalism. *Physica Status Solidi B*, *248*, 1989–2007.

Lodi, P. J., & Knowles, J. R. (1991). Neutral imidazole is the electrophile in the reaction catalyzed by triosephosphate isomerase: Structural origins and catalytic implications. *Biochemistry*, *30*(28), 6948–6956.

Löwdin, P. (1950). On the nonorthogonality problem connected with the use of atomic wave functions in the theory of molecules and crystals. *The Journal of Chemical Physics, 18*, 365.

Martin, R. M. (2004). *Electronic structure*. Cambridge (UK): Cambridge University Press.

Martín-García, F., Mendieta-Moreno, J. I., López-Viñas, E., Gómez-Puertas, P., & Mendieta, J. (2012). The role of Gln61 in HRas GTP hydrolysis: A quantum mechanics/molecular mechanics study. *Biophysical Journal, 102*, 152–157.

Marx, D., & Hutter, J. (2009). *Ab initio molecular dynamics*. Cambridge (UK): Cambridge University Press.

Mendieta-Moreno, J. I., Walker, R., Lewis, J. P., Gomez-Puertas, P., Mendieta, J., & Ortega, J. (2014). FIREBALL/AMBER: An efficient local-orbital DFT QM/MM method for biomolecular systems. *Journal of Chemical Theory and Computation, 10*, 2185–2193.

Park, S., Khalili-Araghi, F., Tajkhorshid, E., & Schulten, K. (2003). Free energy calculation from steered molecular dynamics simulations using Jarzynskis equality. *The Journal of Chemical Physics, 119*, 3559.

Reed, A. E., Weinstock, R. B., & Weinhold, F. (1985). Natural population analysis. *The Journal of Chemical Physics, 83*, 735.

Řezáč, J., & Hobza, P. (2012). Advanced corrections of hydrogen bonding and dispersion for semiempirical quantum mechanical methods. *Journal of Chemical Theory and Computation, 8*, 141–151.

Řezáč, J., Riley, K. E., & Hobza, P. (2011). S66: A well-balanced database of benchmark interaction energies relevant for biomolecular structures. *Journal of Chemical Theory and Computation, 7*, 2427–2438.

Salomon-Ferrer, R., Case, D., & Walker, R. (2012). An overview of the Amber biomolecular simulation package. *WIREs Computational Molecular Science, 3*, 198–210.

Samanta, M., Murthy, M. R. N., Balaram, H., & Balaram, P. (2011). Revisiting the mechanism of the triosephosphate isomerase reaction: The role of the fully conserved glutamic acid 97 residue. *ChemBioChem, 12*(12), 1886–1896.

Sankey, O. F., & Niklewski, D. J. (1989). Ab initio multicenter tight-binding model for molecular-dynamics simulations and other applications in covalent systems. *Physical Review B, 40*, 3979–3995.

Schönhammer, K., Gunnarsson, O., & Noack, R. M. (1995). Density-functional theory on a lattice: Comparison with exact numerical results for a model with strongly correlated electrons. *Physical Review B, 52*, 2504.

Senn, H. M., & Thiel, W. (2009). QM/MM methods for biomolecular systems. *Angewandte Chemie, 48*, 1198–1229.

Stewart, J. J. P. (2004). Optimization of parameters for semiempirical methods IV: Extension of MNDO, AM1, and PM3 to more main group elements. *Journal of Molecular Modeling, 10*(2), 155–164.

Walba, H., & Isensee, R. W. (1961). Acidity constants of some arylimidazoles and their cations. *The Journal of Organic Chemistry, 26*(8), 2789–2791.

Walker, R. C., Crowley, M. F., & Case, D. A. (2008). The implementation of a fast and accurate QM/MM potential method in Amber. *Journal of Computational Chemistry, 29*, 1019–1031.

Warshel, A., & Levitt, M. (1976). Theoretical studies of enzymic reactions: Dielectric, electrostatic and steric stabilization of the carbonium ion in the reaction of lysozyme. *Journal of Molecular Biology, 103*, 227–249.

Zhang, Y. (2005). Pseudobond ab initio QM/MM approach and its applications to enzyme reactions. *Theoretical Chemistry Accounts, 116*, 43–50.

Zwanzig, R. W. (1954). High-temperature equation of state by a perturbation method. I. Nonpolar gases. *The Journal of Chemical Physics, 22*(8), 1420–1426.

CHAPTER FOUR

Explicit Drug Re-positioning: Predicting Novel Drug–Target Interactions of the Shelved Molecules with QM/MM Based Approaches

Ankur Omer*,†,1, Venkatesan Suryanarayanan‡,1, Chandrabose Selvaraj‡, Sanjeev Kumar Singh‡,2, Poonam Singh*,†,2

*Division of Toxicology, CSIR-Central Drug Research Institute, Lucknow, Uttar Pradesh, India
†Academy of Scientific & Innovative Research (AcSIR), New Delhi, India
‡Department of Bioinformatics, Computer Aided Drug Design and Molecular Modeling Lab, Science Block, Alagappa University, Karaikudi, Tamil Nadu, India
[1]Equal contribution.
[2]Corresponding authors: e-mail address: skysanjeev@gmail.com; poonam_singh@cdri.res.in

Contents

Advances in Protein Chemistry and Structural Biology, Volume 100
ISSN 1876-1623
http://dx.doi.org/10.1016/bs.apcsb.2015.07.001

Abstract

With the demand to enhance the speed of the drug discovery process there has been an increased usage of computational approaches in drug discovery studies. However because of their probabilistic outcomes, the challenge is to exactly mimic the natural environment which can provide the exact charge polarization effect while estimating the binding energy between protein and ligand. There has been a large number of scoring functions from simple one to the complex one available for estimating binding energy. The quantum mechanics/molecular mechanics (QM/MM) hybrid approach has been the preferred choice of interest since last decade for modeling reactions in biomolecular systems. The application of QM/MM approach has been expanded right from rescoring the already known complexes and depicting the correct position of some novel molecule to ranking a large number of molecules. It is expected that the application of QM/MM-based scoring will grow in all areas of drug discovery. However, the most promising area will be its application in repositioning, that is, assigning novel functions or targets to the already existing drugs, as this would stop the rising attrition rates as well as reduce the overall time and cost of drug discovery procedure.

1. INTRODUCTION

Due to the increase in number of diseases, there has been a large amount of money spent on drug discovery research. In the year 1975, there was around $4 billion US Dollar (USD) invested on drug discovery while in the year 2009, $40 billion USD was invested on the drug discovery. But the number of drugs approved has remained constant, where in 1976 and 2009 there were 26 new drugs got approved and in 2013 there were only 27 new drugs got approved. According to Pammolli, the average time of drug discovery has been increased to 13.9 years from 2000 which was 9.7 years earlier in 1990s (Pammolli, Magazzini, & Riccaboni, 2011) (Fig. 1).

Drug repositioning means the identification of some novel function or novel target of already existing drug that can be utilized as a very effective tool to find out the treatment of some deadly diseases. Pharmaceutical companies are still reliant on the traditional approach for drug repositioning. Out of the 75 FDA approved molecules from 1999 to 2008, around 50% were discovered by phenotypic-based and target-based screening approaches (Hurle et al., 2013; Swinney & Anthony, 2011). Some successful examples of drug repositioning includes the use of Sildenafil in erectile dysfunction. It received US FDA approval to enter into the United States market in 1998 after it was rejected in the phase II clinical trials for the treatment of angina. Thalidomide developed as a sedative was found to cause severe birth defects

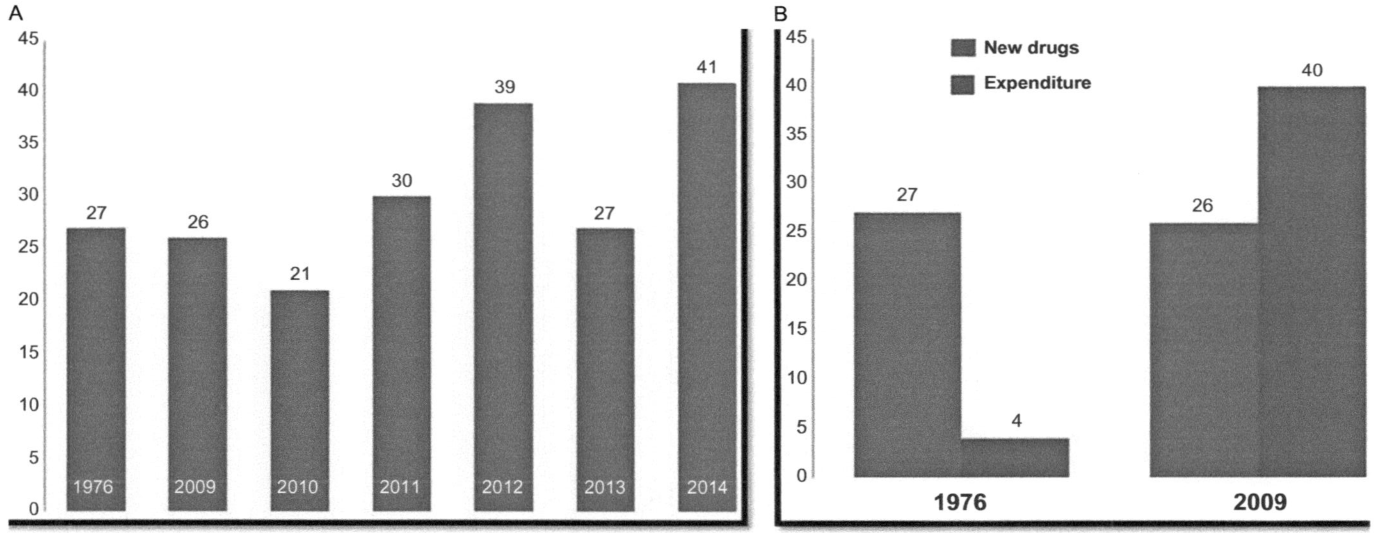

Figure 1 (A) Showing the comparison of number of new drugs approved in 1976 and the drugs approved in past 6 years. (B) comparison between the expenditure on drug discovery projects in 1976 and 2009.

leading to its withdrawal from the market. In 2006, it gained US FDA approval for its anticancer activity to treat multiple myeloma in combination with dexamethasone. Table 1 shows some of the old functions of drugs and their newly assigned functions.

Quantum mechanical/molecular mechanical (QM/MM) methods have its origin in 1970s (Warshel & Levitt, 1976). Before 1990s, the field remained dormant, but in 1990s there was outburst of the articles containing the application of QM/MM-based applications to study biochemical systems (Bruice & Kahn, 2000; Cho, Guallar, Berne, & Friesner, 2005; Field, Bash, & Karplus, 1990). The dramatic acceptance of these methods was well supported by the rapid advancement in the field of computational science which enabled the occurrence of heavy QM/MM calculation in alignment maintaining the accuracy (Claeyssens et al., 2006). Although the hybrid QM/MM approach has become the method of preference to study the biomolecular systems, the complexity associated with the biomolecules due to the presence of a large number of atoms requires exhaustive

Table 1 List of Some Popular Drugs with Comparison to Their Original Disease and the Newly Assigned Repositioned Disease

Drug	Original Disease	New Disease
Arsenic	Tuberculosis and syphilis	Acute promyelocytic leukemia (Zhu, Chen, Lallemand-Breitenbach, & de The, 2002)
Ceftriaxone	Antibiotic	Amyotrophic lateral sclerosis (Rothstein et al., 2005)
Eflornithine	Cancer	African trypanosomes (Marton & Pegg, 1995)
Fosmidomycin	Urinary tract infections	Antimalarial (Borrmann et al., 2004; Jomaa et al., 1999; Missinou et al., 2002)
Fumagillin	Antiamebic	Anticancer angiogenesis inhibitor (Griffith et al., 1997; Kruger & Figg, 2000)
Miltefosine	Cancer	Visceral leishmania (Sundar et al., 2002)
Paromomycin	Amebicide	Visceral leishmaniasis (Nwaka & Ridley, 2003)
Quinacrine	Antimalarial	Prion diseases (Korth, May, Cohen, & Prusiner, 2001; Nakajima et al., 2004)
Thalidomide	Sedative	Cancer (D' Amato, Loughnan, Flynn, & Folkman, 1994)
Retinoic acid	Acne	Acute promyelocytic leukemia (Fang et al., 2002)

sampling. Since last decade, more number of reviews were documented in QM/MM field regarding its application and development (Amara & Field, 1998; Beck & Clark, 1998; Cunningham & Bash, 1997; Friesner & Beachy, 1998; Gao, 1998; Gao & Thompson, 1998; Merz, 1998; Merz, Stanton, & Schreiner, 1998; Mordasini & Thiel, 1998; Ruiz-Lpez & Rivail, 1998).

2. THE PRINCIPLE

The molecular mechanics which is based on empirical potential energy functions has become a widely used concept since last one or two decades. There are a number of reasons to integrate MM with quantum mechanics; as MM is a empirical function based approach, the accuracy of the approach is limited. On the other side, QM functions can potentially generate realistic potential energy surfaces, can include environment-dependent polarization effects and charge transfer interactions. The basic need of integrating both the methods is to account those electronic structure changes which are not considered by MM methods. The main objective of QM/MM-based approach is to describe the interaction between ligand molecule and receptor and the binding ability of the ligand with the receptor. Figure 2 shows a protein receptor molecule bound with its ligand molecule.

The Hamiltonian operator is of fundamental importance for most of the quantum calculations and it corresponds to the total energy of the system. The Hamiltonian of the system according to Fig. 2 can be expressed as:

$$H = H_{\mathbf{QM}} + H_{\mathbf{MM}} + H_{\mathbf{QM/MM}}$$

H_{QM} = Hamiltonian accounting for all QM particles of ligand.

H_{MM} = Hamiltonian accounting for all MM particles of protein.

$H_{\mathrm{QM/MM}}$ = Hamiltonian accounting for the interaction between QM and MM particles within the system.

The van der Waal's interactions at molecular mechanics level can be described by simple functions like Lennard-Jones potential, while electrostatic term enters Fock matrix as self-consistent field (SCF) method. There are a large number of approaches to analyze the interaction between the two QM and MM systems. These approaches can broadly be divided into two categories subtractive QM/MM coupling and additive QM/MM coupling (Menikarachchi & Gascon, 2010).

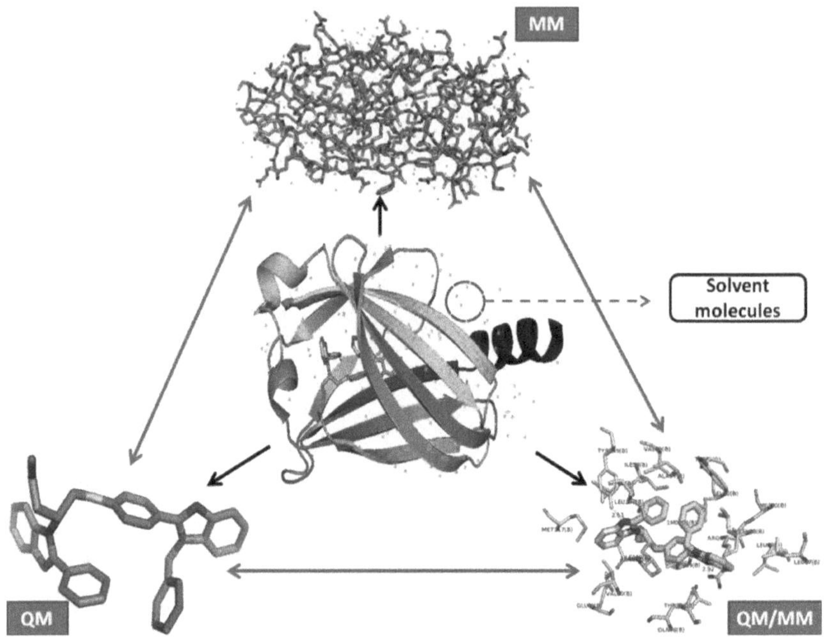

Figure 2 Shows the partitioning of the protein–ligand complex into the QM applied region, MM applied region, and QM/MM applied regions.

3. SUBTRACTIVE QM/MM COUPLING

It utilizes three steps to calculate energy of the system. In first step, the energy of the total system is calculated at MM level. The second step includes the addition of QM energy of the isolated system and the third step includes the calculation of MM energy value of the QM system and finally this value is subtracted. The main advantage of the method is its simplicity. ONIOM method was the most widely used subtractive QM/MM coupling approach, which was developed by Morokuma and colleagues (Maseras & Morokuma, 1995; Svensson et al., 1996).

The subtractive coupling approach can be visualized as MM approach devoid of certain region and replaced by some higher level treatment. The implementation of the ONIOM method is bit straight in comparison to the other existing methods as there is no communication required between QM and MM routines. On the other side, it is limited by the fact that a flexible force field is required to elucidate the effect of chemical changes during any reaction; also it is difficult to model biological charge transfer processes due to absence of polarization by the MM environment on QM electron density.

4. ADDITIVE QM/MM COUPLING

As the QM system is embedded within the MM system, the potential energy of the system is a sum of MM, QM, and QM/MM energy terms which can be represented as:

$$V_{QM/MM} = V_{QM}(QM) + V_{MM}(MM) + V_{QM/MM}(QM + MM)$$

These interactions can be explained at various levels based on the way to treat electrostatic interaction between QM and MM systems (Groenhof, 2013).

4.1 Mechanical Embedding

It is a simple and straight approach, which handles the QM/MM scheme on the same footing as MM–MM scheme, i.e., at MM level.

4.1.1 Drawbacks

The major drawbacks are as follows:

- Requires an accurate set of MM parameters for both QM and MM system.
- It ignores the perturbation of the electronic structure of the MM system due to the electrostatic interaction between the QM and MM systems (Lin & Truhlar, 2005).

The problem lead us to draw attention toward the mechanically embedded three-layered ONIOM method which overcomes the drawback of two-layered ONIOM method. Conceptually, ONIOM can handle any number of layers. Presently, implementation is limited to three layers of (QM/QM/MM) calculations. In the three-layered method, a buffer layer is added as middle layer making it computationally less expensive. One can label such a treatment as QM1/QM2/MM. In a two-layered ONIOM method, the energy of the whole system can be accounted by:

$$E_{ONIOM} = E_{real,MM} + E_{model,QM} - E_{model,MM}$$

The real system contains all the atoms, and calculations are done at MM level. For model system, both QM and MM is carried out (Vreven & Morokuma, 2003).

4.2 Electrostatic Embedding

It does not require the MM electrostatic parameters for the atoms in the MM system, as this method includes certain one electron terms in QM Hamiltonian which allows to deal electrostatic interactions between QM and MM systems at a much advanced level, therefore removing the shortcomings of mechanical embedding approach. Although there has been noticeable development in the electrostatic embedded ONIOM method, it has made implementation more complicated and has also increased computational cost (Vreven & Morokuma, 2003; Vreven, Morokuma, Farkas, Schlegel, & Frisch, 2003).

4.3 Polarization Embedding

It is the next level of sophistication which includes the polarizability of the MM atoms. In this scheme, both the regions can mutually polarize each other (Groenhof, 2013). Thus, MM region atoms polarize the QM region atoms and QM region atoms can also induce polarization on the MM system. Different approaches have been developed to model polarization of MM atoms. The fluctuating charge model (Rappe & Goddard, 1991), the induced dipole model (Warshel et al., 2006), and charge-on-aspiring model (Lamoureux & Roux, 2003) are the models of polarization of MM atoms among the most popular methods developed. Inspite of being the earliest QM/MM approach, the method remained scarce due to the lack of polarizable biomolecular MM force fields (Singh & Kollman, 1986; Warshel & Levitt, 1976). Although a large number of polarizable solvent model for the simulation of liquid water were available, much of the work were under progress for the polarization of protein force fields (Grossfield, Ren, & Ponder, 2003; Ren & Ponder, 2003; Stern, Rittner, Berne, & Friesner, 2001; Yu, Hansson, & Gunsteren, 2003, 2004).

5. LIGAND POLARIZATION

The importance of polarization was known since very long time and there has been a continuous effort to develop suitable force field which can accurately model the protein–ligand interactions. Ligand polarization can be divided into two parts: first, the calculations were done with the ligand during docking procedure when ligand is bound to protein; second, the calculations were done to estimate the role of polarization in protein–ligand binding.

5.1 QM-Polarized Ligand Docking

There are a large number of force fields which quite correctly are able to mimic the partial charges on ligand atoms, however they do not consider the charge polarization induced by protein environment. The QM-polarized ligand docking (QPLD) aims to replace the partial charges on the ligand with quantum mechanical derived charges (Cho, Gullar, Berne, & Friesner, 1995). Ligand is treated as QM region and DFT method is applied for quantum calculations (Shaik et al., 2010). The problem of selection of incorrect pose while applying polarization charge method can be curbed by using an unbiased algorithm without any previous experimental knowledge about the ligand pose (Kaminiski, 2005; Kaminiski, Stern, Berne, & Friesner, 2004; Kaminski et al., 2002). The method utilizes five poses instead of one, charges for the ligands are extracted, redocking is done with each of these set, and then final pose having least interaction energy is selected, thus increasing the docking accuracy. Selvaraj et al. in the year 2014, tried to find out the reason for why HIV-1 PR inhibitors are rejected by HTLV-1 PR structure, inspite of having great similarity in their protein structure. The chemical structures were fetched based upon the similarity criteria with the available HIV-PR inhibitors using QPLD as one of the approach and finally screened around nine compounds showing potential features toward inhibiting HTLV-1 PR (Selvaraj et al., 2014a, 2014b).

6. PROTEIN POLARIZATION

As the protein is a large body in comparison to ligand and is described by MM, the role of polarization of protein in QM/MM calculations has not received the expected level of attention as gained by ligand polarization.

Nevertheless, further development in interaction energies calculation is expected. There have been a large number of works that demonstrate the next generation of potential (Halgren & Damm, 2001; Kaminski et al., 2002; Maple et al., 2005; Patel & Brooks, 2006; Piquemal et al., 2003; Ren & Ponder, 2002, 2003; Stern et al., 1999; Xie & Gao, 2007; Yu et al., 2003).

Recently, Gao et al. (2014) have developed a new strategy called explicit polarization (X-Pol) which is meant for electronic structure calculations based on the partition of a condensed phase or macromolecular system into fragments (Gao et al., 2014).

6.1 Boundary Treatment

In most of the QM/MM-based studies, boundary is assigned and a covalent bond is left dangling. The bond so left must be capped in order to satisfy its valency, and in the case of electrostatic or polarized embedding in case of QM region overpolarization due to MM region should be prevented. There are three different methods to deal with these problems: link atoms, localized orbitals, and special boundary atoms (Thiel, 2009).

In *link atom method*, an additional atom (generally hydrogen) or any monovalent atom is added to the QM frontier atoms. The hydrogen atom is added at the free valency of the atom generated by the separation of QM/MM system. With each atom so linked three degrees of freedom are generated, although there are different methods to handle it, the most common is to fix the position of the link atom so that it lies in the bond being cut and to redistribute the forces to the two atoms of the bond being cut (Eichler, Kolmel, & Sauer, 1997; Thiel, 2009). Although the method is simple in its form, addition of extra atom aids some problems:

- With the addition of each atom three additional degrees of freedom is introduced not originally present in the system. This raises the question about the position of the link atoms to be added.
- The problem of overpolarization due to the point charge on frontier atoms of MM system.
- Introduction of new artificial interactions with other atoms (Senn & Thiel, 2007).

Localized-orbital method places hybrid orbitals at boundaries. This provides a theoretically satisfactory boundary treatment at the QM level. Some of the so-placed orbitals at the boundary are fixed so that they do not participate in SCF iterations.

Boundary atom method replaces the front atoms of the MM systems such that they participate in calculation just like MM atoms, the replaced atoms should also satisfy the valency of the atoms at QM front. The QM features are often parameterized, and such boundary atoms methods are considered to be more accurate than link atom methods, but they are less popular as they require special parameters which are not usually available.

There are a number of studies which stated the merits and demerits of different approaches. Link atoms method is the most widely used approach and there are a number of studies which compare link atom method to localized orbital methods (Hall, Hindle, Burton, & Hillier, 2000; Nicoll, Hindle, MacKenzie, Hillier, & Burton, 2001; Reuter, Dejaegere, Maigret, & Karplus, 2000), while there are a number of studies which have compared the link atom methods on the way the boundary charges are handled. The hybrid orbitals-based approaches are theoretically much fundamental. These methods avoid some of the inherent problems associated with link atom method arising due to addition of atoms, however, technically they are more complicated due to the orthogonality constraints which prevents mixing of frozen orbitals with the active orbitals (Senn & Thiel, 2007).

In general, it can be concluded that all the available methods provide reasonable accuracy when used with care such that there should be least distortions of the charge distributions (Senn & Thiel, 2007).

7. QM/MM MOLECULAR DYNAMICS

Hybrid QM/MM are extensively used local electronic events in MM systems with diverse areas of applications (Bakowies & Thiel, 1996; Field, Bash, & Karplus, 1990; Sherwood, 2000; Singh & Kollman, 1986; Warshel & Levitt, 1976). Biomolecular systems support semiempirical neglect of diatomic differential overlap-type Hamiltonian-based QM/MM-based approaches and density functional tight-binding Hamiltonian (Seabra, Walker, Elstner, Case, & Roitberg, 2007; Walker, Crowley, & Case, 2008). These methods are computationally efficient while on the other side the accuracy of these methods are limited by their approximate nature (Nam, Cui, Gao, & York, 2007; Rossi & Truhlar, 1995). However, QM/MM MD simulations are computationally quite demanding, in order to maintain this, the direct sampling of the QM region can be avoided while sampling the MM region completely (Thiel, 2009). Zhang et al. sampled the MM region while keeping the QM region fixed, for QM atoms the charges were derived using electrostatic potential. There are a number of activities to improve the existing technologies (Hu & Yang, 2008; Senn & Thiel, 2009; Zhang, Liu, & Yang, 2000).

8. GEOMETRY OPTIMIZATION

Optimizing the stationary points on QM/MM, potential energy surface (PES) is the first step to explore the reaction pathway. It usually gives an important view on the mechanism of reaction. Extending the traditional approach to study the reaction under *in silico* environment is quite difficult. The traditional approach has been to characterize stationary points on the potential energy surface. The minima are represented as reactant and products while the transition states are the lowest saddle points connecting the minima. However, to extend this approach to QM/MM potential energy surfaces is quite difficult as the QM/MM system possess much higher dimensionality which demands an efficient geometry optimizer (Groenhof, 2013; Klahn, Braun-Sand, Rosta, & Warshel, 2005).

8.1 QM/MM Exploration of Potential Energy Surfaces

For a molecular system, potential energy surface (PES) arises when the Born–Oppenheimer approximation, that is electronic distribution of the molecule adapts itself quickly in the solution of the Schrödinger equation. The energy of a molecule is by virtue of function of the positions of the nuclei, that is a PES. The computational chemistry is an efficient tool to evaluate the accurate PES to describe the chemistry of interest.

9. APPLICATIONS OF QM/MM METHODS TO STRUCTURE-BASED DRUG DESIGN

9.1 QM/MM Methods to Aid the Understanding of Ligand–Receptor Interactions

Tuttle et al. applied QM/MM-based approaches to understand various factors which can be used to differentiate between the binding of closely related ligand molecules (Tuttle, 2012). The authors demonstrated that how the noncovalent interactions could affect the binding ability of Latrunculin A and its analogues. The study aimed to elucidate the binding mode of Latrunculin A, naturally occurring analogue Latrunculin B and a synthetic analogue L32 computationally to G actin. The study identified one compound (L32) possessing biological activity as good as naturally occurring latrunculins.

In another study, Tuttle et al. explored the role of dispersion forces in the binding site of the antibody 34E4. The group demonstrated dispersion forces to be the reason of stability between the binding site and the hapten. The study has shed light on the importance of dispersion-corrected QM to obtain actual description of the noncovalent interactions which plays the most important role in determining the binding ability of ligands (Tuttle & Thiel, 2008).

9.2 QM/MM Methods in Scoring Refinement

The realistic representation of charge transfer when a ligand binds to a protein is a necessary aspect for estimating the ability of ligand binding. The application of rescoring is widely accepted in the field of ligand binding to metalloprotein receptors. The metalloproteins may contain large charge in the binding site due to polarization of proximal amino acid residues. QM/MM approach provides recalculation of these charges while docking by including the susceptible amino acid residues into the QM region. With the increasing QM size the computational cost will be raised, but the implication of accurate QM methods will not be a limiting part in presence of modern computing architectures (Tuttle, 2012).

Balaz et al. provided a direct application to the use of QM/MM methods to rescore the binding of a ligand to the receptor. The group employed the QM/MM study for 28 hydroxamate inhibitors binding to zinc-dependent matrix metalloproteinase-9 (Khandelwal et al., 2005). Another study performed by Cho and Rinaldo used an "extended QM/MM rescoring method." The glide was used for docking. The method employed the single-point QM/MM calculations to recalculate the charge distribution in metal ion, ligand, and the atoms of surrounding amino acids. The method was tested on eight matrix metalloproteinases, out of which seven showed agreement to the calculated and experimental binding orientation. This method was a direct example of the utility of QM/MM methods in SBDD (Cho & Rinaldo, 2009).

9.3 QM/MM Methods in Drug Repositioning

There are a number of computational approaches available to reposition a drug which includes structural similarity search, ligand similarity search, and many more. The concept of drug repositioning is based upon the principle "Promiscuity of drugs" (Croset, 2014). This fact opens the door for searching new targets for the existing drug candidates. There are a number of tools which can estimate the similarity between existing protein structures and molecules (Fig. 3).

The application of QM/MM in repositioning the already existing molecules toward novel target will really open the efficient way to screen novel targets quickly. There have been few studies which utilizes QM/MM-based approach to assign some novel target or a novel function to the pre-existing drug. These studies will certainly help in understanding the way QM/MM-based approaches be implemented for repositioning the roles of existing drugs. Selvaraj et al. searched for the reason for the rejection of HIV-PR inhibitors from the HTLV-1 PR binding site. Based on the findings, it was concluded that none of the HIV-PR inhibitors were binding with the Met 37 of HTLV-1 PR structure. In order to search new molecules,

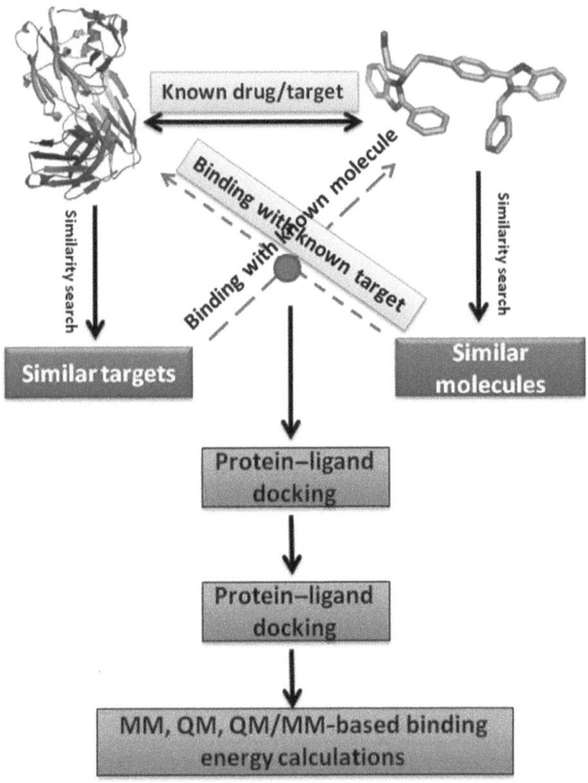

Figure 3 Drug repositioning strategies using QM/MM-based docking approaches.

the already known molecules were taken as reference and a number of analogues were searched having similarity greater than 90%. QM/MM calculation was performed to include polarization effects induced by receptor on the final confirmation of ligand. The study that explored nine compounds from the pubchem showed potential to inhibit HTLV-1 PR as were found to interact specifically with Met 37 (Selvaraj et al., 2014a, 2014b).

In another study, Selvaraj et al. examined the conformation of each HIV-PR drugs with different parameters of interactions. The authors used quantum mechanics-based interaction and binding energy calculation to check whether the HIV-1 PRIs compounds can work as HTLV-1 PR inhibitors or not. QPLD protocol was applied to improve the charges on ligand atoms by replacing them with the QM-derived charges. Protein was treated as MM region and ligand as QM region and DFT method was used. The charges were applied on the system "on the go" rather than using fixed charges assigned by the force-field OPLS-AA. The work suggested that some of

HIV inhibitors such as Darunavir, Amprenavir, Tipranavir, and Indinavir showed good energy with HTLV-1 PR (Selvaraj et al., 2014a, 2014b).

Nelfinavir, a potential HIV-protease inhibitor has shown promising results against cancer. Xie et al. used MM–GBSA method and reverse docking approach to propose multiple protein kinase targets by Nelfinavir. The procedure followed binding site comparison, protein–ligand docking, MD simulations, binding energy calculations by MM–GBSA method, and network construction to find out the putative potential targets. Although the method utilized MM-based method for binding energy calculation, definitely showed the potential of such calculations (Xie, Evangelidis, Xie, & Bourne, 2011).

In a recent study, MM-PB(GB)SA and QM/MM–GBSA methods were used to calculate the binding affinities of the compounds. Wichapong and group used ligand docking and QM- and MM-based scoring functions to distinguish the active and inactive Myt1 kinase inhibitors. The findings were proved experimentally. GOLD version 5.0 (Jones, Willett, & Glen, 1995; Jones, Willett, Glen, Leach, & Taylor, 1997) was used to perform docking. The study concluded that QM/MM–GBSA scoring provides more accurate results to differentiate active and inactive compounds, novel and potent Myt1 kinase inhibitors were identified and also paved the path to identify and screen potential molecules (Wichapong et al., 2014).

Hamza in 2002 performed preliminary docking studies with methotrexate an anticancer drug (Hamza, 2002; Kamal, Nasim, & Al-Jafari, 1996) against 1-Cys peroxiredoxins of *Plasmodium falciparum*. He constructed three-dimensional homology modeled structure and then performed docking studies by different methods MM, MD, QM. The study first performed docking studies on the X-ray crystal structure of human version of the protein, and based upon the knowledge of these interactions further docking was performed with the molecules against the plasmodial 1-Cys peroxiredoxins protein. The study compared methotrexate/Pf-1-Cys peroxiredoxins and mercaptosuccinate/Pf-1-Cys peroxiredoxins interaction energies where, methotrexate clearly showed higher activity. Considering all the known aspects, he suggested the inhibitory role of methotrexate against Pf-1-Cys peroxiredoxins (Hamza, 2002).

Another study used old drugs against the LEDGF/p75-binding pocket of HIV-1 integrase. The authors performed molecular docking experiments using Glide 5.5 Standard Precision method (Friesner et al., 2004; Glide et al., 2009). After this precise ligand–receptor binding, free energy was calculated by using MM–GBSA approach. 26 old drugs were finally selected

Table 2 Some of the Databases and Servers Which can be Helpful in Drug Repositioning Studies (Links for Respective Softwares/Tools Are Provided in the Reference Section)

Database/Resource	Description
PROMISCUOUS database	Protein–protein and drug–protein interactions resource
DrugBank	Detailed drug and drug target information
TherapeuticsTargets Database (TTD)	Provides information about drug targets and respective drugs and corresponding pathway
SuperTarget	Drug-related targets and side effects
Pharmmapper server	Drugs and their target information
Manually Annotated Targets and Drugs Online Resource (MATADOR)	Resource for protein–chemical interactions
Ondex database	Integrated dataset for *in silico* drug discovery
FDA: The Rare Disease Repurposing Database	Includes FDA information about drugs
POSSUM	Pocket similarity search

and purchased for screening assays, out of which 8 drugs were found to be potential inhibitors of LEDGF/p75-HIV-1 integrase interaction (Hu et al., 2012). Some of the important databases, servers, and other online resources which can help in obtaining the existing information about targets, their inhibitors, and interactions between them are provided in Table 2. Table 3 is enlisting a number of standalone softwares which can help to perform QM, MM, or dynamics on the target system along with the information about the availability and platform for which the softwares are available.

10. FIVE YEARS VIEW POINT: FUTURE OF QM/MM-BASED REPOSITIONING

As the number of diseases has been increasing, there is a shift in the research environment in pharmaceutical companies toward the faster screening of drug candidates. Drug repositioning is an effective strategy to fasten the search of potential drug candidates. Although there are many

Table 3 List of Tools and Softwares Available on the Web to Conduct QM, MM Calculations, and Molecular Dynamics Simulation (Links for Respective Softwares/Tools Are Provided in the Reference Section)

Integrated Packages	Availability	Platform
GAMESS (US)	Academic	PC, Macintosh, Linux, Unix, VMS
Jaguar	Commercial	SGI, RS/6000, HP-UX, Cray, Alpha, Linux
Spartan	Commercial	Windows, Linux, Macintosh, SGI, RS/6000, Alpha, HP-UX
ADF	Commercial	Cray, SGI, PC (Pentium Pro or newer with Linux), DEC, Fujitsu, RS/6000, NEC, HP
CRYSTAL	Academic (UK)/ commercial	Linux, Windows, UNIX
Gaussian	Commercial	Windows, Linux, DEC, Cray, Fujitsu, HP-UX, RS/6000, NEC, SGI, Sun
MOLPRO	Commercial	Linux, Alpha, Cray, Fujitsu, AIX, SGI, Sun, HP-UX, NEC
Q-Chem	Commercial	Windows, Linux, DEC, Cray, Fujitsu, RS/6000, SGI, Sun
AMPAC	Commercial	Windows, Linux, SGI, RS6000, Alpha, Sun, HP-UX
MOPAC	Academic/ commercial	UNIX, PC
YAeHMOP	Academic	Linux, Macintosh, RS/6000, HP-UX, SGI
Discovery studio	Commercial	Windows, Linux
Molecular mechanics/molecular dynamics/Monte Carlo software		
MacroModel	Commercial	SGI, RS/6000
MOE	Commercial	Windows, Linux, SGI, Sun, HP-UX
TINKER	Academic	Windows, Linux, SGI, Macintosh, RS/6000, Alpha, HP-UX, Sun
QSite	Commercial	Windows, Linux, Macintosh
Desmond	Commercial/ academic	Linux

computationally robust approaches for drug repositioning, still the screening procedures applied are not able to provide efficient results. QM/MM-based scoring functions have shown a great promise in scoring the novel molecules and rescoring the already known structure but the field is still in its infancy and lot of things needed to be explored. Since last decade there has been tremendous advancements in computational power but high-level descriptions of the QM region is still unachieved as they require exhaustive configurational sampling (Heimdal & Ryde, 2012). Moreover, much robust on the fly approach should be developed (Plotnikov & Warshel, 2012). Finding simple solutions for big problems like dealing with complex biological proteins and macromolecules can be the major challenge for quantum and molecular mechanics-based approaches in the coming years.

11. CONCLUSION

As the industries are looking for faster, cheaper, and efficient drug discovery approaches, the conventional approaches of drug discovery are largely being replaced by newer technologies. Since last few years, there has been a number of drug repositioning approaches, specially computational approach have gained a much consideration. Polarizable force-field-based docking methods (such as QM/MM) can be implemented to obtain much effective and realistic outcomes. Although limitation to include large number of atoms in QM region needs much computational time and architecture to estimate the exact polarizable charge transfer to atoms this price has to be paid. Besides the need for improvements in current QM/MM status, there are a number of successful QM/MM examples in all the area of drug discovery which signifies the ability of current methods to model real-world problems accurately. The utmost requirement is to change the view point of the chemists and biologists so as to integrate classical methods with QM/MM-based approaches which might be one of the way to solve fundamental chemical problems and explore the enigmatic issues.

ACKNOWLEDGMENTS

A. O. thankfully acknowledges UGC for providing Senior Research Fellowship, V. S. acknowledges DST-INSPIRE for providing Senior Research Fellowship (No. DST/INSPIRE Fellowship/2012/482), C. S. gratefully acknowledges CSIR for a Senior Research Fellowship (SRF). P. S. and S. K. S. acknowledge CSIR-CDRI and Alagappa University for signing the Memorandum of Understanding. P. S. and A. O. thankfully acknowledge CSIR-CDRI Director for providing continuous support in research activities. CSIR-CDRI manuscript communication number is 9035.

REFERENCES

Amara, P., & Field, M. J. (1998). In P. v. R. Schleyer (Ed.), *vol. 1. Encyclopedia of computational chemistry* (pp. 431–437). Chichester: Wiley.

Bakowies, D., & Thiel, W. (1996). Hybrid models for combined quantum mechanical and molecular mechanical approaches. *Journal of Physical Chemistry, 100*, 10580–10594.

Beck, B., & Clark, T. (1998). Some biological applications of semiempirical MO theory. *Perspectives in Drug Discovery and Design, 9–11*, 131–159.

Borrmann, S., Adegnika, A. A., Matsiegui, P. B., Issifou, S., Schindler, A., Mawili-Mboumba, D. P., et al. (2004). Fosmidomycin-clindamycin for *Plasmodium falciparum* infections in African children. *The Journal of Infectious Diseases, 189*, 901–908.

Bruice, T. C., & Kahn, K. (2000). Computational enzymology. *Current Opinion in Chemical Biology, 4*, 540–544.

Cho, A. E., Guallar, V., Berne, B. J., & Friesner, R. (2005). Importance of accurate charges in molecular docking: Quantum mechanical/molecular mechanical (QM/MM) approach. *Journal of Computational Chemistry, 26*, 915–931.

Cho, A. E., Gullar, V., Berne, B. J., & Friesner, R. A. (1995). Importance of accurate charges in molecular docking: Quantum mechanical/ molecular mechanical (QM/MM) approach. *Journal of Computational Chemistry, 26*, 915–931.

Cho, A. E., & Rinaldo, D. (2009). Extension of QM/MM docking and its applications to metalloproteins. *Journal of Computational Chemistry, 30*, 2609–2616.

Claeyssens, F., Harvey, J. N., Manby, F. R., Mata, R. A., Mulholland, A. J., Ranaghan, K. E., et al. (2006). High-accuracy computation of reaction barriers in enzymes. *Angewandte Chemie (International Ed. in English), 45*, 6856–6859.

Croset, S. C. J. (2014). *Drug repositioning and indication discovery using description logics*. Doctoral dissertation, England: University of Cambridge.

Cunningham, M. A., & Bash, P. A. (1997). Computational enzymology. *Biochimie, 79*, 687–689.

D' Amato, R. J., Loughnan, M. S., Flynn, E., & Folkman, J. (1994). Thalidomide is an inhibitor of angiogenesis. *Proceedings of the National Academy of Sciences of the United States of America, 91*, 4082–4085.

Eichler, U., Kolmel, C. M., & Sauer, J. (1997). Combining ab initio techniques with analytical potential functions for structure predictions of large systems: Method and application to crystalline silica polymorphs. *Journal of Computational Chemistry, 18*, 463–477.

Fang, J., Chen, S. J., Tong, J. H., Wang, Z. G., Chen, G. Q., & Chen, Z. (2002). Treatment of acute promyelocytic leukemia with ATRA and As2O3: A model of molecular target-based cancer therapy. *Cancer Biology & Therapy, 1*(6), 614–620.

Field, M. J., Bash, P. A., & Karplus, M. (1990). A combined quantum mechanical and molecular mechanical potential for molecular dynamics simulations. *Journal of Computational Chemistry, 11*, 700–733.

Friesner, R. A., Banks, J. L., Murphy, R. B., Halgren, T. A., Klicic, J. J., & Mainz, D. T. (2004). Glide: A new approach for rapid, accurate docking and scoring. 1. Method and assessment of docking accuracy. *Journal of Medicinal Chemistry, 47*, 1739–1749.

Friesner, R. A., & Beachy, M. D. (1998). Quantum mechanical calculations on biological systems. *Current Opinion in Structural Biology, 8*, 257–262.

Gao, J. (1998). In P. v. R. Schleyer, N. L. Allinger, P. A. Kollman, T. Clark, H. F. Schaefer, III, J. Gasteiger, & P. R. Schreiner (Eds.), *vol. 2. Encyclopedia of computational chemistry* (pp. 1257–1263). Chichester: Wiley.

Gao, J., & Thompson, M. A. (1998). Combined quantum mechanical and molecular mechanical methods. *ACS Symp. Ser.: vol. 712*. Washington: American Chemical Society.

Gao, F. J., Truhlar, D. G., Wang, Y., Mazack, M. J. M., Loffler, P., & Provorse, M. R. (2014). Explicit polarization: A quantum mechanical framework for developing next generation force. *Accounts of Chemical Research, 47*, 2837–2845.

Glide, version 5.5, Schrödinger, LLC, New York, NY, 2009.

Griffith, E. C., Su, Z., Turk, B. E., Chen, S., Chang, Y. H., Wu, Z., et al. (1997). Methionine aminopeptidase (type 2) is the common target for angiogenesis inhibitors AGM-1470 and ovalicin. *Chemistry and Biology*, *4*, 461–471.

Groenhof, G. (2013). Biomolecular simulations methods and protocols. In L. Monticelli & E. Salonen (Eds.), *Introduction to QM/MM simulations* (pp. 43–66). New York: Humana.

Grossfield, A., Ren, P., & Ponder, J. W. (2003). Ion solvation thermodynamics from simulation with a polarizable force field. *Journal of the American Chemical Society*, *125*(50), 15671–15682.

Halgren, T. A., & Damm, W. (2001). Polarizable force fields. *Current Opinion in Structural Biology*, *11*, 236–242.

Hall, R. J., Hindle, S. A., Burton, N. A., & Hillier, I. H. (2000). Aspects of hybrid QM/MM calculations: The treatment of the QM/MM interface region and geometry optimization with an application to chorismate mutase. *Journal of Computational Chemistry*, *21*(16), 1433–1441.

Hamza, A. (2002). Homology modeling and docking mechanism of the mercaptosuccinate and methotrexate to P. falciparum 1-Cys peroxiredoxin: A preliminary molecular study. *Journal of Biomolecular Structure and Dynamics*, *20*(1), 7–19.

Heimdal, J., & Ryde, U. (2012). Convergence of QM/MM free-energy perturbations based on molecular-mechanics or semiempirical simulations. *Physical Chemistry Chemical Physics*, *14*, 12592–12604.

Hu, G., Li, X., Sun, X., Lu, W., Liu, G., Huang, J., et al. (2012). Identification of old drugs as potential inhibitors of HIV-1 integrase-human LEDGF/p75 interaction via molecular docking. *Journal of Molecular Modeling*, *18*(12), 4995–5003.

Hu, H., & Yang, W. (2008). Free energies of chemical reactions in solution and in enzymes with ab initio quantum mechanics/molecular mechanics methods. *Annual Review of Physical Chemistry*, *59*, 573–601.

Hurle, M. R., Yang, L., Xie, Q., Rajpal, D. K., Sanseau, P., & Agarwal, P. (2013). Computational drug repositioning: From data to therapeutics. *Clinical Pharmacology and Therapeutics*, *93*, 335–341.

Jomaa, H., Wiesner, J., Sanderbrand, S., Altincicek, B., Weidemeyer, C., Hintz, M., et al. (1999). Inhibitors of the nonmevalonate pathway of isoprenoid biosynthesis as antimalarial drugs. *Science*, *285*, 1573–1576.

Jones, G., Willett, P., & Glen, R. C. (1995). Molecular recognition of receptor sites using a genetic algorithm with a description of desolvation. *Journal of Molecular Biology*, *245*, 43–53.

Jones, G., Willett, P., Glen, R. C., Leach, A. R., & Taylor, R. (1997). Development and validation of a genetic algorithm for flexible docking. *Journal of Molecular Biology*, *267*, 727–748.

Kamal, M. A., Nasim, F. H., & Al-Jafari, A. A. (1996). Investigation of the effect of antineoplastic drugs, cyclophosphamide, cisplatin and methotrexate on the turnover kinetics of human erythrocyte acetylcholinesterase. *Biochemistry and Molecular Biology International*, *39*(2), 293–302.

Kaminiski, G. A. (2005). Accurate prediction of absolute acidity of constants in water with a polarizable force field: Substituted phenols, methanol and imidazole. *The Journal of Physical Chemistry B*, *109*, 5884–5890.

Kaminiski, G. A., Stern, H. A., Berne, B. J., & Friesner, R. A. (2004). Development of an accurate and robust polarizable molecular mechanics force field from quantum chemistry. *The Journal of Physical Chemistry. A*, *108*, 621–627.

Kaminski, G. A., Stern, H. A., Berne, B. J., Friesner, R. A., Cao, Y. X., Murphy, R. B., et al. (2002). Development of a polarizable force field for proteins via ab initio quantum

chemistry: First generation model and gas phase tests. *Journal of Computational Chemistry, 23*, 1515–1531.
Khandelwal, A., Lukacova, V., Comez, D., Kroll, D. M., Raha, S., & Balaz, S. (2005). A combination of docking, QM/MM methods, and MD simulation for binding affinity estimation of metalloprotein ligands. *Journal of Medicinal Chemistry, 48*, 5437–5447.
Klahn, M., Braun-Sand, S., Rosta, E., & Warshel, A. (2005). On possible pitfalls in ab initio quantum mechanics/molecular mechanics minimization approaches for studies of enzymatic reactions. *The Journal of Physical Chemistry B, 109*, 15645–15650.
Korth, C., May, B. C., Cohen, F. E., & Prusiner, S. B. (2001). Acridine and phenothiazine derivatives as pharmacotherapeutics for prion disease. *Proceedings of the National Academy of Sciences of the United States of America, 98*, 9836–9984.
Kruger, E. A., & Figg, W. D. (2000). TNP-470: An angiogenesis inhibitor in clinical development for cancer. *Expert Opinion on Investigational Drugs, 9*, 1383–1396.
Lamoureux, G., & Roux, B. (2003). Modeling induced polarization with classical Drude oscillators: Theory and molecular dynamics simulation algorithm. *The Journal of Chemical Physics, 119*, 3025–3039.
Lin, H., & Truhlar, D. G. (2005). Redistributed charge and dipole schemes for combined quantum mechanical and molecular mechanical calculations. *The Journal of Physical Chemistry. A, 109*(17), 3991–4004.
Maple, J. R., Cao, Y., Damm, W., Halgren, T. A., Kaminski, G. A., Zhang, L. Y., et al. (2005). A polarizable force field and continuum solvation methodology for modeling of protein-ligand interactions. *Journal of Chemical Theory and Computation, 1*, 694–715.
Marton, L. J., & Pegg, A. E. (1995). Polyamines as targets for therapeutic intervention. *Annual Review of Pharmacology and Toxicology, 35*, 55–91.
Maseras, F., & Morokuma, K. (1995). IMOMM—A new integrated ab-initio plus molecular mechanics geometry optimization scheme of equilibrium structures and transition-states. *Journal of Computational Chemistry, 16*, 1170–1179.
Menikarachchi, L. C., & Gascon, J. A. (2010). QM/MM approaches in medicinal chemistry research. *Current Topics in Medicinal Chemistry, 10*, 46–54.
Merz, K. M. Jr. (1998). Combined quantum mechanical and molecular mechanical methods. In J. Gao & M. A. Thompson (Eds.), *ACS Symp. Ser.: vol. 712. Quantum mechanical-molecular mechanical coupled potentials* (pp. 2–15). Washington: American Chemical Society.
Merz, K. M., Jr., Stanton, R. V., & Schreiner, R. (1998). Quantum mechanical/molecular mechanical (QM/MM) coupled potentials. In P. v. R. Schleyer, N. L. Allinger, P. A. Kollman, T. Clark, H. F. Schaefer, III, J. Gasteiger, & P. R. Schreiner (Eds.), *Encyclopedia of computational chemistry: vol. 4.* (pp. 2330–2343). Chichester: Wiley.
Missinou, M. A., Borrmann, S., Schindler, A., Issifou, S., Adegnika, A. A., Matsiegui, P. B., et al. (2002). Fosmidomycin for malaria. *Lancet, 360*, 1941–1942.
Mordasini, T., & Thiel, W. (1998). Combined quantum mechanical and molecular mechanical approaches. *Chimia, 52*, 288–291.
Nakajima, M., Yamada, T., Kusuhara, T., Furukawa, H., Takahashi, M., Yamauchi, A., et al. (2004). Results of quinacrine administration to patients with Creutzfeldt-Jakob disease. *Dementia and Geriatric Cognitive Disorders, 17*, 158–163.
Nam, K., Cui, Q., Gao, J., & York, D. M. (2007). Specific reaction parameterization of the AM1/d Hamiltonian for phosphoryl transfer reactions: H, O, and P atoms. *Journal of Chemical Theory and Computation, 3*, 486–504.
Nicoll, R. M., Hindle, S. A., MacKenzie, G., Hillier, I. H., & Burton, N. A. (2001). Quantum mechanical/molecular mechanical methods and the study of kinetic isotope effects: Modelling the covalent junction region and application to the enzyme xylose isomerase. *Theoretical Chemistry Accounts, 106*(1–2), 105–112.

Nwaka, S., & Ridley, R. G. (2003). Virtual drug discovery and development for neglected diseases through public-private partnerships. *Nature Reviews Drug Discovery*, *2*, 919–928.

Pammolli, F., Magazzini, L., & Riccaboni, M. (2011). The productivity crisis in pharmaceutical R&D. *Nature Reviews Drug Discovery*, *10*, 428–438.

Patel, S., & Brooks, C. L. (2006). Fluctuating charge force fields: Recent developments and applications from small molecules to macromolecular biological systems. *Molecular Simulation*, *32*, 231–249.

Piquemal, J. P., Williams-Hubbard, B., Fey, N., Deeth, R. J., Gresh, N., & Giessner-Prettre, C. (2003). Inclusion of the ligand field contribution in a polarizable molecular mechanics: SIBFA-LF. *Journal of Computational Chemistry*, *24*, 1963–1970.

Plotnikov, N. V., & Warshel, A. (2012). Exploring, refining, and validating the paradynamics QM/MM sampling. *The Journal of Physical Chemistry B*, *116*, 10342–10356.

Rappe, A. K., & Goddard, W. A., III. (1991). Charge equilibration for molecular dynamics simulations. *The Journal of Physical Chemistry*, *95*, 3358–3363.

Ren, P. Y., & Ponder, J. W. (2002). Consistent treatment of inter and intramolecular polarization in molecular mechanics calculations. *Journal of Computational Chemistry*, *23*, 1497–1506.

Ren, P., & Ponder, J. W. (2003). Polarizable atomic multipole water model for molecular mechanics simulation. *The Journal of Physical Chemistry B*, *107*, 5933–5947.

Reuter, N., Dejaegere, A., Maigret, B., & Karplus, M. (2000). Frontier bonds in QM/MM methods: A comparison of different approaches. *The Journal of Physical Chemistry. A*, *104*(8), 1720–1735.

Rossi, I., & Truhlar, D. G. (1995). Parameterization of NDDO wavefunctions using Genetic Algorithms: An evolutionary approach to parameterizing potential energy surfaces and direct dynamics calculations for organic reactions. *Chemical Physics Letters*, *233*, 231.

Rothstein, J. D., Patel, S., Regan, M. R., Haenggeli, C., Huang, Y. H., Bergles, D. E., et al. (2005). Beta-lactam antibiotics offer neuroprotection by increasing glutamate transporter expression. *Nature*, *433*, 73–77.

Ruiz-Lpez, M. F., & Rivail, J.-L. (1998). Combined quantum mechanics and molecular mechanics Approaches to chemical and biochemical reactivity. In P. v. R. Schleyer, N. L. Allinger, P. A. Kollman, T. Clark, H. F. Schaefer, III, J. Gasteiger, & P. R. Schreiner (Eds.), *Encyclopedia of computational chemistry: vol. 1* (pp. 437–448). Chichester: Wiley.

Seabra, G. M., Walker, R. C., Elstner, M., Case, D. A., & Roitberg, A. E. (2007). Implementation of the SCC-DFTB method for hybrid QM/MM simulations within the Amber Molecular Dynamics Package. *Journal of Physical Chemistry A*, *111*, 5655–5664.

Selvaraj, C., Singh, P., & Singh, S. K. (2014a). Molecular insights on analogs of HIV PR inhibitors toward HTLV-1 PR through QM/MM interactions and molecular dynamics studies: Comparative structure analysis of wild and mutant HTLV-1 PR. *Journal of Molecular Recognition*, *27*, 696–706.

Selvaraj, C., Singh, P., & Singh, S. K. (2014b). Molecular modeling studies and comparative analysis on structurally similar HTLV and HIV protease using HIV-PR inhibitors. *Journal of Receptor and Signal Transduction Research*, *34*(5), 361–371.

Senn, H. M., & Thiel, W. (2007). QM/MM methods for biological systems. *Topics in Current Chemistry*, *268*, 173–290.

Senn, H. M., & Thiel, W. (2009). QM/MM methods for biomolecular systems. *Angewandte Chemie International Edition*, *48*, 1198–1229.

Shaik, S., Cohen, S., Wang, Y., Chen, H., Kumar, D., & Thiel, D. (2010). P450 enzymes: Their structure, reactivity and selectivity-modeled by QM/MM calculations. *Chemical Reviews*, *110*, 949–1017.

Sherwood, P. (2000). Hybrid quantum mechanics/molecular mechanics approaches. In J. Grotendorst (Ed.), *NIC series. Modern methods and algorithms of quantum chemistry* (pp. 285–305). Julich: John von Neumann Institute for Computing.

Singh, U. C., & Kollman, P. A. (1986). A combined ab initio quantum mechanical and molecular mechanical method for carrying out simulations on complex molecular systems: Applications to the CH3Cl + Cl– exchange reaction and gas phase protonation of polyethers. *Journal of Computational Chemistry*, 7(6), 718–730.

Stern, H. A., Kaminski, G. A., Banks, J. L., Zhou, R., Berne, B. J., & Friesner, R. A. (1999). Fluctuating charge, polarizable dipole, and combined models: Parameterization from ab initio quantum chemistry. *The Journal of Physical Chemistry B, 103*, 4730–4737.

Stern, H. A., Rittner, F., Berne, B. J., & Friesner, R. A. (2001). Combined fluctuating charge and polarizable dipole models: Application to a five-site water potential function. *The Journal of Chemical Physics, 115*(5), 2237–2251.

Sundar, S., Jha, T. K., Thakur, C. P., Engel, J., Sindermann, H., Fischer, C., et al. (2002). Oral miltefosine for Indian visceral leishmaniasis. *New England Journal of Medicine, 347*, 1739–1746.

Svensson, M., Humbel, S., Froese, R. D. J., Matsubara, T., Sieber, S., & Morokuma, K. (1996). ONIOM: A multilayered integrated MO + MM method for geometry optimizations and single point energy predictions. A test for Diels–Alder reactions and Pt(P(t-Bu)3)2 + H2 oxidative addition. *The Journal of Physical Chemistry, 100*, 19357–19363.

Swinney, D. C., & Anthony, J. (2011). How were new medicines discovered? *Nature Reviews Drug Discovery, 10*, 507–519.

Thiel, W. (2009). QM/MM methodology: Fundamentals, scope, and limitations. In J. Grotendorst, N. Attig, S. Blgel, & D. Marx (Eds.), *Multiscale simulation methods in molecular sciences: vol. 42* (pp. 203–214). Jülich: NIC.

Tuttle, T. (2012). *Drug design strategies: Computational techniques and applications. Quantum mechanical/molecular mechanical approaches in drug design* (pp. 1–26). London: The Royal Society of Chemistry.

Tuttle, T., & Thiel, W. (2008). OM x-D: Semiempirical methods with orthogonalization and dispersion corrections. Implementation and biochemical application. *Physical Chemistry Chemical Physics, 10*, 2159–2166.

Vreven, T., & Morokuma, K. (2003). Investigation of the S0→S1 excitation in bacteriorhodopsin with the ONIOM(MO:MM) hybrid method. *Theoretical Chemistry Accounts, 109*(3), 125–132.

Vreven, T., Morokuma, K., Farkas, O., Schlegel, H. B., & Frisch, M. J. (2003). Geometry optimization with QM/MM, ONIOM, and other combined methods. I. Microiterations and constraints. *Journal of Computational Chemistry, 24*, 760.

Walker, R. C., Crowley, M. F., & Case, D. A. (2008). The implementation of a fast and accurate QM/MM potential method in Amber. *Journal of Computational Chemistry, 29*, 1019.

Warshel, A., & Levitt, M. (1976). Theoretical studies of enzymic reactions: Dielectric, electrostatic and steric stabilization of the carbonium ion in the reaction of lysozyme. *Journal of Molecular Biology, 103*, 227–249.

Warshel, A., Sharma, P. K., Kato, M., Xiang, Y., Liu, H., & Olsson, M. H. M. (2006). Electrostatic basis for enzyme catalysis. *Chemical Reviews, 106*(8), 3210–3235.

Wichapong, K., Rohe, A., Platzer, C., Slynko, I., Erdmann, F., Schmidt, M., et al. (2014). Application of docking and QM/MM-GBSA rescoring to screen for novel Myt1 kinase inhibitors. *Journal of Chemical Information and Modeling, 54*, 881–893.

Xie, L., Evangelidis, T., Xie, L., & Bourne, P. E. (2011). Drug discovery using chemical systems biology: Weak inhibition of multiple kinases may contribute to the anti-cancer effect of Nelfinavir. *PLoS Computational Biology*, 7(4), e1002037.

Xie, W. S., & Gao, J. L. (2007). Design of a next generation force field: The X-POL potential. *Journal of Chemical Theory and Computation, 3*, 1890–1900.
Yu, H., Hansson, T., & Gunsteren, W. F. v. (2003). Development of a simple, self-consistent polarizable model for liquid water. *The Journal of Chemical Physics, 118*(1), 221–234.
Yu, H., Hansson, T., & Gunsteren, W. F. v. (2004). Charge-on-spring polarizable water models revisited: From water clusters to liquid water to ice. *The Journal of Chemical Physics, 121*(19), 9549–9564.
Zhang, Y., Liu, H., & Yang, W. (2000). Free energy calculation on enzyme reactions with an efficient iterative procedure to determine minimum energy paths on a combined ab initio QM/MM potential energy surface. *The Journal of Chemical Physics, 112*, 3483–3492.
Zhu, J., Chen, Z., Lallemand-Breitenbach, V., & de The, H. (2002). How acute promyelocytic leukaemia revived arsenic. *Nature Reviews Cancer, 2*, 705–713.

LINKS

ADF. Available at: http://www.scm.com/.
AMPAC. Available at: http://www.semichem.com/.
CRYSTAL. Available at: http://www.crystal.unito.it/index.php.
Desmond. Available at: http://www.deshawresearch.com/resources.html.
Discovery studio. Available at: http://accelrys.com/products/discovery-studio/.
Drugbank. Available at: http://drugbank.ca/.
FDA: The Rare Disease Repurposing Database. Available at: http://www.fda.gov/Drugs/GuidanceComplianceRegulatoryInformation/Surveillance/AdverseDrugEffects/default.htm.
GAMESS. Available at: http://www.msg.ameslab.gov/GAMESS/GAMESS.html.
Gaussian. Available at: http://www.gaussian.com/.
Jaguar. Available at: http://www.schrodinger.com/.
MacroModel. Available at: http://www.schrodinger.com/.
MATADOR. Available at: http://matador.embl.de/.
MOE. Available at: http://www.chemcomp.com/.
MOLPRO. Available at: https://www.molpro.net/.
MOPAC. Available at: http://www.schrodinger.com/.
Ondex Database. Available at: http://www.ondex.org/.
Pharmmapper. Available at: http://59.78.96.61/pharmmapper/.
POSSUM. Available at: http://possum.cbrc.jp/PoSSuM/.
Promiscuous. Available at: http://bioinformatics.charite.de/promiscuous.
Q-Chem. Available at: http://www.q-chem.com/.
QSite. Available at: www.schrodinger.com/QSite/.
Spartan. Available at: http://www.wavefun.com/products/spartan.html.
SuperTarget. Available at: http://bioinf-apache.charite.de/supertarget/.
TherapeuticsTargets Database. Available at: http://bidd.nus.edu.sg/group/cjttd/ttd.asp.
TINKER. Available at: http://dasher.wustl.edu/tinker/.
YAeHMOP. Available at: http://yaehmop.sourceforge.net/.

CHAPTER FIVE

Enzymatic Halogenases and Haloperoxidases: Computational Studies on Mechanism and Function

Amy Timmins, Sam P. de Visser[1]
Manchester Institute of Biotechnology and School of Chemical Engineering and Analytical Science, The University of Manchester, Manchester M1 7DN, Manchester, United Kingdom
[1]Corresponding author: e-mail address: sam.devisser@manchester.ac.uk

Contents

Abstract

Despite the fact that halogenated compounds are rare in biology, a number of organisms have developed processes to utilize halogens and in recent years, a string of enzymes have been identified that selectively insert halogen atoms into, for instance, a C—H aliphatic bond. Thus, a number of natural products, including antibiotics, contain halogenated functional groups. This unusual process has great relevance to the chemical industry for stereoselective and regiospecific synthesis of haloalkanes. Currently, however, industry utilizes few applications of biological haloperoxidases and halogenases, but efforts are being worked on to understand their catalytic mechanism, so that their catalytic function can be upscaled. In this review, we summarize

Advances in Protein Chemistry and Structural Biology, Volume 100
ISSN 1876-1623
http://dx.doi.org/10.1016/bs.apcsb.2015.06.001

experimental and computational studies on the catalytic mechanism of a range of haloperoxidases and halogenases with structurally very different catalytic features and cofactors. This chapter gives an overview of heme-dependent haloperoxidases, nonheme vanadium-dependent haloperoxidases, and flavin adenine dinucleotide-dependent haloperoxidases. In addition, we discuss the *S*-adenosyl-L-methionine fluoridase and nonheme iron/α-ketoglutarate-dependent halogenases. In particular, computational efforts have been applied extensively for several of these haloperoxidases and halogenases and have given insight into the essential structural features that enable these enzymes to perform the unusual halogen atom transfer to substrates.

1. INTRODUCTION

Living organisms in both marine and terrestrial environments continuously adapt and utilize available resources. One such group of resources are the halogens, namely, fluorine, chlorine, bromine, and iodine. Although their biochemical usage is relatively rare, over the years hundreds of examples have been identified of microorganisms that incorporate halogen atoms into a wide variety of natural product scaffolds, such as olefins, aromatic rings, heterocyclic rings, and aliphatic carbon chains (Gribble, 2003; Vaillancourt, Yeh, Vosburg, Garneaur-Tsodikova, & Walsh, 2006). This functionalization is achieved by a class of enzymes called halogenases and haloperoxidases, which form the topic of this review.

So far, over 4500 halogenated natural products have been identified (Gribble, 2003) and display a wide variety of chemical functions and applications (Laus, 2001; Stonik & Fedorov, 2014; van Pee, 1996). The efficacy of these natural products in biosystems is dependent on the presence of the halogen atom and often they lose activity when the halogen substituent is removed. For example, substitution of the aryl chloride substituents on vancomycin by styryl groups of increasing size resulted in the complete loss of activity against vancomycin-resistant *S. aureus* (Nakama et al., 2010). It is, therefore, important to understand the mechanism of substrate halogenation by halogenases as well as the factors that determine the regioselectivity, the stereoselectivity, and the chemoselectivity of the reaction, as this will enable future biotechnological, biocatalytic, and industrial applications. Since, enzymatic reaction mechanisms proceed extremely fast, often establishing a catalytic cycle is very difficult as intermediate structures do not accumulate long enough to be spectroscopically characterized. Computational tools can then be very helpful and assist experiments with assigning a catalytic cycle and determining the function of active site residues and cofactors. Another common approach is to use so-called biomimetic model complexes that contain the active site features of enzymes but without the protein scaffold (Abu-Omar, Loaiza, & Hontzeas, 2005; Bruijnincx, van Koten, & Klein Gebbink, 2008; Costas, 2011; de Visser, Rohde, Lee, Cho, & Nam, 2013; Kryatov, Rybak-Akimova, & Schindler, 2005). These biomimetic models often react in different ways than their enzymatic counterparts but do establish ligand effects of cofactors.

2. CLASSIFICATION OF HALOGENASES

Over the past 50 years, the classification of halogenases has changed several times as a result of newly discovered enzymes. Currently, halogenases are defined by the nature of the active halogenating agent and are subdivided into families based upon their individual mechanisms, cofactors, and substrates. Below are described the electrophilic halogenases (heme-dependent, vanadium-dependent, and flavin-dependent), nucleophilic halogenases, and the radical halogenases (nonheme iron/α-ketoglutarate (α-KG)-dependent halogenases).

2.1 Heme-Dependent Haloperoxidases

The first halogenase that was discovered was the heme-dependent chloroperoxidase from *Caldariomyces fumago* in 1966 (Hager, Morris, Brown, & Eberwein, 1966; Morris & Hager, 1966; Taurog & Howells, 1966). The studies revealed that it contains a ferriprotoporphyrin IX prosthetic group (Sundaramoorthy, Terner, & Poulos, 1995), and is dependent on hydrogen peroxide. The catalytic cycle of heme-dependent chloroperoxidases is shown in Scheme 1 and starts at the top with the iron(III)-water complex, which is the resting state. Hydrogen peroxide displaces the water molecule and is deprotonated by a distal glutamic acid residue (Glu_{183}) to form an iron(III)-hydroperoxo complex. This structure is protonated on the terminal OH group, possibly by the Glu_{183} amino acid, to form a high-valent ferryl(IV)-oxo heme cation radical species, which is also called Compound I. The latter has been spectroscopically trapped and characterized (Green, Dawson, & Gray, 2004) and has been extensively studied by computational modeling (de Visser, Porro, Quesne, Sainna, & Munro, 2013; Shaik, Kumar, de Visser, Altun, & Thiel, 2005; Vardhaman, Sastri, Kumar, & de Visser, 2011; Vardhaman et al., 2013). Thus, it was shown that Compound I is a versatile oxidant able to react with substrates via hydrogen

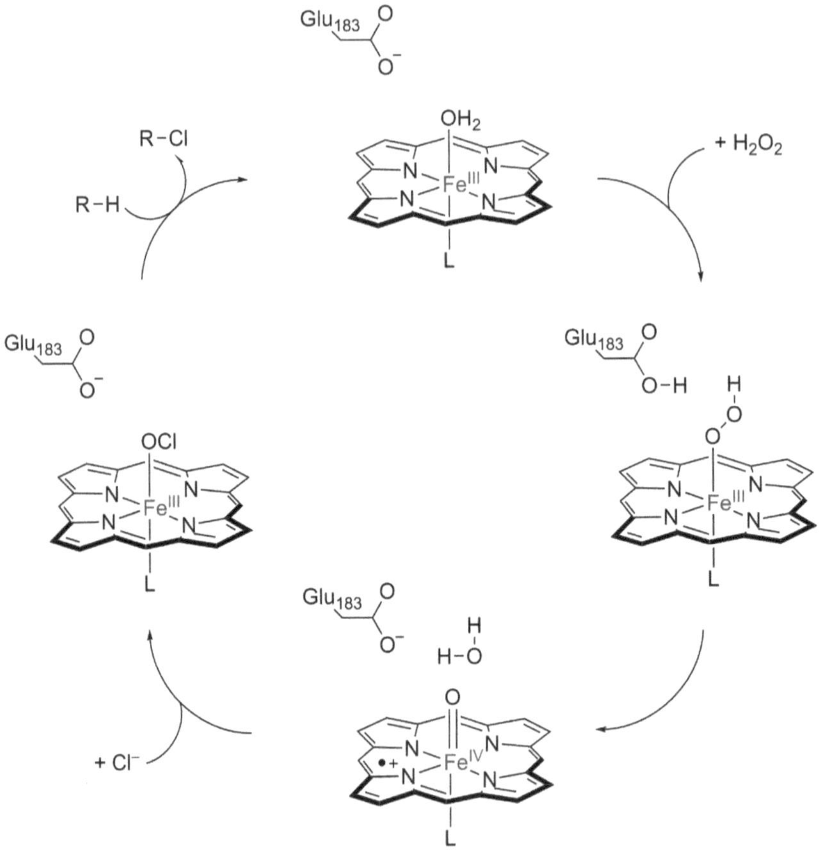

Scheme 1 Catalytic cycle of heme haloperoxidases.

atom abstraction from aliphatic substrates (de Visser, 2010), epoxidation of olefins (Kumar, Karamzadeh, Sastry, & de Visser, 2010; Kumar et al., 2013), sulfoxidation of sulfides (Kumar et al., 2014), and aromatic hydroxylation of arenes (Kumar, Sastry, & de Visser, 2012). These studies showed that Compound I is a triradical species with unpaired electrons in π^*_{FeO} orbitals (π^*_{xz} and π^*_{yz}) that can be either ferromagnetically or antiferromagnetically coupled to a radical on the heme in an a_{2u}-type orbital (Sainna et al., 2015). As such, Compound I has two close-lying spin states (doublet and quartet); on each of these spin states the reaction proceeds, but with different barrier heights due to differences in electron transfer processes. However, reaction mechanisms and rate constants occasionally differ on both spin states and, thereby, leading to so-called spin-state reactivity. For instance, hydrogen atom abstraction takes place as a stepwise mechanism with an initial hydrogen atom abstraction followed by a radical rebound to form alcohol products (de Visser, 2010; Shaik, Kumar, & de Visser, 2008). However, density functional theory (DFT) modeling showed that the lifetime of the radical intermediate is significantly longer in the high-spin state as compared to that in the low-spin state. Because of this, rearrangement patterns and reactions leading to by-products may occur on the quartet spin-state surface, but are unlikely on the doublet spin state (de Visser, Ogliaro, & Shaik, 2001; Kumar, de Visser, & Shaik, 2005), and as a consequence, the two spin-state surfaces lead to different product distributions.

In heme-dependent chloroperoxidases, Compound I reacts with halides (X^-, X = Cl, Br, or I) to form an iron(III)-hypohalide complex. The latter attacks aliphatic substrates (R–H) and converts them to halogenated products (R–X) and returns the catalytic cycle to the resting state (Wagenknecht & Woggon, 1997). Compound I was found to react with substrates with little regioselectivity or stereoselectivity, suggesting a broad substrate specificity that decreases the likelihood of its involvement in the biosynthesis of secondary metabolites (van Pee et al., 2006). Although, Compound I is short-lived, actually it has been spectroscopically characterized with X-ray absorption (Stone, Behan, & Green, 2005), electron-nuclear double-resonance spectroscopy (Kim, Perera, Hager, Dawson, & Hoffman, 2006), and extended X-ray absorption fine structure (EXAFS) studies (Green et al., 2004).

Structural and crystallographic studies on chloroperoxidases revealed no substrate-binding pocket, which seems to implicate that the electrophile, i.e., OCl^-, is released into bulk solution, where it will react with substrates. Under those circumstances, however, the electrophile will react as a normal electrophile in solution, with little regioselectivity and stereoselectivity (Sundaramoorthy, Terner, & Poulos, 1998). These studies contradicted kinetic isotope effect experiments that implicated involvement of the enzyme into the rate-determining hydrogen atom abstraction step (Libby, Beachy, & Phipps, 1996). So far, few applications of chloroperoxidases have been found in industry due to the low stability of the enzyme as well as its inactivation by hydrogen peroxide and the low solubility of many of the organic substrates and products in the reaction process (Spreti, Germani, Icani, & Savelli, 2004). Manoj and Hager (2008) showed that active site mutations changed the reaction mechanism of a two-electron reaction process into a one-electron peroxidation of phenolic substrates. These studies highlighted key amino acids in the reaction mechanism that affected substrate binding as well as the overall catalytic cycle through proton transfer pathways.

The catalytic cycle of chloroperoxidases was studied with DFT methods by Silaghi-Dumitrescu (2008). The calculations focused on the formation energy of the iron(III)-hypochloride from Compound I and a chloride anion and a value of about 15 kcal mol^{-1} in the gas phase with a driving force of 7 kcal mol^{-1} was found. As such, Compound I and the hypochloride-bound structure should exist in rapid equilibrium in the gas phase until the protonation of the latter and expulsion of the HOCl product from the heme.

2.2 Vanadium-Dependent Haloperoxidases

A second group of haloperoxidases was discovered in the 1980s and found to use a vanadium center as catalytic cofactor (Vilter, 1983, 1984) to detoxify hydrogen peroxide to oxidize a halide (de Boer & Wever, 1988; Messerschmidt & Wever, 1996; van Schijndel, Barnett, Roelse, Vollenbroed, & Wever, 1994). In contrast to the heme haloperoxidases, where the oxidation state of the metal ranges from iron(III) to iron(IV), in the vanadium-dependent haloperoxidases the metal acts as a Lewis acid and does not change oxidation state throughout the catalytic cycle (Butler & Sandy, 2009; Vaillancourt et al., 2006). There are three subtypes of vanadium-dependent haloperoxidases linked to their oxidative ability: (i) chloroperoxidases, which oxidize chloride, bromide, and iodide; (ii) bromoperoxidases which oxidize bromide and iodide; and finally (iii) iodoperoxidases that only utilize iodide.

Interestingly, some bromoperoxidases found predominantly in marine algae have demonstrated a preference for indole derivatives even in the presence of other substrates (Martinez et al., 2001), which implicates that they have a certain degree of substrate specificity and may have a function related to marine natural products biosynthesis (Carter-Franklin & Butler, 2004). Bromoperoxidases have also been implicated in natural defence mechanisms of biosystems by reducing the colonization of bacteria at the surface of algae through the disruption of quorum sensing (Borchardt et al., 2001). Chloroperoxidases, on the other hand, are found predominantly in terrestrial fungi and have been associated with the degradation of plant cell walls, thereby facilitating the penetration of fungal hyphae into the host (Wever, 2001).

The catalytic cycle of vanadium-dependent haloperoxidases is currently under debate and little is known on the oxygen-bound intermediates as none have been trapped and characterized experimentally. Akin to the heme-dependent haloperoxidases discussed above, also few industrial applications are known of vanadium-dependent haloperoxidases due to their reduced regioselectivity and stereoselectivity in a reaction with products (Chen & van Pee, 2008). However, a few highly regiospecific and stereospecific vanadium-dependent haloperoxidases have been identified, such as Mcl24 and NapHI, which suggest that the electrophile (HOX) that is formed during the catalytic cycle is controlled through substrate binding and positioning (Bernhardt, Okino, Winter, Miyanaga, & Moore, 2011; Winter & Moore, 2009). Understanding the factors that determine the substrate selectivity will be a key factor to any future applications. Moreover, research is focused to establish what the chemical differences are between chloroperoxidases, on the one hand, with bromoperoxidases and iodoperoxidases, on the other hand.

The proposed catalytic cycle (Butler, 1999; Conte, Coletti, Floris, Licini, & Zonta, 2011) is schematically depicted in Scheme 2 and starts from a vanadium(V)-dioxo-dihydroxo intermediate with an oxo group in the equatorial position. The axial oxo group is linked to the protein via a hydrogen bond with a histidine group (His_{404}). Hydrogen peroxide binding reduces the hydroxo group to an oxo group with the assistance of a distal histidine group (His_{404}), which may relay a proton in the process. The vanadium(V)-oxo group reacts with halide (X^-) to form hypohalide, which reacts with substrates to form halogenated products. The peroxo-bound vanadium(V)-oxo intermediate has been shown to react with aliphatic

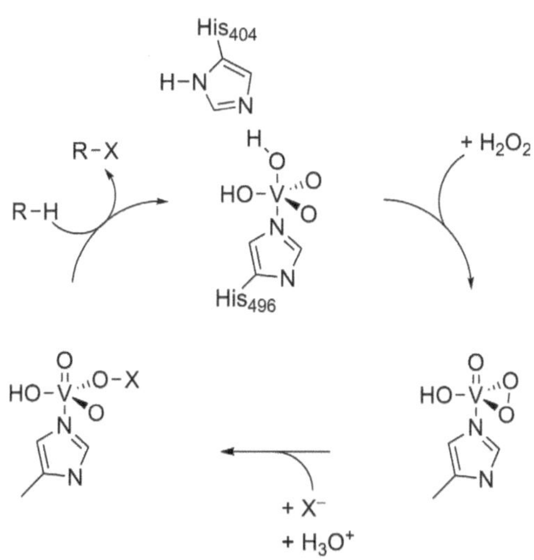

Scheme 2 Catalytic cycle of vanadium-dependent haloperoxidases.

hydrocarbons, sulfides, and olefins to generate alcohols, sulfoxides, and epoxides, respectively (Conte et al., 2011).

A series of computational studies were performed to establish the catalytic cycle of vanadium-dependent haloperoxidases and predict spectroscopic properties of short-lived intermediates. Experimental studies revealed the vanadium to be bound in a trigonal bipyramidal form linked to the protein via one histidine linkage of His_{496}, whereas the other four ligands are oxygen-based. The experimental studies, however, failed to establish whether these oxygen ligands are oxo, hydroxo, or water as monodentate ligands or peroxo as bidentate ligand. Computational studies using DFT model complexes (Schneider, Zampella, Greco, Pecoraro, & De Gioia, 2007; Zampella, Fantucci, Pecoraro, & De Gioia, 2005, 2006) tested possible isomers of peroxo-bound and hydroperoxo-bound isomers of vanadium-dependent haloperoxidases and tested various protonation states of the oxygen ligands. These studies concluded that the most stable resting-state structure is a vanadium(V)-dioxo-dihydroxo structure with the hydroxo *trans* to His_{496}. Hydrogen peroxide binding will then lead to a vanadium(V)-peroxo(oxo)(hydroxo) intermediate, which is attacked by halide to form the vanadium(V)-hypohalide-dioxo(hydroxo) complex. Furthermore, the computational studies identified a key role of a lysine residue in the substrate-binding pocket that polarizes the peroxo cofactor and assists with charge redistribution in the reaction with substrates. More recent quantum mechanics/molecular mechanics (QM/MM) studies (Geethalakshmi, Waller, Thiel, & Bühl, 2009; Waller, Bühl, Geethalakshmi, Wang, & Thiel, 2007; Waller, Geethalakshmi, & Bühl, 2008) focused on characterizing intermediates of the catalytic cycle of vanadium-dependent haloperoxidases and attempted to compare work with experimental ^{51}V nuclear magnetic resonance (NMR) studies. Thus, the resting-state and peroxo-bound structures were investigated with ^{51}V-NMR methods and established isotropic chemical shifts of -931 and -1135 ppm, respectively (Rehder, Casny, & Grosse, 2004). The computational studies, however, predicted different structures than those hypothesized from experiment and come to the conclusion that the peroxo form is unprotonated or alternatively contains one hydroxo group in the equatorial position. In addition to studies on vanadium-dependent chloroperoxidase and bromoperoxidase, the corresponding iodoperoxidase has received somewhat lesser attention. Computational studies, however, revealed large similarities in catalytic cycle and geometries with chloroperoxidases and bromoperoxidases (Pacios & Gálvez, 2010).

2.3 Flavin Adenine Dinucleotide-Dependent Haloperoxidases

The biosynthesis of halogenated natural products is important in a range of biosystems, and, as such these biosystems have developed halogenases or haloperoxidases to perform this unique function. Interestingly, little evidence has emerged of the involvement of heme- and vanadium-dependent haloperoxidases into the biosynthesis of halogenated natural products. In the 1990s, a third class of haloperoxidases was discovered that contains a flavin adenine dinucleotide (FAD) moiety in the active site. The first FAD-dependent halogenase was characterized from the biosynthetic gene cluster for 7-chlorotetracycline (Dairi, Nakano, Aisaka, Katsumata, & Hasegawa, 1995), which implicates that this class of enzymes is responsible for the regioselective and stereospecific substrate halogenation leading to natural products. Subsequently, the protein was identified, expressed, and biochemically and structurally characterized (Chiu et al., 2001; Dong et al., 2005; Eustaquio, Gust, Luft, Chater, & Heide, 2003; Hammer, Hill, Lam, van Pee, & Ligon, 1997; Hohaus et al., 1997; Keller et al., 2000; Nowak-Thompson, Chaney, Wing, Gould, & Loper, 1999; Puk et al., 2002; Sánchez et al., 2002; Seibold et al., 2006; Trefzer et al., 2002; Weitnauer et al., 2001; Wijnands & van Pee, 2004; Yeh, Garneau, & Walsh, 2005; Yu et al., 2002; Zehner et al., 2005). The studies established the existence of two subfamilies of FAD-dependent halogenases, namely, (i) for the halogenation of small free molecules and (ii) enzymes that react with substrates tethered to a thiolate domain in nonribosomal polypeptide synthase system (Wijnands & van Pee, 2004). Despite differences in activity and classification, both halogenase subfamilies require $FADH_2$ and molecular oxygen to oxidize a chloride anion into hypochloride. In addition, these enzymes contain an FAD reductase domain to regenerate the enzyme to the resting state.

There has been controversy and scientific debate for some time regarding the question whether the generated electrophile, i.e., OCl^-, is actually bound to the enzyme or diffuses into the solution as HOCl. Crystal structure coordinates determined of the isozymes PrnA and RebH (Dong, Kotzsch, Dorward, van Pee, & Naismith, 2004; Dong et al., 2005) support the idea that HOCl acts as an electrophile, but is not freely diffusible into bulk solution. Instead, the HOCl appears to be restricted to a channel that directs it toward the substrate (Dong et al., 2005). Active site mutations, and, in particular, the K79A and E346Q mutations in PrnA, confirmed this hypothesis and demonstrated the importance of these residues to catalysis. Thus, Lys_{79} is thought to activate the electrophile either through hydrogen-bonding

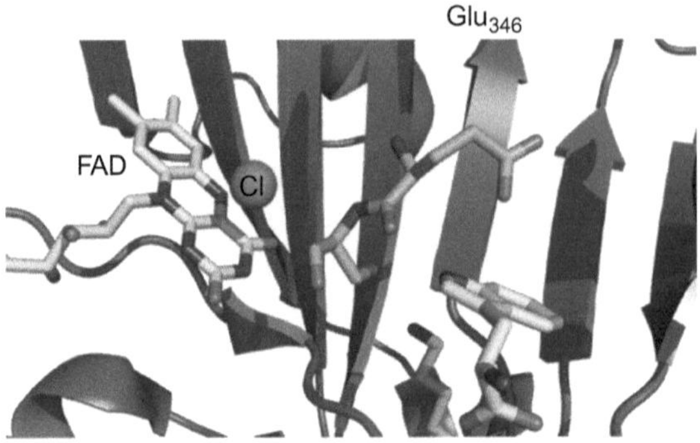

Figure 1 Active site structure of FAD-dependent haloperoxidase.

interactions or alternatively through the formation of a chloroamine (Nightingale et al., 2000; Yeh et al., 2006; Yeh, Blasiak, Koglin, Drennan, & Walsh, 2007). By contrast, the Glu_{346} residue is believed to stabilize a Wheland intermediate, which can be generated after attack of the aromatic substrate (Dong et al., 2005) (Fig. 1). Figure 1 displays the active site structure of a typical FAD-dependent haloperoxidase with FAD and halide (chloride) bound. As can be seen the substrate is anchored through interactions with the Glu_{346} and Lys_{79} amino acid groups that are involved in the haloperoxidation of the substrate.

2.4 *S*-Adenosyl-L-Methionine Fluorinase

Although the heme-, vanadium-, and FAD-dependent haloperoxidases all utilize halogens in their catalytic cycles, as a matter of fact these systems only work with chloride, bromide, and iodide, whereas none of these haloperoxidases can activate fluoride. Part of this is not only due to the low bioavailability of fluoride with respect to chloride, bromide, and iodide but also because of the large oxidation potential of fluoride, which is well higher than that of hydrogen peroxide and therefore prevents haloperoxidase activity with fluoride. The first enzyme that was discovered that reacts as a fluorinase was the *S*-adenosyl-L-methionine (SAM) fluorinase, F1A, from *Streptomyces cattleya* (O'Hagan, Schaffrath, Cobb, Hamilton, & Murphy, 2002; O'Hagan & Deng, 2015). This enzyme catalyzes the reaction of SAM, Scheme 3, with fluoride to form 5′-fluorodeoxyadenosine and L-methionine via a nucleophilic substitution reaction (S_{N2}-type reaction).

In recent years, three more fluorinases have been identified in bacterial strains through genome mining (Deng & Ma, 2014). Alongside the

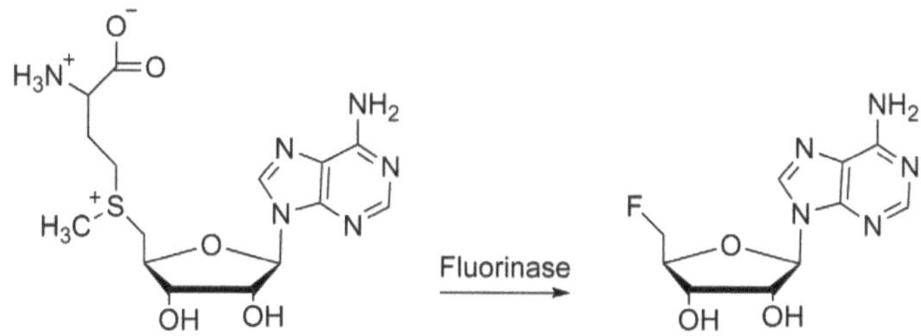

Scheme 3 Chemical reaction catalyzed by SAM-fluorinase enzymes.

structurally and genetically similar chlorinase Sa1L from *Salinispora tropica*, all fluorinases belong to the DUF62 superfamily of which the majority perform hydrolysis of SAM rather than halogenation (Deng & O'Hagan, 2008). Clearly, the mechanistic details of hydrolysis versus halogenation are similar and may result from a bifurcation pathway in the mechanism.

The use of fluoride in biology is hampered by the difficulty of desolvating this highly electronegative ion in the enzyme as well as the potential toxicity of fluorinated metabolites to the host organism (Vaillancourt et al., 2006), all of which have to be dealt with accordingly by the organism itself. Computational and structural studies that compare wild-type and mutant versions of F1A and Sa1L have suggested that specific amino acids in the halogen-binding site have the role of excluding water when the halide is bound (Senn, 2014). In some mutants, space becomes available to hold a water molecule, which results in hydrogen-bonding interactions to the halogen ion, which decreases its nucleophilicity toward substrate SAM, and, thereby reduces its catalytic activity. On top of that, the presence of water has been proposed to disrupt backbone hydrogen bonds, and, consequently may disrupt the local secondary structure of the protein (Senn, 2014). Accordingly, the interaction between the halide and the protein backbone induces conformational changes associated with enzyme activity.

There is still much to be discovered regarding this class of enzymes. For example, what are the contributions of conformational changes to the catalytic cycle associated with halide binding, how are halide ions desolvated by the protein, and what are the chemical and physical properties that determine the bond formation and product release. Clearly, extensive studies will be needed before this class of enzymes can be applied in a biotechnological context.

A combination of quantum mechanical (QM) modeling and QM/MM studies was performed on SAM fluorinase (Senn, O'Hagan, & Thiel, 2005). The calculations started from a reactant complex, whereby an F^- anion was added to the crystal structure coordinates and hydrogen bonded to hydroxyl groups of a Thr and Ser residue and the peptide proton of a Tyr group. A nucleophilic attack of the F^- on the substrate leads to the fluorinated product. The calculations established that the enzyme conformation is important for driving the chemical reaction, as it cannot be performed in a solution phase due to a high-energetic step. Furthermore, the key step that enables the SAM fluorinase to catalyze this energetically difficult reaction is by desolvating fluoride from water and transferring it into the substrate-binding pocket. The low barrier obtained is as a result of ideal positioning of the substrate with respect to the fluoride.

2.5 Nonheme Iron/α-Ketoglutarate-Dependent Halogenases

A final class of halogenases that will be covered in this review is the nonheme iron/α-KG-dependent halogenases. Investigation into the biosynthesis of barbamide from the marine cyanobacterium *Lyngbya majuscula* identified leucine as the precursor for its trichloromethyl product. Labeling studies, however, concluded that another enzyme than an FAD-dependent halogenase was responsible for this halogenation reaction. Genetic analysis of the cloned biosynthetic gene cluster revealed two genes encoding BarB1 and BarB2 that were similar to SyrB2 from *Pseudomonas syringae* and were characterized as originating from a mononuclear nonheme iron/α-KG-dependent halogenase (van Pee et al., 2006). It was found that the halogenase required molecular oxygen, α-KG, iron, a halogen (either Br^- or Cl^-), and an aliphatic substrate.

Since then, several other α-KG-dependent halogenases have been identified, among those the SyrB2, CytC3, and CurA forms have been structurally characterized. Indeed, all three structures have functions associated with the production of halogenated secondary metabolites, or, alternatively, are precursors of substrates that are tethered to a phosphopentathiene (PPT) arm associated with an acyl carrier protein (ACP) (Smith, Gruschow, & Goss, 2013; Wagner, Omari, & Konig, 2009). Consistent with other members of the α-KG-dependent oxidases, such as the hydroxylases, they all have the same structural fold, namely, an eight-stranded antiparallel β-jelly roll motif to which an iron atom is located (Fig. 2).

The resting state of nonheme iron hydroxylases is defined by a facial triad consisting of two histidine and one carboxylic acid (of either Asp or Glu) linkages, the so-called 2His-1Asp/Glu motif (Fig. 2B). However, in nonheme iron/α-KG-dependent halogenases, the carboxylate amino acid in the protein is replaced by an alanine/glycine residue and as a result, a cavity

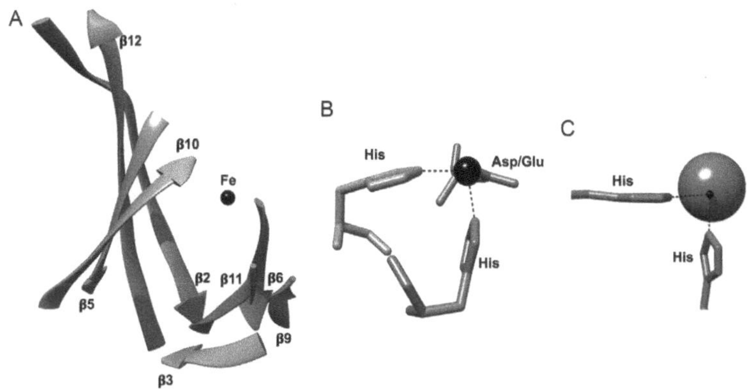

Figure 2 (A) β-jelly roll motif taken from the SyrB2 structure (2FCV.pdb) consisting of eight antiparallel β-strands with an iron atom located at its center. (B) The facial triad found in mononuclear nonheme iron/α-ketoglutarate-dependent dioxygenases (3GZE.pdb). (C) The facial triad found in mononuclear nonheme iron/α-ketoglutarate-dependent halogenases (2FCV.pdb).

is created nearby the metal center that enables binding of a halide ion (Senn, 2014), (Fig. 2C). It was thought that the Asp/Ala substitution defined the difference in activity between nonheme iron/α-KG-dependent hydroxylases and halogenases; however, mutation studies on nonheme iron/α-KG-dependent hydroxylases whereby the Asp residue was replaced by Ala failed to induce any halogenase activity. The same phenomenon was observed for mutations of the nonheme iron/α-KG-dependent halogenase, most probably due to little sequence similarity between the two enzymes. This further implicates that their biochemical function evolved through divergent evolution (Anderson & Chapman, 2006; Senn, 2014).

The nonheme iron/α-KG-dependent halogenases are desirable biotechnological targets due to their ability to introduce a halogen into an aliphatic group. The majority of these enzymes utilize amino acids or derivatives of amino acids as substrates, but a unique α-KG-dependent halogenase, namely, HctB from *L. majuscula* is known to chlorinate a nonamino-acyl substrate (Pratter, Ivkovic, et al., 2014; Pratter, Light, Solomon, & Straganz, 2014).

3. GENERAL MECHANISM OF α-KETOGLUTARATE-DEPENDENT HALOGENASES

The catalytic cycle of nonheme iron/α-KG-dependent hydroxylases and halogenases shows close similarity and is schematically depicted in Scheme 4 (Bollinger, Price, Hoffart, Barr, & Krebs, 2005; Bruijnincx

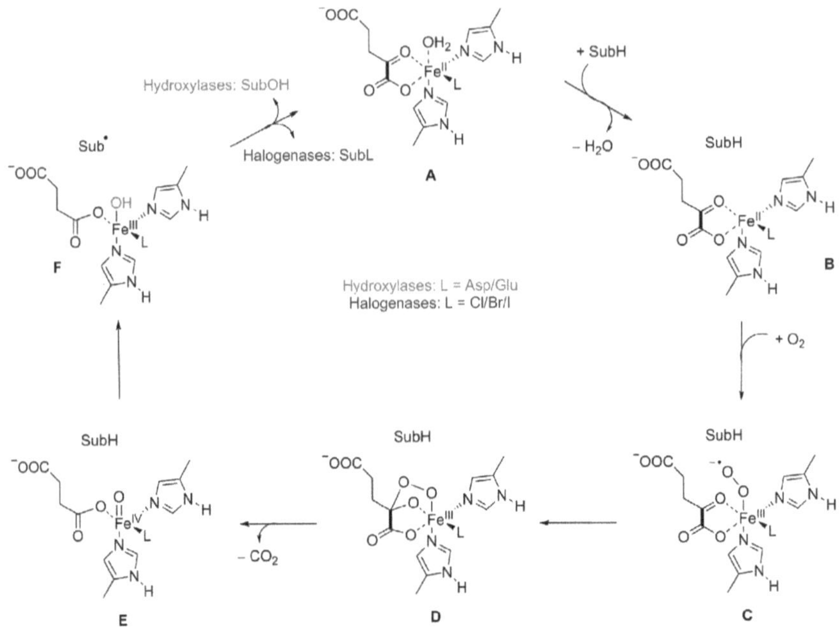

Scheme 4 Catalytic cycle of nonheme iron/α-ketoglutarate-dependent hydroxylases and halogenases.

et al., 2008; Buongiorno & Straganz, 2013; de Visser, 2009; de Visser & Kumar, 2011). Details of the catalytic cycle are still shrouded in mysteries, but in the past decade considerable progress has been made to detect and characterize some of the short-lived intermediates. The cycle starts from the resting state (A), which is an iron(II) center bound to two histidine residues of the protein. In the α-KG-dependent hydroxylases, a third linkage of the protein (labeled as L in Scheme 4) comes from a carboxylate group of either a Glu or an Asp residue. In the α-KG-dependent halogenases, this group is missing and in its position is either a Gly or an Ala residue, which makes space for a halide-binding pocket. The three remaining ligand-binding sites of the iron(II) center are occupied by α-KG as a bidentate ligand through the carboxylate and keto groups, whereas the last binding site holds a water molecule. When substrate binds into the substrate-binding pocket, the axial water molecule is displaced to form the penta-coordinated iron(II) species (B). Molecular oxygen fills the empty binding site and forms the iron(III)-superoxo intermediate (C) and its terminal oxygen atom attacks the keto-position of α-KG to generate a bicyclic ring structure (D). In the latter structure, the dioxygen bond is weakened and subsequently breaks to form an iron(IV)-oxo active species (E). At the same time, carbon dioxide is released from α-KG that is converted into succinate. Iron(IV)-oxo intermediates are efficient oxidants and react with substrates via hydrogen atom abstraction and lead to an iron(III)-hydroxo complex (F) and a substrate radical (de Visser, 2006a, 2006b, 2006c, 2010). At this stage, the mechanism bifurcates between the α-KG-dependent hydroxylases and halogenases (Fig. 3). Thus, rebound of the OH group to the substrate radical leads to alcohol products and returns the system to the resting state by replacing the alcohol with a water molecule. On the other hand, in the halogenases, the halogen binds the substrate radical and leads to halogenated products instead. Indeed, both α-KG-dependent hydroxylases and halogenases were found to react with a rate-determining hydrogen atom abstraction as evidenced from a large kinetic isotope effect for the substitution of the hydrogen with deuterium in the substrate (Grzyska et al., 2005; Price, Barr, Glass, Krebs, & Bollinger, 2003; Price, Barr, Tirupati, Bollinger, & Krebs, 2003). In spite of the differences in active site structure and orientation of the α-KG-dependent hydroxylases and halogenases, there are still questions that remain to be answered regarding the catalytic cycle as a whole and in relation to the halogenases specifically. These questions revolve around two key parts of the mechanism: First, the generation of the iron(IV)-oxo species, and second, the formation of the halogenated product from the iron(III)-hydroxo(halide) complex, which may require an isomerization.

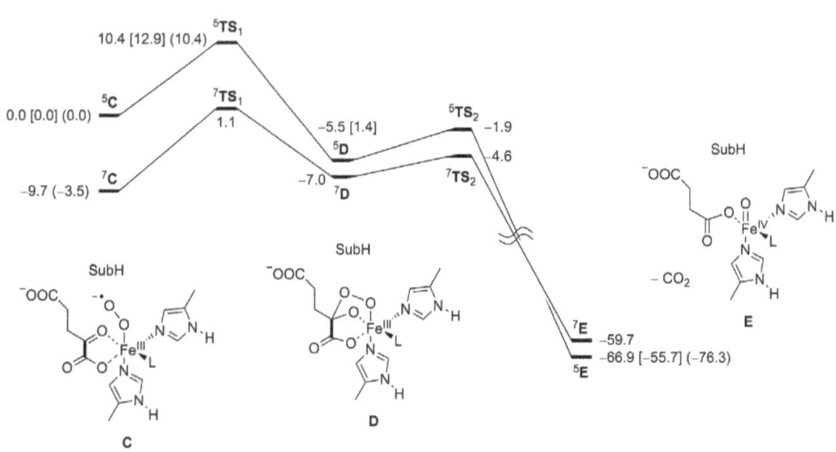

Figure 3 DFT calculated mechanism for the formation of an iron(IV)-oxo species from an iron(III)-superoxo intermediate. All energies are in kcal mol^{-1} relative to ^{5}**C**. Data given are from de Visser (2007) (data out of parenthesis), Topol et al. (2006) (data in square brackets), and Borowski et al. (2004) (data in parenthesis).

At the heart of understanding, the steps leading to the generation of the iron(IV)-oxo species in nonheme iron/α-KG-dependent hydroxylases and halogenases, computational studies have been performed by several groups (Borowski, Bassan, & Siegbahn, 2004; de Visser, 2007; Godfrey, Porro, & de Visser, 2008; Quesne, Latifi, Gonzalez-Ovalle, Kumar, & de Visser, 2014). This was mainly due to experimental difficulties to characterize other short-lived intermediates in the catalytic cycle. The main results of DFT calculations on the oxygen activation step in the catalytic cycle of nonheme iron α-KG-dependent hydroxylases, i.e., the step from (C) to (E) in Scheme 4, are depicted in Fig. 3 as obtained from data from Borowski et al. (2004), Topol et al. (2006), and de Visser (2007). The early work of Borowski et al. and Topol et al. only considered the lowest lying quintet spin state and produced a profile whereby the iron(III)-superoxo is converted into a ring structure (D), which then splits off CO_2 to form a peroxosuccinate-iron(III) complex before a homolytic bond cleavage of the dioxygen bond to form the iron(IV)-oxo and succinate groups (E). All studies implicate a rate-determining initial reaction step, i.e., from (C) to (D) via transition state $\mathbf{TS}_1$. Later studies revealed a low-lying septet spin state (de Visser, 2007), which may be the ground state and which crosses over to the quintet spin state in the peroxosuccinate-iron(III) structure. Nevertheless, all computational and experimental studies agree to a "consensus mechanism" described above in Scheme 4.

3.1 Generation and Characterization of the Iron(IV)-Oxo Species

The majority of the experimental and computational studies reported have been conducted with reference to the nonheme iron/α-KG-dependent hydroxylases. So far, no computational studies have been reported on the generation of the iron(IV)-oxo species in nonheme iron/α-KG-dependent halogenases, and as such, the contribution of the chloride atom on the process of oxygen activation is not known. However, it is assumed that it follows a pathway similar to those found for α-KG-dependent hydroxylases (Borowski et al., 2004; de Visser, 2007).

The activation of oxygen in most nonheme iron/α-KG-dependent hydroxylases and halogenases studied is thought to be linked to the dissociation of the axial water molecule in the resting state and subsequent transition from a hexa-coordinate octahedral geometry to a penta-coordinate trigonal bipyramidal structure. The release of the water molecule in the resting state occurs only in the presence of substrate or cosubstrates, in a process

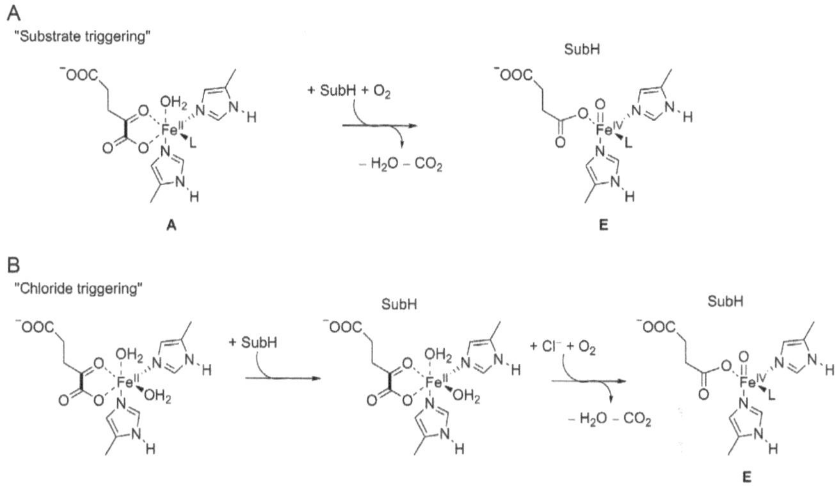

Scheme 5 (A) Substrate triggering mechanism in α-ketoglutarate-dependent halogenases. (B) Chloride triggering mechanism in α-ketoglutarate-dependent halogenases.

coined "substrate triggering" (Scheme 5). Substrates and cosubstrates appear to bind to the protein in a specific order, which may be important to reduce and prevent uncoupling and self-hydroxylation reactions. Oxygen activation, therefore, can only occur when all the necessary substrates and cosubstrates are bound. For example, the activation of oxygen in SyrB2 is 5000 times faster in the presence of substrate (Light, Hangasky, Knapp, & Solomon, 2013). Nevertheless, several nonheme iron/α-KG-dependent hydroxylases are able to react with oxygen in the absence of a substrate. The question, therefore, is what factors are there that control the dissociation of water alongside or apart-from substrate binding? One possible answer may be related to the strength of the interaction between the axial water molecule and the iron(II) center, which may be controlled by primary and secondary sphere interactions.

In the mononuclear nonheme iron/α-KG-dependent hydroxylase CS2, a carboxyl group of one of the amino acids of the facial triad is able to form a hydrogen bond to the axial water molecule, which increases its hydroxyl character and the strength of water—iron bond. This is compensated by the position of the substrate with respect to the axial water molecule: the stronger the water is bound, the closer the substrate has to interact. For example, in CS2, the substrate is known to bind directly over the water molecule, whereas in taurine/α-KG-dependent dioxygenase (TauD) it binds off center (Neidig et al., 2007). Interestingly, in TauD, the orientation of the carboxyl group compared to that in CS2 prevents for the formation of a direct hydrogen bond. Instead an Asp residue in the secondary coordination sphere (not identified in any other hydroxylase) is thought to be involved in a hydrogen-bonding network that holds the water in place albeit more weakly.

The relationship between the strength of water binding and the position of the substrate in α-KG-dependent halogenases is challenging to investigate as there are no substrate-bound crystal structures known. However, Mössbauer spectroscopic studies on CytC3 revealed that in the presence of substrate the structure changes from six-coordination to five-coordination, whereas this happens to a much lesser degree in TauD and CS2 (Neidig et al., 2007). These studies implicate that the nonheme iron/α-KG-dependent halogenases proceed via a "substrate triggering" mechanism (Scheme 5), whereby halide and α-KG bind the iron first and only thereafter substrate enters the pocket, which triggers dioxygen binding and the start of the catalytic cycle. The alternative "chloride triggering" mechanism starts with α-KG binding to the metal center followed by substrate placement into the pocket and lastly chloride binding that triggers dioxygen uptake. The "substrate triggering" versus "chloride triggering" mechanisms depend on the orientation of the substrate with respect to the axial water molecule. Thus, in the "substrate triggering" mechanism, the metal center is in hexacoordination and stays that way upon substrate binding and addition of O_2. Consequently, the substrate will be near the axial water molecule. In the "chloride triggering" mechanism, the substrate is positioned in the pocket before chloride binds, which may result in structural changes to the iron center and give a larger distance between the axial water molecule and the substrate as well as a weakened iron—water bond. This weak bond is thought to result from the hydrogen-bonding network between the chloride, a noncoordinated water molecule, and the bound water (Neidig et al., 2007). It is assumed here that the process of binding chloride has no direct effect on the protein structure.

Substrate placement and positioning in nonheme iron/α-KG-dependent halogenases is different from those observed in the analogous hydroxylases, such as TauD and CS2. The reason may be evolutionary and related to the regioselectivity of halogenation versus hydroxylation reaction. A study on the role of substrate positioning on product distributions in the halogenase SyrB2 found strong differences in the ratio of halogenated versus hydroxylated products upon bringing the substrate closer to the metal center, which led to dominant substrate hydroxylation (Matthews, Krest, et al., 2009; Matthews, Neumann, et al., 2009). Clearly, therefore, the rebound step in the catalytic cycle, i.e., from (F) to (A) in Scheme 4, is highly dependent on the position of the substrate. As a consequence, catalytic cycle steps prior to the radical rebound, such as oxygen activation, may have had to compensate for this.

The recently characterized mononuclear nonheme iron/α-KG-dependent halogenase HctB appears to deviate from the "substrate triggering" phenomenon observed in other halogenases and hydroxylases suggesting instead that chloride binding is the trigger for activating the transition from a six-coordinate to a five-coordinate iron complex and subsequent activation of oxygen. Results from molecular dynamics simulations in the presence or absence of chloride ions concluded that chloride binding induced a conformational change associated with Glu_{224}. This amino acid was originally bound to the iron either directly or indirectly (via a bound water molecule), and chloride binding results in a rearrangement of backbone hydrogen bonds thereby opening a substrate channel that may bring the substrate in closer proximity to the axial water and promoting its dissociation (Pratter, Ivkovic, et al., 2014; Pratter, Light, et al., 2014). These studies assume that substrate binds in its allocated binding site and only after chloride is attached to the metal. However, spectroscopic studies revealed that chloride binding led to a perturbation in the electronic ground state of the iron complex, consistent with a change in the metal geometry. Perturbations in the electronic ground state were also observed upon substrate binding but not to the same extent. This conformational change could simply reflect an alteration in the protein environment surrounding the chloride, which affects how it interacts with the iron and the ligands bound to it, namely, the axial water which would then be influenced to a greater extent by the presence of substrate leading to its dissociation. It has been suggested that the "chloride triggering" mechanism in HctB is associated with its unusual substrate, which is a fatty acyl group rather than an amino acid derivative as in other halogenases (Butler & Sandy, 2009; Neumann, Fujimori, & Walsh, 2009; Pratter, Ivkovic, et al., 2014; Pratter, Light, et al., 2014).

The disruption of hydrogen-bonding interactions can result in the weakening of the iron—water bond and studies on the nonheme iron enzyme factor-inhibiting hypoxia (FIH) implicated a role of the substrate in this process (Light, Hangasky, Knapp, & Solomon, 2014). It is, therefore, possible that chloride binding has a similar effect as substrate binding and can mimic the role of the substrate in hydroxylases. Overall, the role of chloride binding could be to disrupt the hydrogen bond between Glu_{224} and the axial water which would destabilize the iron—water bond with the help of α-KG, and also stabilize the five-coordinate complex, probably via donation of negative charge. On the other hand, a study performed on the halogenase SyrB2 using anions, such as N_3^- and NO_2^-, to replace Cl^- identified a chloride dependence on oxygen activation (Matthews et al., 2014), in the same way to HctB. Therefore, the "chloride triggering" mechanism may be a common mechanism for the whole class of α-KG-dependent halogenases, and it may very well be that halogenases can operate with both the "substrate triggering" and "chloride triggering" mechanisms. Sequence alignments of SyrB2 and HctB show that in SyrB2, a serine residue occupies the position of Glu_{224} in HctB (Pratter, Ivkovic, et al., 2014; Pratter, Light, et al., 2014), which implicates that the role of Glu_{224} is specific to HctB, but the role of chloride in donating negative charge and promoting the formation of a five-coordinate complex is common. These studies further demonstrate the ability of mononuclear nonheme iron/α-KG-dependent halogenases to insert various anions into substrates other than chloride, thereby increasing their potential for biotechnological applications (Matthews et al., 2014). It appears, therefore, that nature has designed a flexible system tailored to each nonheme iron/α-KG-dependent hydroxylase, with substrate and a number of cosubstrates, which operates through the activation of molecular oxygen. In fact, one study proposed three factors necessary to promote the loss of water specific to FIH: (i) steric interactions that destabilize the six-coordinate complex; (ii) hydrogen-bonding interactions from the substrate and protein backbone that stabilize the five-coordinate complex; and (iii) electron donor properties of α-KG, which stabilize the five-coordinate complex (Light et al., 2014).

This "chloride triggering" mechanism proposed for HctB calls into question the importance of the order of substrate binding on the construction of a fully functional active site. SyrB2 can be crystallized in the presence of chloride, but without α-KG, whereas CytC3 can only be crystallized in the presence of chloride when α-KG is present. The comparison of CytC3 and SyrB2 implicates that the chloride-binding pocket is fully constituted only upon conformational changes induced through α-KG binding. The ability of α-KG to induce conformational changes upon binding to the metal is further supported from crystal structures of CurA with and without α-KG bound. Consequently, the SyrB2 and HctB halogenases appear to have a fully constituted chloride-binding pocket, whereas in CurA and CytC3 this is not the case (Buongiorno & Straganz, 2013). These differences may be linked to the activation of oxygen and the regioselectivity of halogenation over hydroxylation and, therefore, it may be essential in SyrB2 and HctB to have halide bound prior to α-KG. Overall, conformational changes upon substrate and cosubstrate binding must be a common factor in the α-KG-dependent halogenases toward to constitution of a complete active site linked to the activation of oxygen, the formation of an iron(IV)-oxo species, and the formation of halogenated products.

3.2 Regioselectivity of Halogenation Versus Hydroxylation

The initial catalytic mechanism for mononuclear nonheme iron/α-KG-dependent hydroxylases was proposed over 30 years ago (Scheme 4), and since then researchers have been trying to trap and characterize each of these intermediates; however, several of these are extremely short-lived and difficult to isolate. One such intermediate is the iron(IV)-oxo species, which was identified for the hydroxylases prolyl-4-hydroxylase (P4H) and TauD using a combination of spectroscopic techniques. Thus, using UV–visible, IR, resonance Raman, EXAFS, EPR, and Mössbauer spectroscopic techniques, the iron(IV)-oxo intermediate was found to have a short Fe–O distance as well as a high-spin state (Bollinger et al., 2005; Matthews, Krest, et al., 2009; Matthews, Neumann, et al., 2009). Kinetic isotope effects then confirmed that the rate-determining step in the reaction mechanism included a hydrogen atom abstraction reaction by the iron(IV)-oxo intermediate and, hence, confirmed it as the active oxidant in nonheme iron/α-KG-dependent hydroxylases and halogenases.

The identification of similar intermediates in the hydroxylases and halogenases provided evidence for the large overlap in the catalytic cycles of these two enzyme classes until the formation of the iron(IV)-oxo intermediate. Only, thereafter, the two mechanisms bifurcate resulting in halogenated products for the α-KG-dependent halogenases, whereas the hydroxylases give alcohol products instead. Interestingly, the α-KG-dependent halogenase SyrB2 has been shown to hydroxylate nonnatural substrates (Pratter, Ivkovic, et al., 2014; Pratter, Light, et al., 2014), which implicates that minor differences to the substrate, its positioning, and the second coordination sphere interactions influence the regioselectivity of hydroxylation over halogenation.

Although nonheme iron/α-KG-dependent halogenases show large structural overlap with the corresponding hydroxylases, the regioselectivity of hydroxylation over halogenation is no solely dependent on the replacement of the Asp/Glu metal-ligand residue with an Ala residue. It does, however, create space for chloride to bind, but site-specific mutations showed that this is not the sole reason as considerable amount of hydroxylated products were observed (Anderson & Chapman, 2006; Borchardt et al., 2001).

3.3 Substrate Placement

An obstacle to studying the effect of substrate placement on product outcome is the unavailability of any crystal structure with the actual substrate bound for all nonheme iron/α-KG-dependent halogenases. Therefore, indirect spectroscopic techniques were applied, such as NMR, matrix-assisted laser desorption/ionization-time of flight mass spectrometry, and Mössbauer spectroscopy. In addition, a series of computational studies were performed to gain insight into the substrate placement. Mononuclear nonheme iron/α-KG-dependent hydroxylases are typically found in nonribosomal peptide synthase and polyketide synthase pathways for which the substrate is delivered by either a peptidyl carrier protein (PCP) or an ACP via PPT arm (Pratter, Ivkovic, et al., 2014; Pratter, Light, et al., 2014). Thus, the substrate as a whole consists of three major components, namely, the ACP/PCP, the PPT arm, and a portion with an aliphatic C—H bond. It is the interaction of these three parts individually and together which help to determine the product outcome. Detection of this product is also difficult relying on the use of labeled substrates for its full characterization. Figure 4 gives two examples of SyrB2 and CmaB that use an ACP with PPT arm linked to an amino acid residue. These acyl carriers are important for substrate activity and it was found that free substrate was 14 times less active than substrate linked to an acyl carrier (Pratter, Ivkovic, et al., 2014; Pratter, Light, et al., 2014).

3.4 Role of the Substrate

In SyrB2, the substrate is linked to an ACP unit called SyrB1 and its binding to the halogenase domain triggers the catalytic cycle and the formation of an iron(IV)-oxo intermediate (Matthews, Krest, et al., 2009). It, therefore,

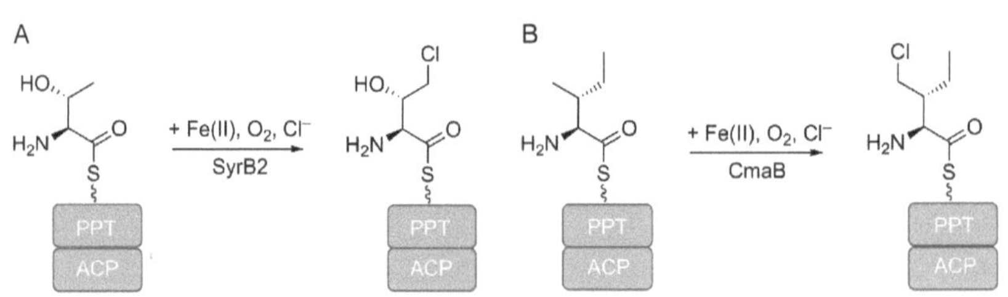

Figure 4 Mechanism of substrate halogenation by (A) SyrB2 and (B) CmaB.

follows that the ACP, PPT arm, and the substrate are all required for adequate activity. The function of the PPT arm is to deliver the substrate into the active site in such a way that it promotes the halogenation activity. As stated previously, the position of the substrate is not only linked to the formation of the iron(IV)-oxo species (Matthews, Krest, et al., 2009) but also the hydrogen abstraction step. Evidence from kinetic isotope effect experiments (KIEs) implicated a rate-determining hydrogen atom abstraction mechanism whereby the iron(IV)-oxo species is converted into an iron(III)-hydroxo group. The hydrogen atom abstraction step likely proceeds with a considerable amount of quantum chemical tunneling as evidenced from the large KIE values obtained (Matthews, Neumann, et al., 2009). Nevertheless, the rate of hydrogen atom abstraction by α-KG-dependent halogenases is much lower than the corresponding hydroxylases, which may be related to the distance of the substrate from the iron(IV)-oxo moiety. This may be an essential evolutionary consequence of favoring halogenation over hydroxylation as the rate of hydrogen atom transfer and hydroxylation of the substrate radical have been found to be directly correlated to the proximity of the substrate to the iron(IV)-oxo species (Buongiorno & Straganz, 2013; Matthews, Neumann, et al., 2009).

One feature associated with the PPT arms is linked with the positioning of the substrate into the ideal conformation in order to promote substrate halogenation and prevent hydroxylation reactions. A study with different PPT chain lengths showed that the rate of hydrogen atom abstraction was unaffected, but, by contrast, dramatic changes in the chemoselectivity of the reaction were observed (Kulik & Drennan, 2013). Furthermore, hydrogen-bonding interactions of either substrate or PPT chain may influence the position of the substrate as well. For instance, the SyrB2 substrate Thr group has an alcohol group that interacts with the protein and stabilizes the structure in a specific orientation, which likely promotes the halogenation reaction. Indeed, conformational changes to the substrate binding were found to lead to substrate hydroxylation (Kulik, Blasiak, Marzari, & Drennan, 2009).

The idea that the substrate influences the product outcome is not new, as the reaction of SyrB2 with nonnatural substrates tethered to the PPT arm covalently bound to an ACP revealed differing ratios of hydroxylation and halogenation compared to the natural substrate, which was attributed to interactions of the substrate with the cofactor (Matthews, Krest, et al., 2009; Matthews, Neumann, et al., 2009). HctB, by contrast, gives a mixture of products when it acts upon the hexanoyl substrate, which is found to be independent of chloride concentration. The products observed are 5-oxo-hexanoic acid (50%), 5,5-dichlorohexanoic acid (25%), and 5-chloro-4-enoic acid (25%) (Pratter, Ivkovic, et al., 2014; Pratter, Light, et al., 2014). Of interest is that this substrate is unlike others identified for known α-KG-dependent halogenases that use 3-hydroxy-3-(-4-pentynyl) glutaryl (JamE), 2-aminobutyric acid (CytC3), leucine-*S*-BarA (BarB1/B2), and isoleucine (CmaB) (Butler & Sandy, 2009; Neumann et al., 2009; Pratter, Ivkovic, et al., 2014; Pratter, Light, et al., 2014). Other α-KG-dependent halogenases such as CS2 are able to perform multiple reactions dependent on the substrate (Neidig et al., 2007). It may very well be linked to the domain structure of HctB (Fig. 5), because unlike known α-KG-dependent halogenases, HctB possesses a three-domain structure. Located at the N-terminus is the halogenase domain, which is linked to an acyl-coenzyme A binding protein (ACBP) domain and an ACP (Neidig et al., 2007) domain. Interestingly, the function of the ACBP that separates the two catalytic domains is unknown for HctB and was previously thought to have a regulatory role, although this could not be confirmed.

The proposed "substrate triggering" and "chloride triggering" mechanisms (Matthews, Neumann, et al., 2009; Pratter, Ivkovic, et al., 2014; Pratter, Light, et al., 2014) that lead up to the binding of molecular oxygen and the biosynthesis of the iron(IV)-oxo species implicate strong influences of the substrate on enzymatic activity. Key to understanding the regioselectivity and chemoselectivity of the reaction process is the relative pathways leading to either the chloride or hydroxyl rebound after the hydrogen atom abstraction step. The orientation of the iron(IV)-oxo species relative to this reactive C—H bond of the substrate may be key in this. In

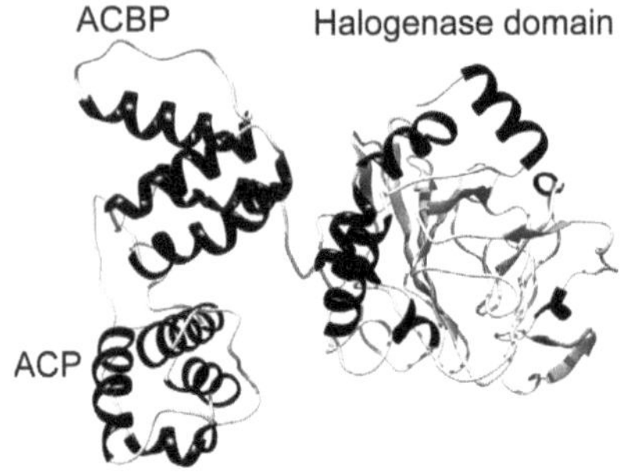

Figure 5 Structure of α-ketoglutarate-dependent halogenase HctB with three domains, namely, the halogenase domain, the acyl-coenzyme A binding protein domain (ACBP), and the acyl carrier protein (ACP). Structure was constructed from the sequence of HctB.

theory, there are many different geometric orientations of the iron(IV)-oxo species possible with various coordination states. Thus, the succinate ligand can bind as a monodentate ligand through the carboxylate anion only or alternatively as a bidentate ligand with both oxygen atoms of the same carboxylate group linked to the metal. As such, the iron(IV)-oxo species can exist in a five-coordinate trigonal bipyramidal or square pyramidal geometries (structures $\mathbf{1}_5$ in Scheme 6) as well as a six-coordinate octahedral geometry (structures $\mathbf{1}_6$ in Scheme 6). In addition, the oxo group can be placed axial or equatorial with respect to the chloride ligand (the atom in axial position is given as a substrate to the structure in Scheme 6). Despite this, only a singular penta-coordinate geometry has been observed in non-heme iron/α-KG-dependent hydroxylases (Grzyska et al., 2005; Hoffart, Barr, Guyer, Bollinger, & Krebs, 2006; Riggs-Gelasco et al., 2004; Price, Barr, Glass, et al., 2003; Price, Barr, Tirupati, et al., 2003), whereas for the corresponding halogenases two geometries were characterized with Mössbauer spectroscopy (Galonic, Barr, Walsh, Bollinger, & Krebs, 2007; Matthews, Neumann, et al., 2009). A further set of combined Mössbauer and DFT calculations on the iron(IV)-oxo species of TauD revealed that the experimentally determined Mössbauer spectra could only be explained with a penta-coordinated iron(IV)-oxo species (Sinnecker et al., 2007).

Computational studies on biomimetic complexes (Quesne & de Visser, 2012) investigated the role of the equilibrium between structures with either

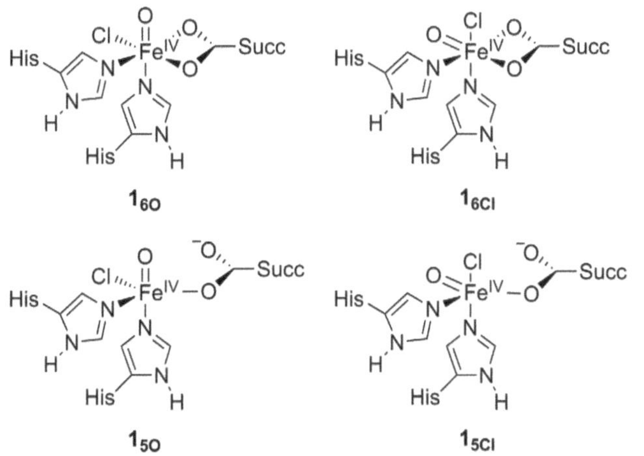

Scheme 6 Hexa-coordinated ($\mathbf{1}_6$) and penta-coordinated ($\mathbf{1}_5$) isomers of the iron(IV)-oxo species of α-ketoglutarate-dependent halogenases.

axial chloride or axial oxo groups. It was found that the product distribution of hydroxylated versus halogenated substrates depends on the atom in the equatorial position as it influenced the bonding energy of the chloride and oxo groups. Both pathways were found to start from an initial and rate-determining hydrogen atom abstraction from substrate to form an iron(III)hydroxo(chloride) complex. They then analyzed the bond dissociation energies of the Fe^{III}—OH and Fe^{III}—Cl bonds in various isomeric structures (see Fig. 6). With the hydroxo group in the axial position and chloride in the equatorial position (top of Fig. 6), the weakest bond is the Fe—Cl bond by 9.2 kcal mol^{-1}. However, interchanging the hydroxo and chloride positions (lower part of Fig. 6) gave a reversal of the bond energy strength, whereby the Fe—OH bond is weaker than the Fe—Cl bond by 13.2 kcal mol^{-1}. These thermochemical analyses implicate that the ligand in the equatorial position will be most likely transferred in the rebound stage of the reaction mechanism as its bond energy to the metal is the weakest. They proposed an analogous process in nonheme iron/α-KG-dependent halogenases, where the binding position of the chloride with respect to the hydroxo group would determine the product distributions. However, an enzymatic system, of course, may have influences from primary and secondary sphere interactions that may affect the product distributions as well.

Indeed, experimental and computational studies of SyrB2 demonstrated that the substrate positioning affected what isomeric iron(IV)-oxo species

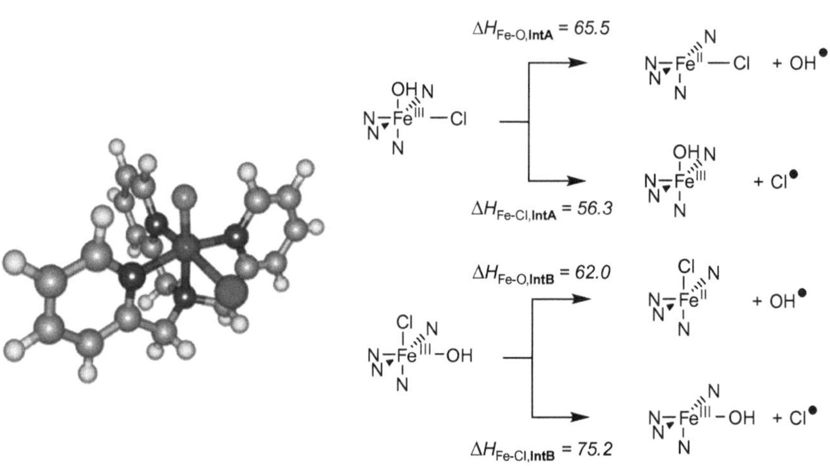

Figure 6 Thermodynamics of Fe^{III}—OH versus Fe^{III}—Cl bond breaking in different isomers of the iron(III)-hydroxo(chloride) complex. All data are in kcal mol^{-1}.

was present and thereby influenced the regioselectivity of substrate hydroxylation over halogenation (Borowski, Noack, Mariusz, Zych, & Siegbahn, 2010; Matthews, Krest, et al., 2009; Matthews, Neumann, et al., 2009). Thus, the computational modeling showed that an essential isomerization step was needed after the hydrogen atom abstraction, which swaps the positions of the hydroxo and chloride ligands and then gives regioselective substrate halogenation.

Orbital interactions often dictate the orientation of substrate and oxidant. The iron(IV)-oxo species has a quintet spin ground state with $\pi^*_{xy}{}^{\uparrow}\pi^*_{xz}{}^{\uparrow}\pi^*_{yz}{}^{\uparrow}\sigma^*_{x^2-y^2}{}^{\uparrow}\sigma^*_{z^2}{}^{0}$ electronic occupation and a slightly higher lying triplet spin state with $\pi^*_{xy}{}^{2}\pi^*_{xz}{}^{\uparrow}\pi^*_{yz}{}^{\uparrow}\sigma^*_{x^2-y^2}{}^{0}\sigma^*_{z^2}{}^{0}$ occupation (de Visser, 2006b). In hydrogen atom abstraction reactions by the iron(IV)-oxo species in the quintet spin state, an electron is transferred from substrate (SubH) into a virtual $\sigma^*_{z^2}$ orbital along the Fe—O bond (de Visser, 2006c), which gives an $[Fe^{III}(OH)\text{-}Sub^{\bullet}]$ complex with orbital occupation $\pi^*_{xy}{}^{\uparrow}\pi^*_{xz}{}^{\uparrow}\pi^*_{yz}{}^{\uparrow}\sigma^*_{x^2-y^2}{}^{\uparrow}\sigma^*_{z^2}{}^{\uparrow}\sigma_{Sub}{}^{\downarrow}$, see Fig. 7. This electron transfer, generally, leads to a linear Fe–O–H–Sub orientation and has been assigned a $^5\sigma$-pathway. On the triplet spin state, by contrast, the hydrogen atom abstraction fills the π^*_{xz} orbital with a second electron and as this is an antibonding orbital between the $3d_{xz}$ on iron with $2p_x$ on oxygen, the substrate orients itself under an angle Fe–O–H of around 120° instead, which has been termed the $^3\pi$-pathway. Over the years, attempts have been made to stabilize the $^3\pi$-pathway to find experimental evidence for this. Addition of bulky substituents on the metal ligands prevented substrate approach along the "ideal" orientation in the $^5\sigma$-pathway, and forced a smaller Fe–O–H angle of 120° that still favored the $^5\sigma$ over the $^3\pi$-pathways, although the energy gap decreased (Latifi, Sainna, Rybak-Akimova, & de Visser, 2013). However, using a penta-coordinated ligand system an aromatic

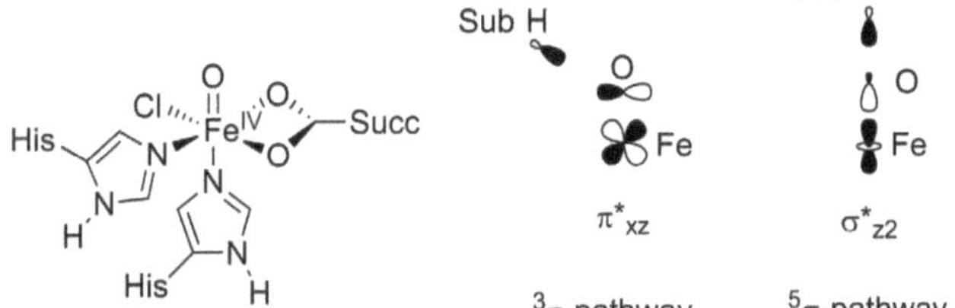

Figure 7 σ- versus π-pathways for hydrogen atom abstraction by iron(IV)-oxo complexes.

hydroxylation pathway could be followed whereby one of the aromatic ligand substituents was hydroxylated via a $^3\pi$-pathway (Sahu et al., 2014).

To establish the catalytic mechanism of nonheme iron/α-KG-dependent halogenases, a range of DFT and QM/MM studies were performed. The initial studies using DFT were performed in the gas phase and did not include the protein and second coordination sphere effects (de Visser & Latifi, 2009; Pandian, Vincent, Hillier, & Burton, 2009). Several suggestions were proposed to block the substrate hydroxylation pathway and give preferential halogenation reactions only, including protonation of the hydroxo group by a neighboring arginine residue (Pandian et al., 2009) or a reaction of hydroxo with carbon dioxide (de Visser & Latifi, 2009). However, detailed QM/MM studies (Borowski et al., 2010; Quesne, Faponle, Goldberg, & de Visser, 2015) implicated important effects of the protein and substrate orientation and proposed an iron(III)-hydroxo isomerization pathway akin to that found for the α-KG-dependent hydroxylase AlkB (Quesne et al., 2014). Figure 8 summarizes the key results of the isomerization pathway and the proposed products obtained for SyrB2. Thus, hydrogen atom abstraction gives an iron(III)-hydroxo(chloride) complex with the hydroxo in the axial position. This complex can rebound either the OH or the Cl group to form hydroxylated or halogenated products with barriers of 2.4 and 6.5 kcal mol^{-1}, respectively, and hence, should give preferential hydroxylation over halogenation. Isomerization of this complex give the iron(III)-chloride(hydroxo) isomer with the chloride in the axial position. Calculations of the radical rebound step gives dramatic changes and a high hydroxo-rebound barrier of 11.1 kcal mol^{-1} and virtually a negligible barrier for halogenation. Therefore, the isomerization step is essential to provide halogenation products and exclude the formation of hydroxylated

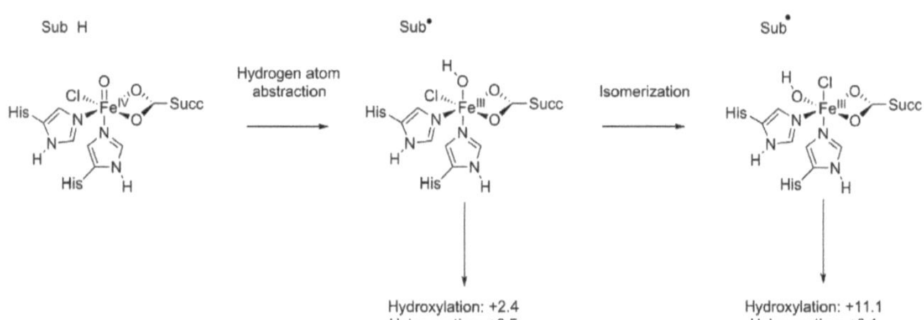

Figure 8 QM/MM calculated reaction mechanism of substrate halogenation versus hydroxylation by SyrB2 enzymes. Values in kcal mol^{-1}.

products. This mechanism was rationalized by calculations of bond dissociation energies, similar to those reported above in Fig. 6, with weaker Fe—ligand bonds along the axial axis than along the Fe-equatorial axis.

3.5 QM/MM Studies of HctB Halogenases

In order to gain insight into the mechanism of HctB and its regioselectivity preference of halogenation over hydroxylation, a detailed QM/MM study was performed (Timmins & de Visser, 2015). The general protocol for QM/MM studies is schematically depicted in Scheme 7. Thus, the initial structure is an available set of crystal structure coordinates from the protein databank (pdb). Usually, these structures refer to long-lived intermediates of the catalytic cycle of an enzyme, such as a resting state or a substrate-bound structure. There can be major changes in the protein upon substrate binding due to refolding of the protein and/or the closure of the substrate-binding pocket, so a substrate-bound structure is always preferable over a resting-state structure. In the case of HctB, however, no crystal structure coordinates were available, so the homology model (Pratter, Ivkovic, et al., 2014) was used. This system had the ACP attached to the protein and substrate located

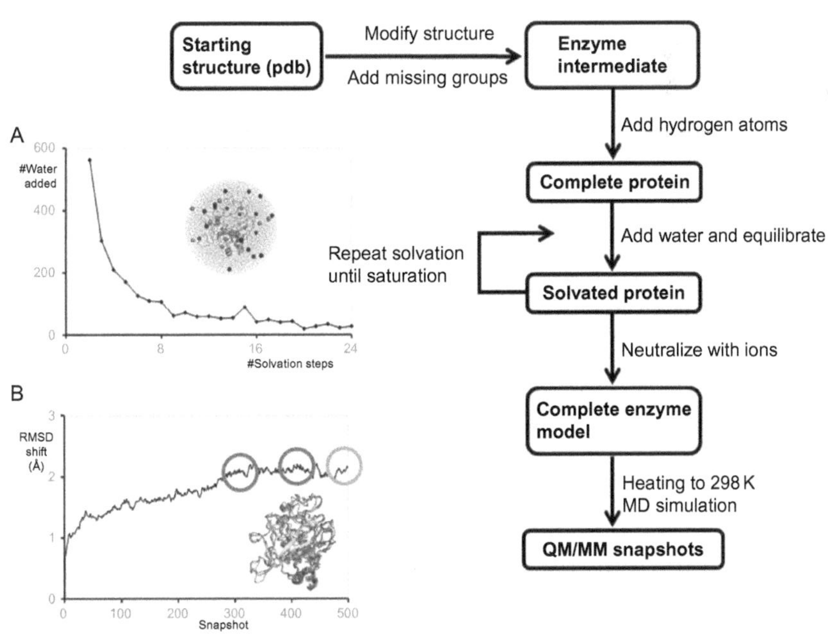

Scheme 7 QM/MM setup and protocols. Insets show the iterative solvation procedure and MD simulation.

in one specific position. However, we manually moved the substrate to alternative binding positions as described below.

Normally, a QM/MM calculation starts from an experimentally determined crystal structure and the available pdb is analyzed in detail and checked for completeness (missing groups and mutations identified). Often these are resting or substrate-bound structures. However, the resting-state or substrate-bound structures are not the required QM/MM starting structures but a structure well later into the catalytic cycle, namely, the iron(IV)-oxo species in our case. Thus, the pdb file needs active site modifications, whereby in our case, α-KG is replaced by succinate and the iron(II)-water complex by an iron(IV)-oxo group. These changes are made to the pdb file manually and subsequently a docking program is utilized to find the most suitable positions of the substrate, which in HctB is the ACP linked to substrate, i.e., hexanoic acid. In this particular system, we located five possible substrate-binding positions, which we used as starting points of the calculations.

After the initial creation of the enzyme intermediate, hydrogen atoms are added to the structure to complete atomic valencies. This is not straightforward work, although part of it can be done with computational software packages, such as PropKa, which generally assumes neutral pH with all arginine and lysine residues protonated and all aspartate and glutamate residues deprotonated. After addition of hydrogen atoms, all charged groups and cysteine bridges will need to be manually checked for their correct protonation states. Thus, a histidine residue has a neutral pK_a and can be singly protonated, doubly protonated, or fully deprotonated in the protein, but this is dependent on the local environment of the histidine residue. Therefore, manual analysis of each individual histidine side chain for neighboring hydrogen-bonding donor and acceptor groups gives an indication of its protonation state. For instance, histidine residues located on the surface of the enzyme tend to be doubly protonated, whereas those in the inner parts of the protein are often singly protonated. In our case, His_8 was chosen to be doubly protonated; His_{67}, His_{111}, His_{157}, His_{185}, His_{227}, and His_{256} were chosen to be singly protonated on Nδ; and finally His_{64}, His_{163}, His_{268}, and His_{273} were chosen also to be singly protonated but on Nε.

The next stage is the addition of solvent, i.e., water, to the protein. We use an iterative process, where a sphere with radius of 8 Å is added around a protein, which is then minimized (the protein coordinates are kept fixed). After minimization, more water molecules are added and the solvation procedure is repeated until few (usually a cutoff of 20 water molecules is used) water molecules can be added. The inset (A) in Scheme 7 shows this process where the number of added water molecules is given as a function of the number of steps. Finally, the complete enzyme model is formed after ions are added to neutralize the overall charge: Mg^{2+}/Ca^{2+} or Cl^{-} to balance the negative or positive protein charges.

The complete enzyme model is then subjected to a molecular dynamics simulation, whereby the system is initially heated up to room temperature (298 K) for 500–1000 ns. An example of an MD simulation for HctB is given in (B) in Scheme 7. As can be seen, the root-mean-square deviation of the bond length changes goes up for the first 200–300 ns, but then stabilizes. The structures after stabilization are considered useful starting points for the QM/MM calculations and our usual approach is to select a number of different snapshots at different time intervals of the MD simulation (Kumar, Thiel, & de Visser, 2011; Porro, Sutcliffe, & de Visser, 2009; Quesne et al., 2014).

Using the snapshots at 300, 400, and 500 ns, we started a QM/MM optimization of the iron(IV)-oxo species of HctB, whereby we use Turbomole to describe the QM region and Charmm linked via ChemShell for the MM partition. The link-atom approach is used to describe the borders between the QM and MM regions, whereby each bond between the QM and MM areas is replaced by a C–H group. The obtained structures show good agreement with small model complexes with similar bond lengths and bond angles (de Visser, 2006a, 2007; de Visser & Latifi, 2009).

3.6 Summary

Enzymatic haloperoxidases and halogenases are unique enzymes in biology that catalyze the formation of C—X bonds. This is an important reaction in organic chemistry and as such has great relevance to the chemical industry. Currently, few applications of haloperoxidases and halogenases are being used in industry, which is mainly due to the lack of knowledge of their catalytic mechanism and function. Computation has assisted experiment extensively and helped to gain insight into the mechanism of halogenation of organic substrates and the nature of the active complex. There are still many questions that have to be answered before commercialization can be made. For instance, most halogenases appear to be highly substrate specific and often give considerable amount of by-products, usually alcohols. More research will be needed to improve substrate selectivity, substrate scope, as well as regioselectivity enhancement.

ACKNOWLEDGMENTS

The BBSRC is thanked for a studentship to A. T. and the National Service of Computational Chemistry Software for CPU time provided for this work.

REFERENCES

Abu-Omar, M. M., Loaiza, A., & Hontzeas, N. (2005). Reaction mechanisms of mononuclear non-heme iron oxygenases. *Chemical Reviews*, *105*, 2227–2252.

Anderson, R. J. L., & Chapman, S. K. (2006). Molecular mechanisms of enzyme-catalysed halogenation. *Molecular BioSystems*, *2*, 350–357.

Bernhardt, P., Okino, T., Winter, J. M., Miyanaga, A., & Moore, B. S. (2011). A stereoselective vanadium-dependent chloroperoxidase in bacterial antibiotic biosynthesis. *Journal of the American Chemical Society*, *133*, 4268–4270.

Bollinger, J. M., Jr., Price, J. C., Hoffart, L. M., Barr, E. W., & Krebs, C. (2005). Mechanism of taurine: α-Ketoglutarate dioxygenase (TauD) from *Escherichia coli*. *European Journal of Inorganic Chemistry*, *2005*, 4245–4254.

Borchardt, S. A., Allain, E. J., Michels, J. J., Stearns, G. W., Kelly, R. F., & McCoy, W. F. (2001). Reaction of acylated homoserine lactone bacterial signalling molecules with oxidised halogen antimicrobials. *Applied and Environmental Microbiology*, *67*, 3174–3179.

Borowski, T., Bassan, A., & Siegbahn, P. E. M. (2004). Mechanism of dioxygen activation in 2-oxoglutarate-dependent enzymes. A hybrid DFT study. *Chemistry A European Journal*, *10*, 1031–1041.

Borowski, T., Noack, H., Mariusz, R., Zych, K., & Siegbahn, P. E. M. (2010). Mechanism of selective halogenation by SyrB2: A computational study. *Journal of the American Chemical Society*, *132*, 12887–12898.

Bruijnincx, P. C. A., van Koten, G., & Klein Gebbink, R. J. M. (2008). Mononuclear non-heme iron enzymes with the 2-His-1-carboxylate facial triad: Recent developments in enzymology and modeling studies. *Chemical Society Reviews*, *37*, 2716–2744.

Buongiorno, D., & Straganz, G. D. (2013). Structure and function of atypically coordinated enzymatic mononuclear non-heme-Fe(II) centers. *Coordination Chemistry Reviews*, *257*, 541–563.

Butler, A. (1999). Mechanistic considerations of the vanadium haloperoxidases. *Coordination Chemistry Reviews*, *187*, 17–35.

Butler, A., & Sandy, M. (2009). Mechanistic considerations of halogenating enzymes. *Nature*, *460*, 848–854.

Carter-Franklin, J. N., & Butler, A. (2004). The reactivity of vanadium bromoperoxidase in the biosynthesis of halogenated marine natural products. *Journal of the American Chemical Society*, *126*, 15060–15066.

Chen, X., & van Pee, K. H. (2008). Catalytic mechanisms, basic roles and biotechnological and environmental significance of halogenating enzymes. *Acta Biochimica et Biophysica Sinica*, *40*, 183–193.

Chiu, H. T., Hubbard, B. K., Shah, A. N., Eide, J., Fredenburg, R. A., Walsh, C. T., et al. (2001). Molecular cloning and sequence analysis of the complestatin biosynthetic gene cluster. *Proceedings of the National Academy of Sciences of the United States of America*, *98*, 8548–8553.

Conte, V., Coletti, A., Floris, B., Licini, G., & Zonta, C. (2011). Mechanistic aspects of vanadium catalysed oxidations with peroxides. *Coordination Chemistry Reviews*, *255*, 2165–2177.

Costas, M. (2011). Selective C-H oxidation catalysed by metalloporphyrins. *Coordination Chemistry Reviews*, *255*, 2912–2932.

Dairi, T., Nakano, T., Aisaka, K., Katsumata, R., & Hasegawa, M. (1995). Cloning and nucleotide sequence of the gene responsible for chlorination of tetracycline. *Bioscience Biotechnology and Biochemistry*, *59*, 1099–1106.
de Boer, E., & Wever, R. (1988). The reaction mechanism of the novel vanadium-bromoperoxidase. A steady-state kinetic analysis. *The Journal of Biological Chemistry*, *263*, 12326–12332.
de Visser, S. P. (2006–1a). Propene activation by the oxo-iron active species of taurine/α-ketoglutarate dioxygenase (TauD) enzyme. How does the catalysis compare to heme-enzymes? *Journal of the American Chemical Society*, *128*, 9813–9824.
de Visser, S. P. (2006–2b). Differences in and comparison of the catalytic properties of heme and non-heme enzymes with a central oxo-iron group. *Angewandte Chemie International Edition*, *45*, 1790–1793.
de Visser, S. P. (2006–3c). What factors influence the ratio of C–H hydroxylation versus C=C epoxidation by a nonheme cytochrome P450 biomimetic? *Journal of the American Chemical Society*, *128*, 15809–15818.
de Visser, S. P. (2007). Can the peroxosuccinate complex in the catalytic cycle of taurine/α-ketoglutarate dioxygenase (TauD) act as an alternative oxidant? *Chemical Communications*, 171–173.
de Visser, S. P. (2009). Elucidating enzyme mechanism and intrinsic chemical properties of short-lived intermediates in the catalytic cycles of cysteine dioxygenase and taurine/α-ketoglutarate dioxygenase. *Coordination Chemistry Reviews*, *253*, 754–768.
de Visser, S. P. (2010). Trends in substrate hydroxylation reactions by heme and nonheme iron(IV)-oxo oxidants give correlations between intrinsic properties of the oxidant with barrier height. *Journal of the American Chemical Society*, *132*, 1087–1097.
de Visser, S. P., & Kumar, D. (Eds.). (2011). *Iron-containing enzymes: Versatile catalysts of hydroxylation reactions in nature*. Cambridge, UK: Royal Society of Chemistry Publishing.
de Visser, S. P., & Latifi, R. (2009). Carbon dioxide: A waste product in the catalytic cycle of alpha-ketoglutarate dependent halogenases prevents the formation of hydroxylated by-products. *The Journal of Physical Chemistry B*, *113*, 12–14.
de Visser, S. P., Ogliaro, F., & Shaik, S. (2001). How does ethene inactivate cytochrome P450 en route to its epoxidation? A density functional study. *Angewandte Chemie International Edition*, *40*, 2871–2874.
de Visser, S. P., Porro, C. S., Quesne, M. G., Sainna, M. A., & Munro, A. W. (2013). Overview on recent theoretical studies discriminating the two-oxidant versus two-state-reactivity models for substrate monoxygenation by cytochrome P450 enzymes. *Current Topics in Medicinal Chemistry*, *13*, 2218–2232.
de Visser, S. P., Rohde, J.-U., Lee, Y.-M., Cho, J., & Nam, W. (2013). Intrinsic properties and reactivities of mononuclear nonheme iron-oxygen complexes bearing the tetramethylcyclam ligand. *Coordination Chemistry Reviews*, *257*, 381–393.
Deng, H., & Ma, L. (2014). Identification of fluorinases from *Streptomyces sp* MA37, *Norcardia brasiliensis*, and *Actinoplanes sp* N902-109 by genome mining. *Chembiochem*, *15*, 364–368.
Deng, H., & O'Hagan, D. (2008). The fluorinase, the chlorinase and the duf-62 enzymes. *Current Opinion in Chemical Biology*, *12*, 582–592.
Dong, C., Flecks, S., Unversucht, S., Haupt, C., van Pée, K. H., & Naismith, J. H. (2005). Tryptophan 7-halogenase (PrnA) structure suggests a mechanism for regioselective chlorination. *Science*, *309*, 2216–2219.
Dong, C., Kotzsch, A., Dorward, M., van Pee, K. H., & Naismith, J. H. (2004). Crystallization and X-ray diffraction of a halogenating enzyme, tryptophan 7-halogenase, from *Pseudomonas fluorescens*. *Acta Crystallographica Section D: Biological Crystallography*, *60*, 1438–1440.
Eustaquio, A. S., Gust, B., Luft, T., Chater, K. F., & Heide, L. (2003). Clorobiocin biosynthesis in Streptomyces: Identification of the halogenase and generation of structural analogs. *Chemistry & Biology*, *10*, 279–288.

Galonic, D. P., Barr, E. W., Walsh, C. T., Bollinger, J. M., Jr., & Krebs, C. (2007). Two interconverting Fe(IV) intermediates in aliphatic chlorination by the halogenase CytC3. *Nature Chemical Biology*, *3*, 113–116.

Geethalakshmi, K. R., Waller, M. P., Thiel, W., & Bühl, M. (2009). ^{51}V NMR chemical shifts calculated from QM/MM models of peroxo forms of vanadium haloperoxidases. *The Journal of Physical Chemistry B*, *113*, 4456–4465.

Godfrey, E., Porro, C. S., & de Visser, S. P. (2008). Comparative quantum mechanics/molecular mechanics (QM/MM) and density functional theory calculations on the oxo-iron species of taurine/α-ketoglutarate dioxygenase. *The Journal of Physical Chemistry. A*, *112*, 2464–2468.

Green, M. T., Dawson, J. H., & Gray, H. B. (2004). Oxoiron(IV) in chloroperoxidase compound II is basic: Implications for P450 chemistry. *Science*, *304*, 1653–1656.

Gribble, G. W. (2003). The diversity of naturally produced organohalogens. *Chemosphere*, *52*, 289–297.

Grzyska, P. K., Ryle, M. J., Monterosso, G. R., Liu, J., Ballou, D. P., & Hausinger, R. P. (2005). Steady-state and transient kinetic analysis of taurine/α-ketoglutarate dioxygenase: Effects of oxygen concentration, alternative sulfonates, and active-site variants on the Fe^{IV}-oxo intermediate. *Biochemistry*, *44*, 3845–3855.

Hager, L. P., Morris, D. R., Brown, F. S., & Eberwein, H. (1966). Chloroperoxidase: II. Utilisation of halogen anions. *The Journal of Biological Chemistry*, *241*, 1769–1777.

Hammer, P. E., Hill, D. S., Lam, S. T., van Pee, K. H., & Ligon, J. M. (1997). Four genes from Pseudomonas fluorescens that encode the biosynthesis of pyrrolnitrin. *Applied and Environmental Microbiology*, *63*, 2147–2154.

Hoffart, L. M., Barr, E. W., Guyer, R. B., Bollinger, J. M., Jr., & Krebs, C. (2006). Direct spectroscopic detection of a C–H-cleaving high-spin Fe(IV) complex in a prolyl-4-hydroxylase. *Proceedings of the National Academy of Sciences of the United States of America*, *103*, 14738–14743.

Hohaus, K., Altmann, A., Burd, W., Fischer, I., Hammer, P. E., Hill, D. S., et al. (1997). NADH-dependent halogenases are more likely to be involved in halometabolite biosynthesis than haloperoxidases. *Angewandte Chemie (International Ed in English)*, *36*, 2012–2013.

Keller, S., Wage, T., Hohaus, K., Hölzer, M., Eichhorn, E., & van Pée, K. H. (2000). Purification and partial characterization of tryptophan 7-halogenase(PrnA) from Pseudomonas fluorescens. *Angewandte Chemie (International Ed in English)*, *39*, 2300–2302.

Kim, S. H., Perera, R., Hager, L. P., Dawson, J. H., & Hoffman, B. M. (2006). Rapid freeze-quench ENDOR study of chloroperoxidase compound I: The site of the radical. *Journal of the American Chemical Society*, *128*, 5598–5599.

Kryatov, S. V., Rybak-Akimova, E. V., & Schindler, S. (2005). Kinetics and mechanisms of formation and reactivity of non-heme iron oxygen intermediates. *Chemical Reviews*, *105*, 2175–2226.

Kulik, H. J., Blasiak, L. C., Marzari, N., & Drennan, C. L. (2009). First-principles study of non-heme Fe(II) halogenase SyrB2 reactivity. *Journal of the American Chemical Society*, *131*, 14426–14433.

Kulik, H. J., & Drennan, C. L. (2013). Substrate placement influences reactivity in non-heme Fe(II) halogenases and hyroxylases. *The Journal of Biological Chemistry*, *288*, 11233–11241.

Kumar, D., de Visser, S. P., & Shaik, S. (2005). Multistate reactivity in styrene epoxidation by Compound I of cytochrome P450: Mechanisms of products and side products formation. *Chemistry A European Journal*, *11*, 2825–2835.

Kumar, S., Faponle, A. S., Barman, P., Vardhaman, A. K., Sastri, C. V., Kumar, D., et al. (2014). Long-range electron transfer triggers mechanistic differences between iron(IV)-oxo and iron(IV)-imido oxidants. *Journal of the American Chemical Society*, *136*, 17102–17115.

Kumar, D., Karamzadeh, B., Sastry, G. N., & de Visser, S. P. (2010). What factors influence the rate constant of substrate epoxidation by Compound I of cytochrome P450 and analogous iron(IV)-oxo oxidants. *Journal of the American Chemical Society, 132*, 7656–7667.

Kumar, D., Latifi, R., Kumar, S., Rybak-Akimova, E. V., Sainna, M. A., & de Visser, S. P. (2013). Rationalization of the barrier height for para-Z-styrene epoxidation by iron(IV)-oxo porphyrins with variable axial ligands. *Inorganic Chemistry, 52*, 7968–7979.

Kumar, D., Sastry, G. N., & de Visser, S. P. (2012). Axial ligand effect on the rate constant of aromatic hydroxylation by iron(IV)-oxo complexes mimicking cytochrome P450 enzymes. *The Journal of Physical Chemistry B, 116*, 718–730.

Kumar, D., Thiel, W., & de Visser, S. P. (2011). Theoretical study on the mechanism of the oxygen activation process in cysteine dioxygenase enzymes. *Journal of the American Chemical Society, 133*, 3869–3882.

Latifi, R., Sainna, M. A., Rybak-Akimova, E. V., & de Visser, S. P. (2013). Does hydrogen bonding-donation to manganese(IV)-oxo and iron(IV)-oxo oxidants affect the oxygen atom transfer ability? A computational study. *Chemistry – A European Journal, 19*, 4058–4068.

Laus, G. (2001). Biological activities of natural halogen compounds. *Studies in Natural Products Chemistry, 25*, 757–809.

Libby, R. D., Beachy, T. M., & Phipps, K. A. (1996). Quantitating direct chlorine transfer from enzyme to substrate in chloroperoxidase-catalysed reactions. *The Journal of Biological Chemistry, 271*, 21820–21872.

Light, K. M., Hangasky, J. A., Knapp, M. J., & Solomon, E. I. (2013). Spectroscopic studies of the mononuclear non-heme Fe^{II} Enzyme FIH: Second-sphere contributions to reactivity. *Journal of the American Chemical Society, 135*, 9665–9674.

Light, K. M., Hangasky, J. A., Knapp, M. J., & Solomon, E. I. (2014). First- and second-sphere contributions to Fe(II) site activation by cosubstrate binding in non-heme Fe enzymes. *Dalton Transactions, 43*, 1505–1508.

Manoj, K. M., & Hager, L. P. (2008). Chloroperoxidase, a Janus enzyme. *Biochemistry, 47*, 2997–3003.

Martinez, J. S., Carroll, G. L., Tschirret-Guth, R. A., Altenhoff, G., Little, D. R., & Butler, A. (2001). On the regiospecificity of vanadium bromoperoxidase. *Journal of the American Chemical Society, 123*, 3289–3294.

Matthews, M. L., Chang, W.-C., Layne, A. P., Miles, L. A., Krebs, C., & Bollinger, J. M., Jr. (2014). Direct nitration and azidation of aliphatic carbons by an iron-dependent halogenase. *Nature Chemical Biology, 10*, 209–217.

Matthews, M. L., Krest, C. M., Barr, E. W., Vaillancourt, F. H., Walsh, C. T., Green, M. T., et al. (2009). Substrate-triggered formation and remarkable stability of the C-H bond-cleaving chloroferryl intermediate in the aliphatic halogenase, SyrB2. *Biochemistry, 48*, 4331–4343.

Matthews, M. L., Neumann, C. S., Miles, L. A., Grove, T. L., Booker, S. J., Krebs, C., et al. (2009). Substrate positioning controls the partition between halogenation and hydroxylation in the aliphatic halogenase, SyrB2. *Proceedings of the National Academy of Sciences of the United States of America, 106*, 17723–17728.

Messerschmidt, A., & Wever, R. (1996). X-ray structure of a vanadium-containing enzyme: Chloroperoxidase from the fungus Curvularia inaequalis. *Proceedings of the National Academy of Sciences of the United States of America, 93*, 392–396.

Morris, D. R., & Hager, L. P. (1966). Chloroperoxidase: I. Isolation and properties of the crystalline glycoprotein. *The Journal of Biological Chemistry, 241*, 1763–1768.

Nakama, Y., Yoshida, O., Yoda, M., Araki, K., Sawada, Y., Nakamura, J., et al. (2010). Discovery of a novel series of semisynthetic vancomycin derivatives effective against vancomycin-resistant bacteria. *Journal of Medicinal Chemistry, 53*, 2528–2533.

Neidig, M. L., Brown, C. D., Light, K. M., Fujimori, D. G., Nolan, E. M., Price, J. C., et al. (2007). CD and MCD of CytC3 and taurine dioxygenase: Role of the facial triad in α-KG-dependent oxygenases. *Journal of the American Chemical Society*, *129*, 14224–14231.

Neumann, C. S., Fujimori, D. G., & Walsh, C. T. (2009). Halogenation strategies in natural product biosynthesis. *Chemistry & Biology*, *15*, 99–109.

Nightingale, Z. D., Lancha, A. H., Handelman, S. K., Dolnikowski, G. G., Busse, S. C., Dratz, E. A., et al. (2000). Relative reactivity of lysine and other peptide-bound amino acids to oxidation by hypochlorite. *Free Radical Biology & Medicine*, *29*, 425–433.

Nowak-Thompson, B., Chaney, N., Wing, J. S., Gould, S. J., & Loper, J. E. (1999). Characterization of the pyoluteorin biosynthetic gene cluster of Pseudomonas fluorescens Pf-5. *Journal of Bacteriology*, *181*, 2166–2174.

O'Hagan, D., & Deng, H. (2015). Enzymatic fluorination and biotechnological developments of the fluorinase. *Chemical Reviews*, *115*, 634–649.

O'Hagan, D., Schaffrath, C., Cobb, S. L., Hamilton, J. T. G., & Murphy, C. D. (2002). Biochemistry: Biosynthesis of an organofluorine molecule. *Nature*, *416*, 279.

Pacios, L. F., & Gálvez, O. (2010). Active site, catalytic cycle, and iodination reactions of vanadium iodoperoxidase: A computational study. *Journal of Chemical Theory and Computation*, *6*, 1738–1752.

Pandian, S., Vincent, M. A., Hillier, I. H., & Burton, N. A. (2009). Why does the enzyme SyrB2 chlorinate, but does not hydroxylate, saturated hydrocarbons? A density functional theory (DFT) study. *Dalton Transactions*, *31*, 6201–6207.

Porro, C. S., Sutcliffe, M. J., & de Visser, S. P. (2009). Quantum mechanics/molecular mechanics studies on the sulfoxidation of dimethyl sulfide by compound I and compound 0 of cytochrome P450: Which is the better oxidant? *The Journal of Physical Chemistry. A*, *113*, 11635–11642.

Pratter, S. M., Ivkovic, J., Birner-Gruienberger, R., Breinbauer, R., Zangger, K., & Stranganz, G. D. (2014). More than just a halogenase: Modification of fatty acyl moieties by a trifunctional metal enzyme. *Chembiochem*, *15*, 567–574.

Pratter, S. M., Light, K. M., Solomon, E. I., & Straganz, G. D. (2014). The role of chloride in the mechanism of O_2 activation at the mononuclear nonheme Fe(II) center of the halogenase HctB. *Journal of the American Chemical Society*, *136*, 9385–9395.

Price, J. C., Barr, E. W., Glass, T. E., Krebs, C., & Bollinger, J. M., Jr. (2003). The first direct characterization of a high-valent iron intermediate in the reaction of an α-ketoglutarate-dependent dioxygenase: A high-spin Fe(IV) complex in taurine/α-ketoglutarate dioxygenase (TauD) from *Escherichia coli*. *Journal of the American Chemical Society*, *125*, 13008–13009.

Price, J. C., Barr, E. W., Tirupati, B., Bollinger, J. M., Jr., & Krebs, C. (2003). The first direct characterization of a high-valent iron intermediate in the reaction of an alpha-ketoglutarate-dependent dioxygenase: A high-spin Fe^{IV} complex in taurine/alpha-ketoglutarate dioxygenase (TauD) from *Escherichia coli*. *Biochemistry*, *42*, 7497–7508.

Puk, O., Huber, P., Bischoff, D., Recktenwald, J., Jung, G., Süßmuth, R. D., et al. (2002). Glycoprotein biosynthesis in Amycalotopsis mediterranei DSM5908: Function of a halogenase and a haloperoxidase/perhydrolase. *Chemistry & Biology*, *9*, 225–235.

Quesne, M. G., & de Visser, S. P. (2012). Regioselectivity of substrate hydroxylation versus halogenation by a nonheme iron(IV)–oxo complex: Possibility of rearrangement pathways. *Journal of Biological Inorganic Chemistry*, *17*, 841–852.

Quesne, M. G., Faponle, A. S., Goldberg, D. P., & de Visser, S. P. (2015). Catalytic function and mechanism of heme and nonheme iron(IV)-oxo complexes in nature. In M. Costas, & M. Swart (Eds.), Germany: Wiley. in press.

Quesne, M. G., Latifi, R., Gonzalez-Ovalle, L. E., Kumar, D., & de Visser, S. P. (2014). Quantum mechanics/molecular mechanics study on the oxygen binding and substrate hydroxylation step in AlkB repair enzymes. *Chemistry—A European Journal*, *20*, 435–446.

Rehder, D., Casny, M., & Grosse, R. (2004). A vanadium-51 NMR study of the binding of vanadate and peroxovanadate to proteins. *Magnetic Resonance in Chemistry, 42*, 745–749.

Riggs-Gelasco, P. J., Price, J. C., Guyer, R. B., Brehm, J. H., Barr, E. W., Bollinger, J. M., Jr., et al. (2004). EXAFS spectroscopic evidence for an FeO unit in the Fe(IV) intermediate observed during oxygen activation by taurine: α-Ketoglutarate dioxygenase. *Journal of the American Chemical Society, 126*, 8108–8109.

Sahu, S., Quesne, M. G., Davies, C. G., Dürr, M., Ivanović-Burmazović, I., Siegler, M. A., et al. (2014). Direct observation of a non-heme iron(IV)-oxo complex that mediates aromatic C–F hydroxylation. *Journal of the American Chemical Society, 136*, 13542–13545.

Sainna, M. A., Kumar, S., Kumar, D., Fornarini, S., Crestoni, M. E., & de Visser, S. P. (2015). A comprehensive test set of epoxidation rate constants by iron(IV)-oxo porphyrin complexes. *Chemical Science, 6*, 1516–1529.

Sánchez, C., Butovich, I. A., Brana, A. F., Rohr, J., Méndez, C., & Salas, J. A. (2002). The biosynthetic gene cluster for the antitumor rebeccamycin. Characterization and generation of indolocarbazole derivatives. *Chemistry & Biology, 9*, 519–531.

Schneider, C. J., Zampella, G., Greco, C., Pecoraro, V. L., & De Gioia, L. (2007). Mechanistic analysis of nucleophilic substrates oxidation by functional models of vanadium-dependent haloperoxidases: A density functional theory study. *European Journal of Inorganic Chemistry, 2007*, 515–523.

Seibold, C., Schnerr, H., Rumpf, J., Kunzendorf, A., Hatscher, C., Wage, T., et al. (2006). A flavin-dependent tryptophan 6-halogenase and its use in modification of pyrrolnitrin biosynthesis. *Biocatalysis and Biotransformation, 24*, 401–408.

Senn, H. M. (2014). Insights into enzymatic halogenatin from computational studies. *Frontiers in Chemistry, 2*, 1–15.

Senn, H. M., O'Hagan, D., & Thiel, W. (2005). Insight into enzymatic C–F bond formation from QM and QM/MM calculations. *Journal of the American Chemical Society, 127*, 13643–13655.

Shaik, S., Kumar, D., & de Visser, S. P. (2008). A valence bond modeling of trends in hydrogen abstraction barriers and transition states of hydroxylation reactions catalyzed by cytochrome P450 enzymes. *Journal of the American Chemical Society, 130*, 10128–10140.

Shaik, S., Kumar, D., de Visser, S. P., Altun, A., & Thiel, W. (2005). Theoretical perspective on the structure and mechanism of cytochrome P450 enzymes. *Chemical Reviews, 105*, 2279–2328.

Silaghi-Dumitrescu, R. (2008). Halide activation by heme peroxidases: Theoretical predictions on putative adducts of halides with compound I. *European Journal of Inorganic Chemistry, 2008*, 5404–5407.

Sinnecker, S., Svensen, N., Barr, E. W., Ye, S., Bollinger, J. M., Jr., Neese, F., et al. (2007). Spectroscopic and computational evaluation of the structure of the high-spin Fe(IV)-oxo intermediates in taurine: α-Ketoglutarate dioxygenase from *Escherichia coli* and its His99-Ala ligand variant. *Journal of the American Chemical Society, 129*, 6168–6179.

Smith, D. R. M., Gruschow, S., & Goss, R. J. M. (2013). Scope and potential of halogenases in biosynthetic applications. *Current Opinion in Chemical Biology, 17*, 276–283.

Spreti, N., Germani, R., Icani, A., & Savelli, G. (2004). Stabilization of chloroperoxidase by polyethylene glycols in aqueous media: Kinetic studies and synthetic applications. *Biotechnology Progress, 20*, 96–101.

Stone, K. L., Behan, R. K., & Green, M. T. (2005). X-ray absorption spectroscopy of chloroperoxidase compound I: Insight into the reactive intermediate of P450 chemistry. *Proceedings of the National Academy of Sciences of the United States of America, 102*, 16563–16565.

Stonik, V. A., & Fedorov, S. N. (2014). Marine low molecular weight natural products as potential cancer preventative compounds. *Marine Drugs, 12*, 636–671.

Sundaramoorthy, M., Terner, J., & Poulos, T. L. (1995). The crystal structure of chloroperoxidase: A heme peroxidase-cytochrome P450 functional hybrid. *Structure, 3*, 1367–1377.

Sundaramoorthy, M., Terner, J., & Poulos, T. L. (1998). Stereochemistry of the chloroperoxidase active site: Crystallography and molecular modelling studies. *Chemistry & Biology, 5*, 461–473.

Taurog, A., & Howells, E. M. (1966). Enzymatic iodination of tyrosine and thyroglobulin with chloroperoxidase. *The Journal of Biological Chemistry, 241*, 1329–1339.

Timmins, A., & de Visser, S. P. (2015). Regioselectivity of substrate halogenation versus hydroxylation by HctB halogenases: A quantum mechanics/molecular mechanics study. Manuscript in preparation. Unpublished work.

Topol, I. A., Nemukhin, A. V., Salnikow, K., Cachau, R. E., Abashkin, Y. G., Kasprzak, K. S., et al. (2006). Quantum chemical modeling of reaction mechanism for 2-glutarate dependent enzymes: Effect of substitution of iron by nickel and cobalt. *The Journal of Physical Chemistry. A, 110*, 4223–4228.

Trefzer, A., Pelzer, S., Schimana, J., Stockert, S., Bihlmaier, C., Fiedler, H. P., et al. (2002). Biosynthetic gene cluster of simocyclinone, a natural multihybrid antibiotic. *Antimicrobial Agents and Chemotherapy, 46*, 1174–1182.

Vaillancourt, F. H., Yeh, E., Vosburg, D. A., Garneaur-Tsodikova, S., & Walsh, C. T. (2006). Nature's inventory of halogenation catalysts: Oxidative strategies predominate. *Chemical Reviews, 106*, 3364–3378.

van Pee, K.-H. (1996). Biosynthesis of halogenated metabolited by bacteria. *Annual Review of Microbiology, 50*, 375–399.

van Pee, K.-H., Dong, C., Flecks, S., Naismith, J., Patallo, E. P., & Wage, T. (2006). Biological halogenation has moved far beyond haloperoxidases. *Advances in Applied Microbiology, 59*, 127–157.

van Schijndel, J. W. P. M., Barnett, P., Roelse, J., Vollenbroed, E. G. M., & Wever, R. (1994). The stability and steady state kinetics of vanadium chloroperoxidase from the fungus Curvularia inequalis. *European Journal of Biochemistry, 225*, 151–157.

Vardhaman, A. K., Barman, P., Kumar, S., Sastri, C. V., Kumar, D., & de Visser, S. P. (2013). Mechanistic insight into halide oxidation by non-heme iron complexes. Haloperoxidase versus halogenase activity. *Chemical Communications, 49*, 10926–10928.

Vardhaman, A. K., Sastri, C. V., Kumar, D., & de Visser, S. P. (2011). Nonheme ferric hydroperoxo intermediates are efficient oxidants of bromide oxidation. *Chemical Communications, 47*, 11044–11046.

Vilter, H. (1983). Peroxidases from Phaeophyceae. *Botanica Marina, 26*, 429–435.

Vilter, H. (1984). Peroxidases from Phaeophycea: A vanadium(V)-dependent peroxidase from Ascophyllum nodosum. *Phytochemistry, 23*, 1387–1390.

Wagenknecht, H.-A., & Woggon, W.-D. (1997). Identification of intermediates in the catalytic cycle of chloroperoxidase. *Chemistry & Biology, 4*, 367–372.

Wagner, C., Omari, M. E., & Konig, G. M. (2009). Biohalogenation: Nature's way to synthesize halogenated metabolites. *Journal of Natural Products, 72*, 540–553.

Waller, M. P., Bühl, M., Geethalakshmi, K. R., Wang, D., & Thiel, W. (2007). ^{51}V NMR chemical shifts calculated from QM/MM models of vanadium chloroperoxidase. *Chemistry—A European Journal, 13*, 4723–4732.

Waller, M. P., Geethalakshmi, K. R., & Bühl, M. (2008). ^{51}V NMR chemical shifts from quantum-mechanical/molecular-mechanical models of vanadium bromoperoxidase. *The Journal of Physical Chemistry B, 112*, 5813–5823.

Weitnauer, G., Mühlenweg, A., Trefzer, A., Hoffmeister, D., Süßmuth, R. D., Jung, G., et al. (2001). Biosynthesis of the orthosomycin antibiotic avilamycin A: Deductions from the molecular analysis of the avi biosynthetic gene cluster of Streptomyces viridochromogenes Tü57 and production of new antibiotics. *Chemistry & Biology, 8*, 569–581.

Wever, R. (2001). Vanadium dependent haloperoxidases. In A. Messerschmidt, R. Humber, K. Wieghardt, & T. Poulos (Eds.), *Handbook of metalloproteins* (pp. 1417–1428). Chichester: Wiley.

Wijnands, I., & van Pee, K. H. (2004). A novel halogenase gene from the pentachloropseudilin producer Actinoplanes sp. ATCC 33002 and detection of in vitro halogenase activity. *FEMS Microbiology Letters, 237*, 363–367.

Winter, J. M., & Moore, B. S. (2009). Exploring the chemistry and biology of vanadium-dependent haloperoxidases. *The Journal of Biological Chemistry, 284*, 18577–18581.

Yeh, E., Blasiak, L. C., Koglin, A., Drennan, C. L., & Walsh, C. T. (2007). Chlorination by a long-lived intermediate in the mechanism of flavin-dependent halogenases. *Biochemistry, 46*, 1284–1292.

Yeh, E., Cole, L. J., Barr, E. W., Bollinger, J. M., Jr., Ballou, D. P., & Walsh, C. T. (2006). Flavin redox chemistry precedes substrate chlorination during the reaction of the flavin-dependent halogenase RebH. *Biochemistry, 45*, 7904–7912.

Yeh, E., Garneau, S., & Walsh, C. T. (2005). Robust in vitro activity of RebF and RebH, a two-component reductase/halogenase, generating 7-chlorotryptophan during rebeccamycin biosynthesis. *Proceedings of the National Academy of Sciences of the United States of America, 102*, 3960–3965.

Yu, T. W., Bai, L., Clade, D., Hoffmann, D., Toelzer, S., Trinh, K. Q., et al. (2002). The biosynthetic gene cluster of the maytansinoid antitumor agent ansamitocin from Actinosynnema pretiosum. *Proceedings of the National Academy of Sciences of the United States of America, 99*, 7968–7973.

Zampella, G., Fantucci, P., Pecoraro, V. L., & De Gioia, L. (2005). Reactivity of peroxo forms of the vanadium haloperoxidase cofactor. A DFT investigation. *Journal of the American Chemical Society, 127*, 953–960.

Zampella, G., Fantucci, P., Pecoraro, V. L., & De Gioia, L. (2006). Insight into the catalytic mechanism of vanadium haloperoxidases. DFT investigation of vanadium cofactor reactivity. *Inorganic Chemistry, 45*, 7133–7143.

Zehner, S., Bister, B., Sussmuth, R. D., Mendez, C., Salas, J. A., & van Pee, K. H. (2005). A regioselective tryptophan 5-halogenase is involved in pyrroindomycin biosynthesis in Streptomyces rugosporus LL-42D005. *Chemistry & Biology, 12*, 445–452.

CHAPTER SIX

The Importance of the MM Environment and the Selection of the QM Method in QM/MM Calculations: Applications to Enzymatic Reactions

Eric André C. Bushnell*,‡,1, Victoria Erica J. Berryman*, James W. Gauld†, Russell J. Boyd*

*Department of Chemistry, Dalhousie University, Halifax, Nova Scotia, Canada
†Department of Chemistry and Biochemistry, University of Windsor, Windsor, Ontario, Canada
‡Department of Chemistry, Brandon University, Brandon, Manitoba, Canada
[1]Corresponding author: e-mail address: Eric.Bushnell@Dal.Ca

Contents

Abstract

In this chapter, we discuss the influence of an anisotropic protein environment on the reaction mechanisms of saccharopine reductase and uroporphyrinogen decarboxylase, respectively, via the use of a quantum mechanical and molecular mechanical (QM/MM) approach. In addition, we discuss the importance of selecting a suitable DFT functional to be used in a QM/MM study of a key intermediate in the mechanism of 8*R*-lipoxygenase, a nonheme iron enzyme.

In the case of saccharopine reductase, while the enzyme utilizes a substrate-assisted catalytic pathway, it was found that only through treating the polarizing effect of the active site, via the use of an electronic embedding formalism, was agreement with experimental kinetic data obtained. Similarly, in the case of uroporphyrinogen decarboxylase, the effect of the protein environment on the catalytic mechanism was found

Advances in Protein Chemistry and Structural Biology, Volume 100
ISSN 1876-1623
http://dx.doi.org/10.1016/bs.apcsb.2015.06.008

to be such that the calculated rate-limiting barrier is in good agreement with related experimentally determined values for the first decarboxylation of the substrate.

For 8*R*-lipoxygenase, it was found that the geometries and energies of the multi-centered open-shell intermediate complexes formed during the mechanism are quite sensitive to the choice of the density functional theory method. Thus, while density functional theory has become the method of choice in QM/MM studies, care must be taken in the selection of a particular high-level method.

1. INTRODUCTION

Many of the chemical reactions necessary for life proceed too slowly in solution in the absence of a catalyst. Thus, enzymes play a crucial role by accelerating these processes to life-sustaining rates (Kamerlin & Warshel, 2010; Liu & Gauld, 2008). For instance, the hydrolysis of a glycosidic bond has a half-life (under standard conditions in solution) of approximately 5 million years (Huang, Llano, & Gauld, 2010). Yet the enzymes that catalyze such reactions (i.e., glycoside hydrolases) can do so with rates of 1000 times per second (Huang et al., 2010). Uroporphyrinogen decarboxylase (UROD) is yet another example of catalytic rate enhancement. This enzyme performs the quadruple decarboxylation of its substrate URO-III in mere seconds (Bushnell, Erdtman, Llano, Eriksson, & Gauld, 2011). In the absence of this enzyme, the half-life for this same reaction (under standard conditions in solution) is 2.3 billion years (Lewis & Wolfenden, 2008).

During the 1950s and 1960s, the catalytic efficiency of enzymes was believed to be due to enhanced concentration and closeness of catalytic residues near substrates (Hammes, Benkovic, & Hammes-Schiffer, 2011). Since then, however, many ways in which an enzyme may enhance chemical rates have been proposed. However, it should be noted that no single consensus currently exists to explain the source of their rate enhancement abilities (Hammes et al., 2011; Kamerlin & Warshel, 2010; Llano & Gauld, 2010). For instance, transition state stabilization (Llano & Gauld, 2010; Pauling, 1948; Schramm, 2012), preorganization of the active site (Schutz & Warshel, 2001; Warshel, 1978, 1987, 2002, 2003; Warshel & Levitt, 1976; Warshel & Papazyan, 1998; Warshel, Sharma, Kato, & Parson, 2006; Warshel, Sharma, Kato, Xiang, et al., 2006), near attack conformation (Bruice, 2002; Hur & Bruice, 2003; Kollman, Kuhn, & Perakyla, 2002; Lau & Bruice, 1998; Page & Jencks, 1971), entropic effects upon substrate binding (Bruice, 1976; Jencks, 1986; Villa et al., 2000; Villa & Warshel, 2001; Warshel, 1998), desolvation of the substrate (Crosby,

Stone, & Lienhard, 1970; Devi-Kesavan & Gao, 2003; Lee & Houk, 1997; Shurki, Strajbl, Villa, & Warshel, 2002; Warshel, Aqvist, & Creighton, 1989), enzyme conformational dynamics (Careri, Fasella, & Gratton, 1979; Gavish & Werber, 1979; Hammes et al., 2011; Hammes-Schiffer & Benkovic, 2006; Henzler-Wildman & Kern, 2007; Henzler-Wildman et al., 2007; Kamerlin & Warshel, 2010; Kurplus & McCammon, 1983; McCammon, Wolynes, & Karplus, 1979), quantum mechanical tunnelling (Nagel & Klinman, 2006, 2010; Olsson, Mavri, & Warshel, 2006), and low barrier hydrogen bonds (Tuckerman, Marx, Klein, & Parrinello, 1997; Warshel, Papazyan, & Kollman, 1995) have all been proposed as the source or contributing factors to enzymes' rate-enhancing power. A more complete discussion of each of these factors can be found elsewhere (Llano & Gauld, 2010).

Better understanding of the means by which an enzyme catalyzes a chemical reaction may provide valuable information that may lead, for instance, to the design of new therapeutic agents, or new biomimetic catalysts. Yet, understanding how enzymes catalyze chemical reactions remains a challenge for both experimental and computational biochemistry (Lodola, Woods, & Mulholland, 2008). For instance, the catalytic power of an enzyme may be a combination of two or more of the effects listed above; thus, it is very challenging to dissect the significance of each contribution to the overall catalytic power of an enzyme (Warshel, Sharma, Chu, & Aqvist, 2007). However, regardless of the means by which an enzyme may catalyze a reaction, the structure of an enzyme is critical to its function.

The two most common methods to study biochemical reactions are quantum mechanical and molecular mechanical (QM/MM) and QM/cluster approaches. A QM/cluster approach ignores a majority of the protein environment while the remaining truncated system is modeled using an *ab initio* method, such as density functional theory (DFT). The protein environment is often implicitly modeled using a polarizable continuum model with, for instance, a dielectric value of 4 to reflect its slight polarity (Himo, 2006; Hu, Eliasson, Heimdal, & Ryde, 2009). A QM/MM approach on the other hand combines two or more computational methods and allows for the precise investigation of considerably larger systems.

For the QM/cluster approach, there are several examples that discuss the advantages of using such a method (Blomberg, Borowski, Himo, Liao, & Siegbahn, 2014; Himo, 2006; Siegbahn & Himo, 2009). To briefly summarize, the models used in QM/cluster studies are generally very small (in relation to a complete protein) due to the sizeable computational costs incurred (Hu et al., 2009). Indeed, QM systems typically have on average 20–200 atoms (Siegbahn & Himo, 2009). However, systems of 400 or more atoms have been studied (Hu et al., 2009). An advantage to using these smaller models is that it easier to ensure that the states along the reaction coordinates remain in a similar local minimum (Sumner, Söderhjelm, & Ryde, 2013). This is generally not the case for QM/MM approaches (Shaik et al., 2010; Siegbahn & Himo, 2009). It has been suggested that the reason why a QM/cluster approach has shown such good agreement with experiment is because the energies involved in the reaction are usually much larger than long-range interaction energies (Himo, 2006). That is, the long-range electrostatics have a minimal effect on the reaction barriers. Thus, the effect of catalysis within an active site is to a large extent local where the environmental effects are usually of lower order (Himo, 2006).

Generally, the errors in reaction energies for a QM/cluster approach are estimated to be around 12–20 kJ/mol (Himo, 2006; Siegbahn & Blomberg, 1999). However, because of the limited number of atoms in a QM/cluster approach, care in the choice of groups included in the model is needed (Hu, Soderhjelm, & Ryde, 2011). For instance, a previous study on a proton-transfer step in [Ni,Fe]-hydrogenase investigated the effects on relative energies by increasing a QM model based on two different criteria (Hu et al., 2009). More specifically, the QM/cluster model was enlarged based on either: (i) distance to the reactive center or (ii) the effect on the reaction energies (Hu et al., 2009). Using the two different methods, it was observed that the results obtained were quite sensitive to the selection scheme (Hu et al., 2009). Furthermore, even with a model of 400 atoms in the QM model, the wrong selection of atoms included in the model could have dramatic effects (Hu et al., 2009). Indeed, it was found that even after adding 40 amino acid residues to the QM model, a difference of 60 kJ mol^{-1} can be observed in reaction energies depending on the selection scheme. Therefore, it was suggested to only use the residues involved in the mechanism with a PCM approach with dielectric of 4 to model the environment.

Additionally, due to the neglect of the protein environment detailed understanding of the interactions of the environment on the reaction is unknown. For instance, if we consider the QM/MM investigation into the reaction of 2-methyl-3-hydroxypyridine-5-carboxylic acid oxygenase via the use of an ONIOM approach, it was found that the energetics of the reaction pathway are sensitive to the choice of residues in the QM region (i.e., the different QM/MM partitioning schemes) (Tian, Strid, & Eriksson, 2011). In particular, from an investigation of various partitioning schemes, residues that are important in affecting reaction energies and residues that are critical for substrate binding were identified (Tian et al., 2011). Hence, from this work, important insight into the effect of the protein environment on the reaction of 2-methyl-3-hydroxypyridine-5-carboxylic acid oxygenase was gained.

It was in 1976 that Warshel and Levitt introduced the concept of QM/MM; the ideas presented are still considered essential in the field of QM/MM (Senn & Thiel, 2009; Warshel & Levitt, 1976). However, the work by Singh and Kollman (1986) and Field, Bash, and Karplus (1990) are also credited with establishing what QM/MM is today (Vreven & Morokuma, 2006). Several reviews have been published that discuss the details of the QM/MM (see for instance Lodola et al., 2008; Senn & Thiel, 2009; Vreven & Morokuma, 2006). Hence, herein we summarize key points to highlight the advantages of a QM/MM approach.

The QM/MM approach has been established as a valuable tool to investigate inorganic/organometallic systems and has been shown to be a state-of-the-art computational technique for studying biochemical systems (Senn & Thiel, 2009). The use of QM/MM allows for the combination of two (or more in the case of ONIOM) computational techniques and allows for the precise investigation of large systems. For instance, models containing 20,000–30,000 atoms are now quite common (Shaik et al., 2010). The advantage of using QM/MM is that, while MM cannot describe chemical reactions, the typical MM force fields used to model proteins (e.g., AMBER and CHARM22) are extremely good at describing the energetic differences in conformational changes as well as describing nonbonded interactions in large systems such as proteins (Adcock & McCammon, 2006; Lodola et al., 2008).

For QM/MM calculations, there are two different schemes in which the SCF calculation is performed for the QM region. In the mechanical embedding (ME) scheme, the energy of the QM region is calculated in the absence of partial charges of the MM region. Thus, the polarizing effect of the anisotropic protein environment has no impact on the QM calculations. Because the interaction energies between layers are calculated at the MM level of theory, care must be taken in the parameterization of the QM and MM regions (Hu et al., 2011). In the electronic embedding (EE) scheme, the energy of the QM region is calculated in the presence of the point charges of the MM outer layer. Thus, the QM region is polarized in the presence of the anisotropic environment. However, it should be noted that for the EE scheme, the partial charges included near truncation points might lead to over-polarization of the QM region during the SCF procedure

(Hu et al., 2011). Thus, care must be taken in selecting the QM region for the calculations. Interestingly, while it might be expected that the use of an EE scheme would provide a more accurate description of the system, the results of a study by Hu et al. (2011) suggest that the ME scheme is a better choice. However, in their approach, the charges for a majority of the QM layer remained consistent, but for a small section of the reacting core, this was not the case.

When choosing a QM region, it is important to ensure that the QM/MM energy is independent of the connectivity; that is, the truncation between QM and MM is not too close to the reaction center. For instance, if the truncation is too close to the site of reaction, then prior to the reaction there might be no bond angle or torsional term in the MM energy, whereas after bond formation an angle or torsional term would now exist in the MM energy term resulting in the MM energy to be discontinuous. Thus, it is recommended to have the site of truncation at least three bonds away from bond making and breaking processes (Vreven et al., 2006).

Comparison of the QM/cluster and QM/MM approaches generally shows that the energies of the latter approach converge faster as the number of atoms in the QM region increases (Hu et al., 2011).

It should be noted that regardless of the choice of a QM/cluster or QM/MM approach, a key component is the selection of a particular *ab initio* or DFT method. In the case of biochemical and bioinorganic systems, DFT has become the method of choice (Neese, 2006; Ryde, 2007). Indeed, in the case of metal complexes, errors in M–L bond lengths are typically less than 0.07 Å (Neese, 2006; Olsson & Ryde, 2001; Ryde, 2003, 2007; Shen & Ryde, 2004; Sigfridsson, Olsson, & Ryde, 2001; Sigfridsson & Ryde, 2002). Notably, this compares favorably with errors of 0.1 Å and even 0.3 Å that have been seen in crystallographic methods (Cruickshank, 1999; Fields et al., 1994; Nilsson, Lecerof, Sigfridsson, & Ryde, 2003). Indeed, hybrid DFT methods, in particular B3LYP, have become the standard tools of choice for computational investigations on enzymes and metalloenzymes (Becke, 1993a, 1993b; Lee, Yang, & Parr, 1988; Siegbahn, 2006). With respect to metalloenzymes, the good performance of the B3LYP functional is surprising given that the functional was parameterized based on reference molecules that do not include metals (Reiher, Salomon, & Hess, 2001). However, because of this, it has been found that for some metal-containing systems, the use of B3LYP does not lead to reasonable agreement with experiment (Ghosh, Persson, & Taylor, 2003; Kepenekian et al., 2011). For instance, it was found by

Reiher et al. (2001) that in the case of Fe-containing systems with near-degenerate states, such methods are often unable to give a qualitatively or quantitatively correct ordering of states. Moreover, in the case of Fe^{III}(porphyrin)Cl, the B3LYP method predicted that a quartet and not the experimentally observed sextet was the ground state (Ghosh et al., 2003). This poses problems given that accurate determination of the relative energies of near-degenerate states can be essential in elucidating a given systems chemistry (Reiher et al., 2001; Siegbahn & Himo, 2009). From the past studies, it appears that the amount of exact Hartree–Fock (HF) exchange included in DFT methods has a direct consequence on the reliability of the result (Reiher et al., 2001). For metal-containing systems that contain unpaired electrons, it has been proposed that a contribution of 15% HF offers better accuracy (Lundberg & Morokuma, 2007; Reiher, 2002; Reiher et al., 2001; Schenk, Pau, & Solomon, 2004; Siegbahn, Blomberg, & Chen, 2010). Indeed, this has been shown for the calculation of the energetics of Fe^{II}–S complexes (Reiher et al., 2001).

In this chapter, we discuss two case studies (the reaction mechanism of saccharopine reductase and UROD) that focus on the influence of the anisotropic protein environment on reaction mechanisms. In the final case study, we discuss the reliability of several commonly used DFT functionals in the prediction of relative energies of a key intermediate in the mechanism of 8*R*-lipoxygenase.

2. CASE STUDIES

2.1 Saccharopine Reductase

Saccharopine reductase is a key enzyme in the synthesis of lysine. The enzyme catalyzes the condensation reaction of α-aminoadipate-δ-semialdehyde (AASA) with glutamic acid with subsequent reduction by NADPH to give the L-lysine precursor saccharopine (Vashishtha, West, & Cook, 2009). It is noted that while organisms generally use 20 "standard" α-amino acids for the synthesis of proteins, not all organisms are able to synthesize these 20 amino acids. For instance, only certain green plants, bacteria, fungi, and cyanobacteria are able to synthesize L-lysine (Vashishtha et al., 2009; Zabriskie & Jackson, 2000). For the synthesis of L-lysine two distinct pathways exist: the diaminopimelate pathway (green plants, bacteria, and lower fungi), and the α-aminoadipate pathway (cyanobacteria and higher fungi) (Xu, Andi, Qian, West, & Cook, 2006). Fungal species that utilize the α-aminoadipate pathway are *Aspergillus fumigatus*,

Saccharomyces cerevisiae, and *Magnaporthe grisea* (Vashishtha et al., 2009). Notably, the two former species are human fungal agents that pose a risk to those suffering from AIDS, cancer, or recent tissue transplant (Johansson, Steffens, Lindqvist, & Schneider, 2000). *M. grisea* on the other hand is best known for causing rice blast disease and poses a major risk to rice producing countries (Talbot, 1995). Thus, the α-aminoadipate pathway is an attractive target for the development of new therapeutic agents and fungicides (Ribot et al., 2008).

While two possible mechanisms have been previously suggested for saccharopine reductase, neither mechanism identifies the acidic or basic groups that catalyze the reaction (Johansson et al., 2000; Vashishtha et al., 2009). However, one study does suggest (Vashishtha et al., 2009) possible active site residues that may act to catalyze the reaction, whereas the other study does not suggest (Johansson et al., 2000) any suitable active site acid/base residues. Because of this uncertainty with the use of QM/MM, we elucidated the mechanism for saccharopine reductase (Almasi, Bushnell, & Gauld, 2011). Notably, this study (Almasi et al., 2011) identified that saccharopine reductase utilizes a substrate-assisted catalytic pathway and that no active site residues are directly involved in the mechanism. In particular, the acidic and basic groups that catalyze the reaction are functional groups within AASA and glutamate. Hence, the enzyme binds the three required reactant molecules AASA, glutamate, and NADPH in an orientation and polar environment conducive to reaction.

The QM/MM model (shown in Fig. 1) for this study was taken from an X-ray crystal structure (PDB: 1E5Q) of saccharopine reductase from *M. grisea*, which was co-crystallized with the saccharopine product. To generate the reactive complex, the C–N bond of the product molecule was cleaved and hydrogens added to the nitrogen to regenerate the initial glutamate substrate, while an oxygen was added to the carbon forming the initial α-aminoadipate-δ-semialdehyde substrate. Hydrogens were added to the active site model with all ionizable functional groups being modeled in their most likely protonation state at pH 7.

As seen in Fig. 1, part of the glutamate and AASA as well as the nicotinamide moiety of the NADPH cofactor was included in the QM layer. All residues and waters within 15 Å of the substrates were modeled in the MM layer either in their entirety or only as their peptide backbone component (i.e., the residue was modeled as $-NHCH_2CO-$). To ensure the structural integrity of the model, the α-carbon of each residue was held fixed at its crystal structure position.

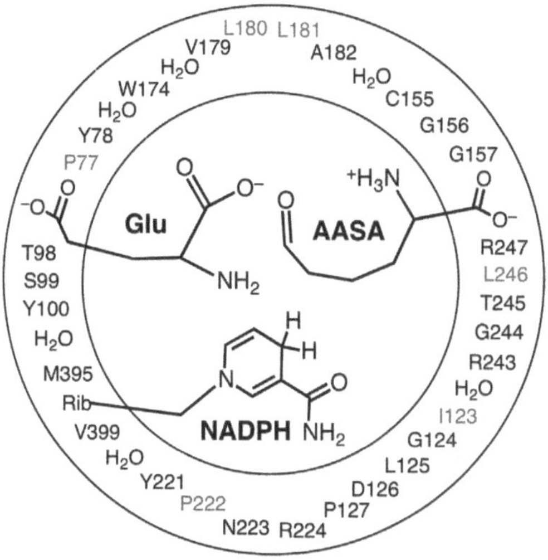

Figure 1 Schematic representation of the QM/MM model: groups in the inner and outer circles are modeled at the QM and MM levels of theory (see below), respectively. Note that residues, waters, and functional groups in black have been included in the model in their entirety, while residues in red (gray in the print version) have only had their peptide backbone included (Almasi et al., 2011).

For the calculations, the combined quantum mechanical and molecular mechanical (QM/MM) method in the ONIOM (Bearpark et al., 2007; Dapprich, Komaromi, Byun, Morokuma, & Frisch, 1999; Humbel, Sieber, & Morokuma, 1996; Maseras & Morokuma, 1995; Morokuma et al., 2001; Svensson et al., 1996; Vreven et al., 2006; Vreven & Morokuma, 2000; Vreven, Morokuma, Farkas, Schlegel, & Frisch, 2003) formalism was used as implemented within the Gaussian 03 program suite (Frisch et al., 2004). The geometries of the stationary points on the PES were obtained at the ONIOM(HF/6-31G(d):AMBER94) level of theory within the ME formalism. Harmonic vibrational frequency calculations were performed at the same level of theory in order to characterize the stationary points as minima or transition structures and to calculate Gibbs energy corrections at standard ambient temperature and pressure.

The relative energies discussed herein were obtained at higher levels of theory based on the optimized geometries, with inclusion of the appropriate free energy correction. Specifically, SP calculations were performed at the: (i) ONIOM(MP2/6-31G(d)//HF/6-31G(d):AMBER94)-ME; (ii) ONIOM(MP2/6-31G(d)//HF/6-31G(d):AMBER94) within the EE formalism.

It is noted that for this chapter, only relative energies are discussed. However, detailed discussion of the optimized geometries of the reactant, product, intermediate, and transition structures can be found in our previous (Almasi et al., 2011) work. The overall potential energy surfaces (PESs) obtained for the catalytic mechanism of saccharopine reductase at the ONIOM(MP2/6-31G(d)//HF/6-31G(d):AMBER94)-ME and ONIOM(MP2/6-31G(d)//HF/6-31G(d):AMBER94)-EE levels of theory are presented in Fig. 2. Notably, as can be seen, inclusion of the polarizing effects of the anisotropic protein environment (i.e., use of the EE formalism) has a large impact on the relative energies of the reaction. In particular, for almost all intermediate, transition state, and the product complexes, their relative energies with respect to **RC** are raised by 27.7–161.4 kJ mol^{-1}. The details of the overall mechanism are discussed below.

The first step in the overall pathway is nucleophilic attack of the glutamates α-amino nitrogen at the R-group carbonyl carbon center of the cosubstrate AASA. At the ONIOM(MP2/6-31G(d)//HF/6-31G(d):AMBER94)-ME level of theory, this occurs with a barrier of 60.0 kJ mol^{-1} (**TS1**). The use of the EE formulism has a marginal effect and results in a reduction of only 9.8 kJ mol^{-1} in the barrier height with respect to **RC**.

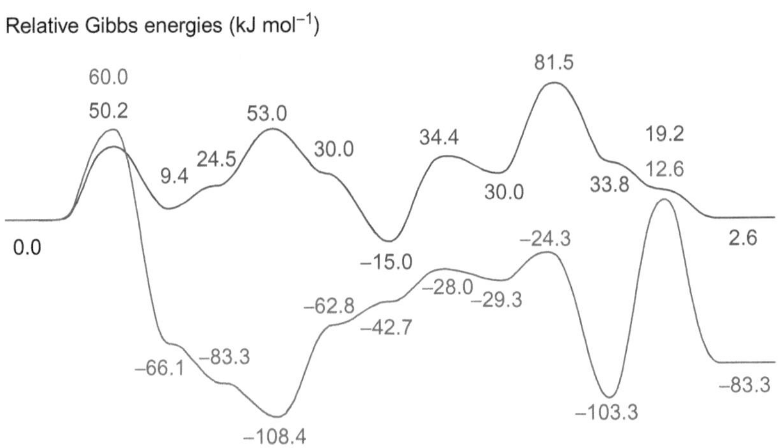

Figure 2 Overall PESs for the catalytic mechanism of saccharopine reductase obtained at the (i) ONIOM(MP2/6-31G(d)//HF/6-31G(d):AMBER94)-ME (red (gray in the print version) line) and (ii) ONIOM(MP2/6-31G(d)//HF/6-31G(d):AMBER94)-EE (blue (dark gray in the print version) line) level of theory. Gibbs energy corrections obtained at the ONIOM(HF/6-31G(d):AMBER94) level of theory have been added to the calculated energies (Almasi et al., 2011).

Thus, it seems that the anisotropic environment has little effect on the formation of the C–N bond. This is unlike the other stationary points for the mechanism.

With formation of the "N-protonated" carbinolamine **I1**, it must first undergo deprotonation of the bridging $-NH_2^+-$ moiety and inversion of the resulting –NH– group to allow for the loss of H_2O resulting in the formation of a Schiff base (Erdtman, Bushnell, Gauld, & Eriksson, 2011). In saccharopine reductase, this proceeds in a stepwise manner from **I1** to **I4** (Almasi et al., 2011). The first step is a transfer of the proton from the bridging $-{}^{Glu}NH_2^+-$ group to the glutamate's side-chain carboxylate group (via **TS2**) resulting in the formation of **I2** (Almasi et al., 2011). In **TS3**, this proton transfers to the oxygen of the carbinolamine's ^{AASA}C–OH hydroxyl group (**I3**) providing the necessary protons for the loss of the water. The final step prior to the formation of the Schiff base is inversion of the –NH– group, which occurs via **TS4** to form **I4**.

At the ONIOM(MP2/6-31G(d)//HF/6-31G(d):AMBER94)-ME level of theory, we can see that both **I1** and **I3** are kinetically unstable with respect to **I2**. Indeed, **TS2** and **TS3** are 17.2 and 20.1 kJ mol^{-1} lower in energy than **I1** and **I3**, respectively. Moreover, **I1** and **I3** are thermodynamically unfavorable lying 42.3 and 65.7 kJ mol^{-1} higher in energy than **I2**, respectively. Thus, at this level of theory, it appears that once **TS1** is crossed, the system quickly rearranges to **I2**. Moreover, **I2** then rearranges directly to **I4** via the unstable **I3** with an overall barrier of 80.4 kJ mol^{-1} (i.e., **TS3** + **TS4**) (Almasi et al., 2011).

As seen in Fig. 2, inclusion of the polarizing effects of the anisotropic protein environment (via the use of an EE formalism) significantly alters the relative energies of **I1**, **I2**, **I3**, **I4**, **TS2**, **TS3** and **TS4**. Indeed, with the exception of **I3**, all lie higher in energy than **RC**. In particular, **I1** now lies 9.4 kJ mol^{-1} higher in energy than **RC** and corresponds to a destabilization of 75.5 kJ mol^{-1}. In the case of **I2**, a destabilization of 161.4 kJ mol^{-1} is observed. If we consider **I1**, **I2**, **I3**, and **I4**, species **I2** now has the highest relative energy with respect to **RC** (i.e., 53.0 kJ mol^{-1}), whereas it was the lowest energy intermediate when an ME formalism was used. Moreover, it is now thermodynamically unfavorable and kinetically unstable with respect to rearrangement back to **I1** or to the subsequent **I3**. Indeed, **TS2** and **TS3** are now 24.5 and 30.0 kJ mol^{-1} lower in energy than **I2**, respectively (Fig. 2). In the case of **I3**, while it remains lower in energy than **RC**, it is now the lowest energy intermediate along the PES. In the case of **TS4**, its relative energy with respect to **RC** has increased by 62.4 kJ mol^{-1}, whereas in the case of **I4** its relative energy has increased by 59.3 kJ mol^{-1}.

As discussed previously (Almasi et al., 2011), a likely reason for these drastic changes can be explained by considering the interactions that exist between the glutamate's carboxylate (modeled in the QM region) and the active site residues Arg224 and Tyr78 (modeled in the MM region). At the ONIOM(MP2/6-31G(d)//HF/6-31G(d):AMBER94)-ME level of theory, the interaction energies between these groups are calculated at the MM level of theory. Thus, in the absence of the polarizing effect of the protein environment, **I2** is most stable due to the neutralization of charges in the QM region. However, at the ONIOM(MP2/6-31G(d)//HF/6-31G(d):AMBER94)-EE, the polarizing effects of these functional groups are included in the QM SCF calculation. Thus, the presence of the charged guanidinium and polar phenol functional groups greater stabilize the anionic Glu–COO^- group in **I1** and **I3**. Moreover, the neutralization of the anionic Glu–COO^- in **I2** would be unflavored. Hence, this instability of **I2** at the ONIOM(MP2/6-31G(d)//HF/6-31G(d):AMBER94)-EE level of theory suggests that it may instead represent a transition structure for Glu–COO^--assisted proton transfer from $-^{Glu}NH_2^+-$ to the carbinolamine hydroxyl oxygen, which would otherwise have required an unfavorable four-membered ring transition structure (Almasi et al., 2011; Erdtman et al., 2011; Hall & Smith, 1998a). Moreover, the involvement of the carboxylate group is comparable to previous studies (Hall & Smith, 1998b; Rosenberg, Silver, Sayer, & Jencks, 1974; Williams, 1987) in which water assists the proton transfer from the bridging $-NH_2^+-$ to the carbinolamine oxygen resulting in lower barriers.

Once the bridging $-^{Glu}NH-$ has inverted, loss of the carbinolamine –OH as water can occur, thus forming the Schiff base. At the ONIOM(MP2/6-31G(d)//HF/6-31G(d):AMBER94)-ME level of theory, this occurs with a very low barrier of 5.0 kJ mol^{-1} (i.e., **TS5**). The resulting imine intermediate **I5** lies markedly lower in energy than **I4** by 74.0 kJ mol^{-1}. The final step in the catalytic mechanism is reduction of the Schiff base by NADPH resulting in the formation of the product, saccharopine. As seen in Fig. 2, at the ONIOM(MP2/6-31G(d)//HF/6-31G(d):AMBER94)-ME level of theory, this occurs with a very high barrier of 115.9 kJ mol^{-1} (i.e., **TS6**) with respect to **I5**. However, with respect to **I2** (the lowest energy intermediate), the rate-limiting barrier is in fact 121.0 kJ mol^{-1}. The product complex formed (**PC**) is -83.3 kJ mol^{-1}

lower in energy than the initial active site-bound reactant complex **RC**. Thus, overall, the pathway is calculated to be exothermic and thus thermodynamically favored at the ONIOM(HF/6-31G(d):AMBER94)–ME level of theory (Fig. 2). Thus, overall at the ONIOM(MP2/6-31G(d)//HF/6-31G(d):AMBER94)-ME level of theory, the rate-limiting step is reduction of the Schiff base. Finite solvent isotope effects indicate that proton(s) are in flight in the rate-limiting step(s) and that this step is limiting under both limiting and saturating substrate concentrations (Vashishtha et al., 2009). This could very well correspond to the PCET process that occurs during the final reduction step. However, at a pH of 7, the measured k_{cat} is 1.06 ± 0.02 which corresponds to an estimated barrier of 72.9 kJ mol^{-1} (Himo, 2006; Vashishtha et al., 2009). Thus, the calculated rate-limiting barrier of 121.0 kJ mol^{-1} is in poor agreement (see above).

For the PES obtained at the ONIOM(MP2/6-31G(d)//HF/6-31G(d):AMBER94)-EE level of theory, considerable changes are observed for the subsequent loss of water via **TS5** to give the Schiff base intermediate **I5**. The barrier for this step is now 51.5 kJ mol^{-1} with respect to **I4** and now represents the overall rate-limiting step along the catalytic pathway. Furthermore, **I5** lies higher in energy than **RC** by 33.8 kJ mol^{-1} corresponding to a destabilization of 137.1 kJ mol^{-1}. However, **I5** is thermodynamically and kinetically unstable with respect to further reaction via **TS6** to give the final product complex **PC** (Fig. 2). Therefore, hydride transfer from the NADPH cofactor to the ^{AASA}C center of the imine now essentially occurs without a barrier. The complex **PC** is calculated to be marginally endothermic compared to **RC** by 2.6 kJ mol^{-1}.

If we consider again the estimated experimental (Himo, 2006; Vashishtha et al., 2009) barrier of 72.9 kJ mol^{-1}, slightly better agreement is seen with an overall barrier of 96.5 kJ mol^{-1} (ΔG(**TS5**) – ΔG(**I3**)). Moreover, as noted above, finite solvent isotope effects indicate that proton(s) are in flight in the rate-limiting step(s) (Vashishtha et al., 2009). In the case of this rate-limiting step, it involves two steps **TS4** and **TS5**. The former is inversion of the NH group, whereas the latter is loss of water via a proton transfer, both of which involve the transfer of a proton.

In summary, saccharopine reductase utilizes a substrate-assisted catalytic pathway in which acid/base groups within the substrates themselves facilitate the mechanistically required proton transfers. While the use of an ME formalism correctly predicts the mechanism, it is only through the use of an EE formalism that we get good agreement with experimental values.

2.2 Uroporphyrinogen Decarboxylase

UROD III catalyzes the fifth step of a multistage multienzymatic process in the synthesis of heme. Heme is an important biomolecule for many organisms that is involved in ligand transport, electron transfer, light-harvesting, and redox mechanisms (Fan, Liu, Hao, Teng, & Niu, 2007; Heinemann, Jahn, & Jahn, 2008; Martins, Grimm, Mock, Huber, & Messerschmidt, 2001; Mock, Trainotti, Kruse, & Grimm, 1995). Specifically, UROD catalyzes the sequential nonsymmetric decarboxylation of the four acetates of uroporphyrinogen III (URO-III) to give coproporphyrinogen III (CP-III) (Fan et al., 2007; Heinemann et al., 2008; Jackson et al., 1976; Martins et al., 2001).

The proposed general mechanism for decarboxylation of the pyrrole-acetate is given in Fig. 3 (Akhtar, 1994; Barnard & Akhtar, 1975, 1979). The initial step involves a proton transfer from a general acid (HA) to C2 of the pyrrole ring. This destabilizes the C3′—C3″OO$^-$ bond resulting in its cleavage to form CO_2 and a basic C3=C3′ moiety. A second general acid (HB) then protonates the methylene C3′ carbon, while the first acid, now in its conjugate base form (A$^-$), abstracts the proton from the –C2H$_2$$^+$– group.

Within the active site, three electrostatic regions have been identified: a negative, a polar-positive, and a noncharged region. The negatively charged region contains an invariant aspartyl residue which has been proposed to orientate the substrate for catalysis, as well as stabilizes various mechanistic intermediates (Fan et al., 2007; Phillips, Whitby, Kushner & Hill, 2003). The polar-positive region contains several residues (such as argininyl and protonated histidyl residues) which interact with the carboxylates of

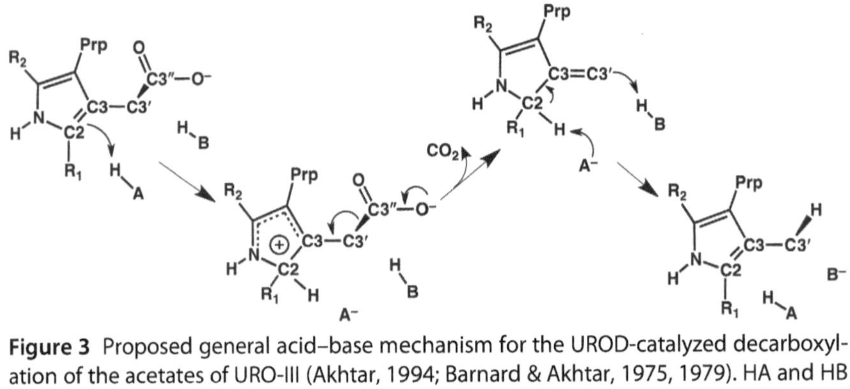

Figure 3 Proposed general acid–base mechanism for the UROD-catalyzed decarboxylation of the acetates of URO-III (Akhtar, 1994; Barnard & Akhtar, 1975, 1979). HA and HB are general acids.

URO-III serving to bind and recognize the substrate (Chaufan, de Molina, & de Viale, 2001; Phillips et al., 2003). Notably, of the polar-positive residues within the active site, one or more arginyl residues (hUROD: Arg37, Arg41, and Arg50) have been found to be catalytically important.

Previously, Silva and Ramos (2005) performed a computational investigation on the first acetate decarboxylation of URO-III as catalyzed by UROD. A cluster/DFT-type approach was used, which as discussed in Section 1 ignores the anisotropic protein environment. Rather an implicit solvation approach is used to model the environment. The chemical model was constructed from a crystal structure of a UROD...product complex (Phillips et al., 2003) and consisted of the side chains of Asp86 and Arg37 and 1,3,4-methyl-2-acetyl pyrrole to model URO-III.

Using this computational model, it was concluded that decarboxylation could proceed in accordance with the general mechanism (Silva & Ramos, 2005). In particular, the initial step is a proton transfer from Arg37 onto C2 of the substrate pyrrole ring with a barrier of 89.5 kJ mol^{-1} (Fig. 4).

With the protonation of C2, the resulting intermediate (**I1**) lies higher in energy than **RC** by 47.3 kJ mol^{-1}. Decarboxylation of the acetate moiety then occurs via **TS2** (Fig. 5) with a barrier 82.4 kJ mol^{-1} (Silva & Ramos, 2005). It is noted that for the DFT/cluster model, release of the CO_2 was in fact hindered by the strong hydrogen bond interaction between the carboxylate of Asp86 and the pyrrole ring's HN moiety.

The third step is protonation of the newly formed methylene carbon from which the CO_2 was lost. Unfortunately, due to the lack of a general acid HB in the DFT/cluster model, direct comparison of the results obtained for this step between the two studies (Bushnell et al., 2011; Silva & Ramos, 2005) is not possible. However, for completeness, we briefly discuss the

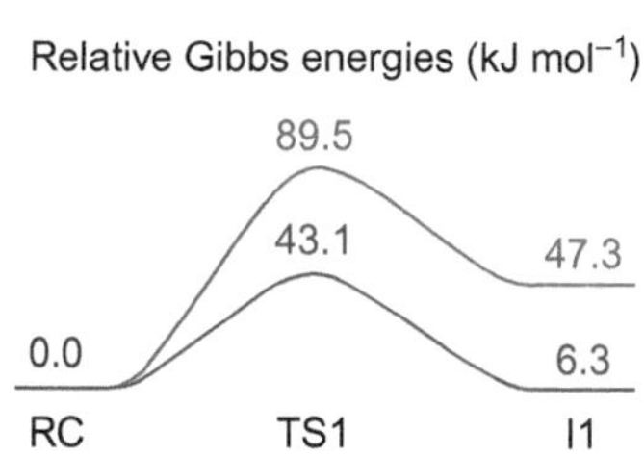

Figure 4 Gibbs energy PES for the initial proton transfer from Arg37 to ring **D** obtained using a QM/MM model (blue (dark gray in the print version) surface) and cluster/DFT model (red (gray in the print version) surface) (Bushnell et al., 2011; Silva & Ramos, 2005).

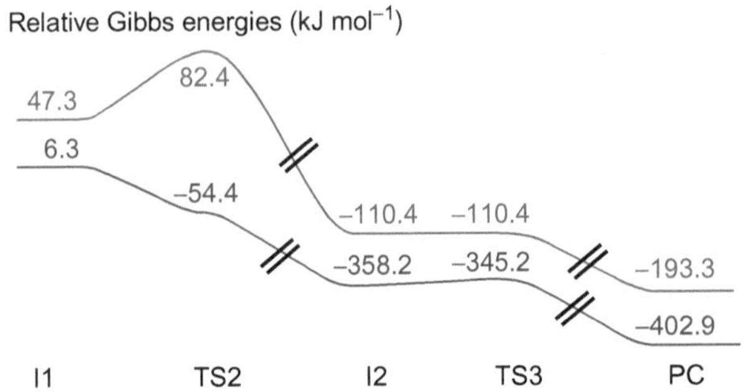

Figure 5 Gibbs energy PES for the remainder of the mechanism using a QM/MM model (blue (dark gray in the print version) surface) and cluster/DFT model (red (gray in the print version) surface) (Bushnell et al., 2011; Silva & Ramos, 2005).

results of both studies. In the DFT/cluster model with a proton transferred to C3′ from the solvent, **I2** is found to lie 157.7 kJ mol^{-1} lower in energy than **I1** (Fig. 5). Subsequently, deprotonation of the C2 carbon of the pyrrole ring occurs with little to no barrier and results in a product complex (**PC**) that lies 193.3 kJ mol^{-1} lower in energy than RC (Silva & Ramos, 2005).

Overall, from the work of Silva and Ramos (2005), it was concluded that the mechanism is thermodynamically and enzymatically feasible with an overall barrier of 89.5 kJ mol^{-1}. Notably, this rate-limiting barrier corresponds to the initial proton transfer from Arg50 to C2. Human UROD has been experimentally measured to have a k_{cat} value of 0.16 s^{-1} (de Verneuil, Sassa, & Kappas, 1983; Himo, 2006; Lewis & Wolfenden, 2008). Using TST, this corresponds to an overall barrier of 77.4 kJ mol^{-1} (Himo, 2006). Hence, it appears that good agreement is reached between the cluster/DFT model and experiment.

However, while agreement was reached in regard to energetics, a key point ignored in the small DFT/cluster approach is that, from experimental observations, the stereochemistry at the C3′ center is retained; in particular, HB must be an active site residue (Akhtar, 1991; Barnard & Akhtar, 1975, 1979). In the small DFT/cluster study, it was assumed that solvent donates the second proton to the C3′ center. In addition, as noted above, key electrostatic regions within the active site that have been identified are not modeled in a DFT/cluster-based approach.

Thus, we (Bushnell et al., 2011) reexamined the catalytic mechanism of UROD using a QM/MM approach in the ONIOM formalism with ME (Bearpark et al., 2007; Dapprich et al., 1999; Humbel et al., 1996;

Maseras & Morokuma, 1995; Morokuma et al., 2001; Svensson et al., 1996; Vreven et al., 2006; Vreven & Morokuma, 2000; Vreven et al., 2003) as implemented in the Gaussian 03 program suite (Frisch et al., 2004). This allowed us to consider the role of a second arginyl (Arg50) residue, found to interact with the leaving carboxylate, as well as to take into account the anisotropic environment around the active site.

The QM/MM model (shown in Fig. 6) for this study was taken from an X-ray crystal structure (PDB: 1R3Y) of human UROD which was co-crystallized with the CP-III product. The details of the docking and MD simulation prior to the QM/MM calculations can be found in our original work (Bushnell et al., 2011). As seen in Fig. 6, the QM/MM model included the substrate URO-III and all active site residues immediately surrounding it, i.e., first-shell residues (Fig. 6). In addition, for those portions of the substrate exposed to solvent, the first solvation shell was retained. A subset of the complete model centered on the reactive region of the active site was then selected for the high-level QM treatment consisting of two arginyl (Arg37 and Arg50) and an aspartyl (Asp86) residue and the first

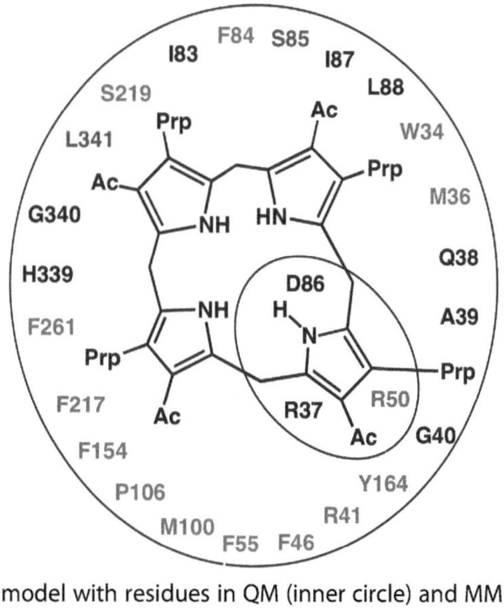

Figure 6 QM/MM model with residues in QM (inner circle) and MM layers. Residues in red (gray in the print version) modeled as side chains with atoms fixed at the truncation position. Remainder of residues modeled as side chain and backbone with Cα positions fixed. Residues in blue (dark gray in the print version) represent side chains modeled as hydrogen (S85 and L341) (Bushnell et al., 2011).

substrate pyrrole that is decarboxylated (Bushnell et al., 2011). Optimized geometries were obtained at the ONIOM(B3LYP/6-31G(d):AMBER94) level of theory (Case et al., 2005). However, in order to obtain more reliable calculated relative Gibbs energies, single-point energy calculations on the above optimized structures were performed at the ONIOM(B3LYP/6-311+G(2df,p):AMBER94) level of theory. Gibbs energies were obtained by adding the Gibbs energy corrections.

Using this larger and more complete model, the initial step is again proton transfer from the guanidinium of Arg37 onto the C2 center of the substrate pyrrole. However, this step now occurs via **TS1** at a cost of 43.1 kJ mol^{-1} relative to the initial substrate-bound active site complex **RC** (Fig. 4). This is approximately half the size of the barrier for this initial step obtained by Silva and Ramos (2005) using a cluster/DFT approach.

The resulting C2-protonated intermediate (**I1**) lies higher in energy than **RC** by just 6.3 kJ mol^{-1}. Thus, inclusion of the anisotropic protein environment causes a stabilizing effect which reduces the relative energy of **I1** by 41.0 kJ mol^{-1}. However, most notably is that with the use of a QM/MM-based approach, the Gibbs energy barrier for decarboxylation is *lower* than that of **I1** (Bushnell et al., 2011). In particular, **TS2** now lies 60.7 kJ mol^{-1} lower in energy than **I1**. This considerable difference in energetic cost for decarboxylation was a result of the explicit inclusion of the protein environment in the QM/MM model which provides an environment favorable for the lengthening of the Asp86···C2-protonated pyrrole interaction enhancing the rate of decarboxylation unlike that seen in DFT/cluster approach (Bushnell et al., 2011; Silva & Ramos, 2005).

For the QM/MM-based study, a second arginyl active site residue (Arg50) was modeled in the QM region which was believed to act as HB. Indeed, Arg50 was found to protonate the methylene carbon (C3′) concomitantly with loss of CO_2. The resulting intermediate **I2** lies 364.5 kJ mol^{-1} lower in energy than **I1** (Fig. 5) (Bushnell et al., 2011). Notably, the identification of Arg50 as HB is supported by experimental conclusions based on the observed retention of stereochemistry at the pyrrole–CH_3 group formed (Akhtar, 1994; Barnard & Akhtar, 1975, 1979). The final step of the reaction involves deprotonation of C2 by Arg37. For the QM/MM-based study, the process was found to occur with a barrier of 13.0 kJ mol^{-1} where the PC lies 402.9 kJ mol^{-1} lower in energy than RC. Thus, like the study by Silva and Ramos, the QM/MM-based study predicts a thermodynamically favorable and kinetically feasible process.

As noted above, human UROD has been experimentally measured to have a k_{cat} value of 0.16 s^{-1} which corresponds to an overall barrier of 77.4 kJ mol^{-1}; yet in the QM/MM study, a rate-limiting barrier of 43.1 kJ mol^{-1} was obtained (de Verneuil et al., 1983; Himo, 2006; Lewis & Wolfenden, 2008). But, it has been suggested (Juaréz, Aldonatti, Vigna, & Ríos de Molina, 2007) that decarboxylation of URO-III generating the 7-carboxylate intermediate is not likely the rate-limiting step. Indeed, for several variants of UROD, the rate-limiting step appears to be decarboxylation of the 7-carboxylate intermediate. For UROD from various species, the experimentally determined barriers for the first decarboxylation lie in the range of 8.4–51.5 kJ mol^{-1} (Chaufan et al., 2001; Juaréz et al., 2007). Thus, the QM/MM calculated barrier is in fact in better agreement with related experimentally reported values for the first decarboxylation of ring D. Moreover, the explicit treatment of the protein environment provided details that were not observed in the DFT/cluster-based calculations.

2.3 8*R*-Lipoxygenase

Lipoxygenases (LOXs) are a ubiquitous family of nonheme iron enzymes involved in the stereo- and regiospecific peroxidation of arachidonic acid (AA) or linoleic acid (LA), in the presence of molecular oxygen (Borowski & Broclawik, 2003; Ivanov et al., 2010; Minor, Steczko, Bolin, Otwinowski, & Axelrod, 1993; Neau et al., 2009; Zhang, Gan, Pavel, Sigal, & Solomon, 1995). The generally accepted mechanism for peroxidation is consistent for *all* LOXs and is as shown in Fig. 7.

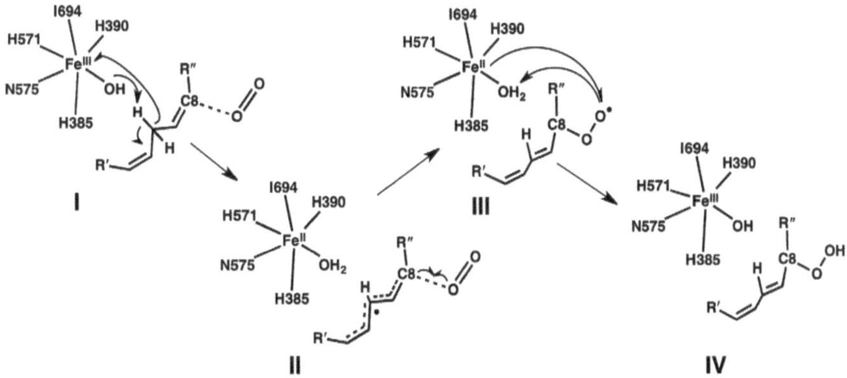

Figure 7 The general mechanism of LOXs.

As seen in Fig. 7, the initial step is a proton-coupled electron transfer from the substrate to the Fe^{III}–OH moiety forming a pentadienyl-type radical (**I**) (Gaffney, 1996). The O_2 within the active site then attacks the pentadienyl radical antarafacial to the Fe center generating a peroxy radical (**II**). This mechanism of substrate activation making it vulnerable to O_2 attack differs with that generally observed in most heme-containing enzymes or the family of nonheme α-ketoglutarate dependent dioxygenases that utilize O_2 to oxidize organic substrates (Liu, Llano, & Gauld, 2009; Shaik et al., 2010).

If we consider intermediate **II** for a moment, we see three centers of unpaired electron density that exist with a total number of seven unpaired electrons: a triplet O_2, the substrate-derived radical, and the high-spin Fe^{II} center. Given the high-spin state of the Fe center and the triplet O_2, there are only two possible electronic configurations that allow for spin conservation upon formation of the peroxyl radical. These configurations have total spins of 3/2 or 5/2 (i.e., quartet and sextet states). The intermediate complexes corresponding to the states are hereafter labeled as ^{4}**II** and ^{6}**II**, respectively. Both these intermediates represent a clear example of a multicentered open-shell system. While DFT methods have become the standard tools of choice for computational investigations of metalloenzymes, such as those that contain iron (Becke, 1993a, 1993b; Lee et al., 1988; Siegbahn, 2006), as noted above, however, the reliability of the calculated results can be quite sensitive to the amount of exact HF exchange included DFT methods (Reiher et al., 2001). To fix such a problem, it has been proposed that a contribution of 15% HF offers better accuracy (Lundberg & Morokuma, 2007; Reiher, 2002; Reiher et al., 2001; Schenk et al., 2004; Siegbahn et al., 2010).

Thus, given the complex nature of ^{4}**II** and ^{6}**II**, we examined the performance of a range of hybrid, meta, and hybrid-meta GGA density functionals to reliably determine the energetics of these mechanistic intermediates of 8*R*-LOX within an ONIOM-type QM/MM model. For purposes of brevity, only a summary of the methods will be discussed herein. However, the details can be found in reference (Bushnell & Gauld, 2013).

For the calculations, the combined QM/MM method in the ONIOM formalism (Bearpark et al., 2007; Dapprich et al., 1999; Humbel et al., 1996; Maseras & Morokuma, 1995; Morokuma et al., 2001; Svensson et al., 1996; Vreven et al., 2006; Vreven & Morokuma, 2000; Vreven et al., 2003) as implemented in the Gaussian 09 program suite was used for the calculations (Frisch et al., 2010). The structures for ^{4}**II** and ^{6}**II** were obtained at the

ONIOM(DFT_i/BS1:AMBER) level of theory in the ME formulism (DFT_i=BP86, B3LYP$^{\pm}$, B3LYP*, B3LYP, M06, and M06-L) (Becke, 1993a, 1993b; Cornell et al., 1995; Lee et al., 1988; Reiher et al., 2001; Zhao & Truhlar, 2008a, 2008b). The basis functions chosen were the 6-31G(d) basis set on all atoms but Fe, where the LANL2DZ+ECPs basis set was used. Single point energies were obtained with 6-311G(2df,p) functions plus the LANL2DZ ECPs for the iron.

The QM/MM model (shown in Fig. 8) for this study was taken from an X-ray crystal structure (PDB: 3FG1) of 8*R*-LOX found in *Plexaura homomalla.* As seen, the QM/MM model constructed for this study included a truncated form of the substrate and the active site residues immediately surrounding it; i.e., first-shell residues were included (Bushnell & Gauld, 2013). Also, second-shell residues that interacted with the Fe center were kept. The reactive region of the active site was modeled using a high-level QM

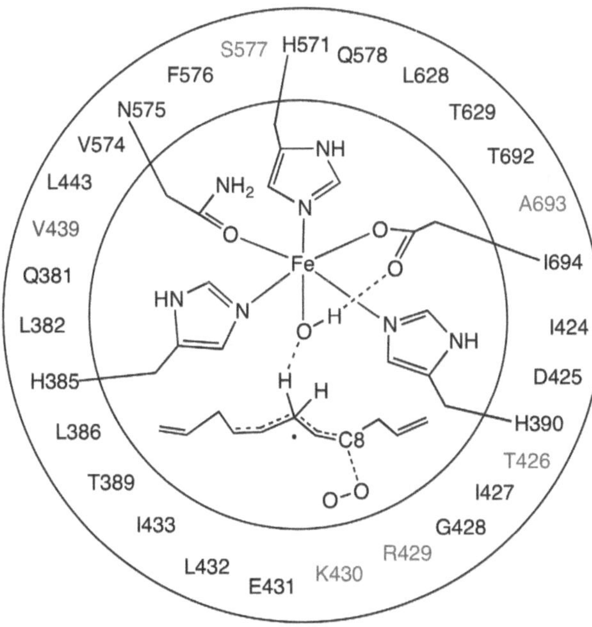

Figure 8 Schematic representation of the QM/MM model used. Groups in the inner circle have been modeled at the DFTi level of theory, while the residues in the outer circle have been modeled at the AMBER level of theory. Color code for residues: included in their entirety (black); modeled as Gly, i.e., only the backbone was included with R-groups replaced by a hydrogen (red; gray in the print version); modeled as only their R-group, i.e., only their C_α and side chains included (blue; dark gray in the print version) (Bushnell & Gauld, 2013).

Table 1 The Relative Energies of the Intermediate Complexes **^{4}II** and **^{6}II** obtained at the ONIOM(DFT$_i$/BS2//DFT$_i$/BS1:AMBER94)-EE Levels of Theory (Bushnell & Gauld, 2013)

	BP86	M06	B3LYP$^{\pm}$	B3LYP*	B3LYP	M06-L
^{4}II	45.6	46	−1.7	0	−2.5	45.2
^{6}II	0	0	0	0	0	0

treatment (Fig. 8, inner circle). This inner region contained the truncated pentadienyl radical substrate; the side chains of His385, His390, His571, and Asn575; the carboxylate of the terminal Ile694; the attacking O_2; and the Fe^{II}–OH_2 center.

Due to the noninteracting nature of the three centers of unpaired electron density, the difference in relative energy between **^{4}II** and **^{6}II** is expected to be marginal. As seen in Table 1 for the various ONIOM(DFT$_i$/BS2//DFT$_i$/BS1:AMBER)-EE levels of theory, this is not always the case. With the use of B3LYP$^{\pm}$, B3LYP*, and B3LYP functionals, the energy differences between the states are within 1 kcal mol^{-1} with the largest difference (0.6 kcal mol^{-1}) occurring for B3LYP. Both the B3LYP$^{\pm}$ and B3LYP functionals predict the quartet (**^{4}II**) to lie marginally lower in energy than the sextet (**^{6}II**).

However, with the use of the BP86 functional, the quartet is predicted to be 45.6 kJ mol^{-1} higher in energy than the sextet. Moreover, the M06 and M06-L functionals predict differences very similar to those obtained with the BP86 functional. In particular, the M06 functional predicts the quartet to be 46.0 kJ mol^{-1} higher in energy than the sextet, whereas the M06-L functional calculates the difference to be 45.2 kcal mol^{-1} higher in energy. This over-favoring of the high-spin state was previously seen in the work of Vancoillie, Zhao, Radon, and Pierloot (2010), which showed that for several heme models, the M06 and M06-L functionals overstabilize the high-spin state with respect to the low-spin state.

Similarly, in regard to the six metal–ligand (M–L) interactions (Fig. 9) as well as the O–O bond length of O_2, the various ONIOM(DFT$_i$/BS2//DFT$_i$/BS1:AMBER)-EE levels of theory do not show equal reliability. Given that the Fe^{III} center exists in the high-spin state for both **^{4}II** and **^{6}II**, the DFT$_i$ methods should give the same or very similar M–L distances for both states. Moreover, given the triplet nature of O_2, the O–O bond length is expected to be the same for both **^{4}II** and **^{6}II**.

From the calculated bond lengths given in Table 2, it is clear that depending on the functional, the M–L bond lengths differ for the quartet

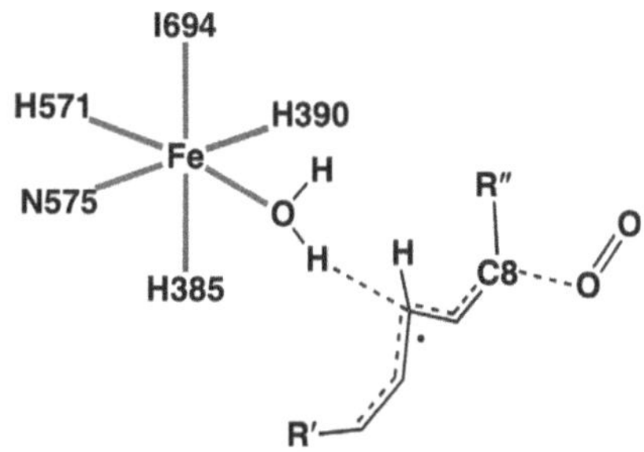

Figure 9 The QM region model used for the active site of 8*R*-LOX with the pentadienyl intermediate (AA$^{\bullet}$) bound. The bond lengths considered in calculating Avg$^{|\Delta r|}$ below are labeled in red (gray in the print version) (Bushnell & Gauld, 2013).

Table 2 Absolute Differences ($|\Delta r|$) Between Optimized **^{4}II** and **^{6}II** Structures, Obtained at the ONIOM(DFT$_i$/BS1:AMBER94) Level of Theory for Selected Interactions (Bushnell & Gauld, 2013)

	Fe-H385	Fe-H390	Fe-H571	Fe-N575	Fe-I694	Fe-OW	O–O	Avg$^{|\Delta r|}$
BP86	0.109	0.091	0.034	0.050*	0.100*	0.072	0.068*	0.076[a]
M06-L	0.042	0.015	0.032	0.160	0.028	0.050*	0.057*	0.054[a]
B3LYP$^{\pm}$	0.061	0.035	0.021*	0.053*	0.064*	0.035	0.007	0.045[a]
B3LYP*	0.012	0.002	0.005*	0.004*	0.021*	0.010	0.000	0.009[a]
B3LYP	0.047	0.036	0.013*	0.056*	0.051*	0.031	0.000	0.039[a]
M06	0.068	0.064	0.041	0.008	0.026*	0.021*	0.047*	0.038[a]

*Complexes in which the quartet bond length is longer.
[a]Averages were calculated only for the M–L interactions, that is the O–O bond length was ignored.
All absolute differences are in Å.

(**^{4}II**) and sextet (**^{6}II**) states. The amount of the differences between the two states is given as the absolute value of Δr between complexes (Table 2). However, for brevity, we will only discuss the average absolute difference (i.e., Avg$^{|\Delta r}$) of the bonds in red (gray in the print version) (Fig. 9) and the O_2 bond (not shown in red (gray in the print version)). For the calculations performed with BP86, B3LYP$^{\pm}$, B3LYP*, B3LYP, M06, and M06-L functionals, Avg$^{|\Delta r|}$ was found to be 0.076, 0.045, 0.009, 0.039, 0.038, and 0.054 Å, respectively. Thus, B3LYP* gives a markedly smaller Avg$^{|\Delta r|}$ value than the other functionals considered. In contrast, the BP86 and M06-L functionals, i.e., those with a 0% HF contribution, have the greatest differences in bond lengths between **^{4}II** and **^{6}II**. While M06-L and BP86 contain 0% HF, the better performance of the former method is somewhat expected given that the functional was trained with compounds that contain transition metals (Zhao & Truhlar, 2008a, 2008b). It should be noted that while B3LYP (20% HF) and M06 (27% HF) have similar $Avg^{|\Delta r|}$ values, there were significant differences in the individual M–L interactions (Table 2) (Bushnell & Gauld, 2013).

In the case of the L_n–M–L_m (where $m \neq n$) bond angles (not shown), differences between **^{4}II** and **^{6}II** also existed depending on the functional used. However, the magnitudes were not as significant as seen in the bond lengths. In particular, for BP86, M06, M06-L, B3LYP$^{\pm}$, B3LYP*, and B3LYP, differences of 2.8°, 1.8°, 2.4°, 1.7°, 0.4°, and 1.4° were observed, respectively (Bushnell & Gauld, 2013).

For the triplet O_2, BP86, M06, and M06-L gave different values for *r*(O–O) between **^{4}II** and **^{6}II** of 0.068, 0.057, and 0.047 Å, respectively (Table 2). In the case of B3LYP$^{\pm}$, B3LYP*, and B3LYP, calculate the expected difference in *r*(O–O) of 0.007, 0.000, and 0.000 Å, respectively.

For the substrate radical itself (AA$^{\bullet}$), modeled as a pentadienyl radical, the BP86, B3LYP$^{\pm}$, B3LYP*, and B3LYP methods all gave very similar results for the two states. In particular, the differences in backbone dihedral angles between **^{4}II** and **^{6}II** were less than 2.0° (not shown). In contrast, M06 and M06-L methods gave differences of 14.1° and 12.3° for these same backbone dihedral angles. This considerable difference was due to AA$^{\bullet}$ shifting considerably in the **^{4}II** complex. In particular, AA$^{\bullet}$ adopted a significantly more planar geometry than in the corresponding **^{6}II** complexes.

Thus, as seen in this case study, care must be taken in the selection of a DFT functional to be used in a QM/MM investigation. This is especially true for systems that contain transition metals.

3. CONCLUSIONS

The effect of the anisotropic polar protein environment on the mechanism of saccharopine reductase was discussed. While saccharopine reductase utilizes a substrate-assisted catalytic pathway (in which acid/base groups within the cosubstrates themselves facilitate the mechanistically required proton transfers), it was found that relative to **RC** almost all intermediates, transition structures, and product complexes were destabilized by as much as 161.4 kJ mol^{-1} upon explicit inclusion of the polarizing effect of the protein environment on the QM region. Importantly, with the polarizing effect of the active site was agreement with experimental kinetic data obtained. Hence, saccharopine reductase binds glutamate, α-aminoadipate-δ-semialdehyde, and NADPH in a manner and polar environment conducive to reaction.

Similarly, the effect of the protein environment on the catalytic mechanism of UROD was discussed. In particular, a comparison was made of its catalytic mechanism as elucidated using a QM/MM and a cluster/DFT approach (Bushnell et al., 2011; Silva & Ramos, 2005). For the QM/MM study, it was found that the stabilizing effect of the surrounding environment provided a significant reduction in both the calculated rate-limiting proton transfer and the barrier for acetate decarboxylation. Notably, inclusion of the protein environment surrounding the active site provided a rate-limiting barrier in good agreement with related experimentally determined values for the first decarboxylation.

Regardless of using either a QM/cluster or QM/MM approach, a key component is the selection of a particular *ab initio* or DFT method. The careful selection of a DFT method was clearly demonstrated for 8*R*-LOX wherein we discussed the performance of a range of hybrid, meta, and hybrid-meta GGA density functionals. In particular, the results presented suggest that, at least for ONIOM QM/MM calculations, the reliable description of geometries and energies of bioinorganic systems is quite sensitive to choice of functional.

4. FUTURE DIRECTIONS

While the QM/MM approach has become an important tool for calculating protein function, such an approach is generally not sufficient for the accurate calculation of reaction Gibbs energies. In particular, QM/MM calculated energy profiles of an enzymatic reaction are typically done with only one starting conformation of a protein. However, chemical reactions proceed with the environment (or active site) fluctuating over many local minima (Hammes et al., 2011). By accounting for such fluctuations, the energetics of a chemical reaction is characterized as a thermal average over the local conformations of the system. That is, an enzymatic reaction is better thought of as a process where the chemical steps are correlated with protein conformational changes (Kosugi & Hayashi, 2011). However, to account for such motion of the protein extensive configurational sampling is needed (Kamerlin & Warshel, 2010).

To determine a statistical average of possible alternative pathways of an enzymatic reaction, there exist several computational techniques (Acevedo & Jorgensen, 2010; Cramer, 2002; Hu et al., 2011; Young, 2001).

One such technique involves the use of a semiempirical (SE) method to describe the QM region, while an MM method describes the environment. The SE method is used due to the extensive number of conformations of the protein, substrate, and solvent that must be performed for calculating the Gibbs energies. As demonstrated in the 8*R*-LOX given the complex nature of the system, SE methods would likely not be suitable. This is due to the large number of approximations made in SE methods to reduce the number two-electron integrals. Thus, as an alternative, a QM/MM MD method can be used where the inner layer is described using a DFT method. However, due to the computational costs incurred in DFT calculations simulation timescales are limited to 0.1 ns, significantly shorter than those using SE methods (Hagiwara, Field, Nureki, & Tateno, 2010). Another choice to determine reaction free energies is the use of a CPMD approach. Similar to QM/MM MD, CPMD uses DFT methods to describe the inner layer where the complete QM/MM model is simulated for a period of time (Dal Peraro, Ruggerone, Raugei, Gervasio, & Carloni, 2007).

However, it is noted that as time goes on computers will advance to a point where accurate *ab initio* or DFT methods can be used in conjunction with a QM/MM MD approach.

ACKNOWLEDGMENTS

We thank the Natural Sciences and Engineering Research Council of Canada (NSERC), the Canada Foundation for Innovation (CFI), and the Ontario Innovation Trust (OIT) for financial support, and SHARCNET, ACEnet, and Compute Canada for computational resources. E.A.C.B. thanks the Killam Trusts for a Killam Post-Doctoral Fellowship.

REFERENCES

Acevedo, O., & Jorgensen, W. L. (2010). Advances in quantum and molecular mechanical (QM/MM) simulations for organic and enzymatic reactions. *Accounts of Chemical Research*, *43*(1), 142–151.

Adcock, S. A., & McCammon, J. A. (2006). Molecular dynamics: Survey of methods for simulating the activity of proteins. *Chemical Reviews*, *106*(5), 1589–1615.

Akhtar, M. (1991). *New comprehensive biochemistry: Biosynthesis of tetrapyrroles*: *Vol. 19*. London: Elsevier.

Akhtar, M. (1994). The modification of acetate and propionate side chains during the biosynthesis of haem and chlorophylls: Mechanistic and stereochemical studies. *Ciba Foundation Symposium*, *180*, 131–152.

Almasi, J. N., Bushnell, E. A. C., & Gauld, J. W. (2011). A QM/MM-based computational investigation on the catalytic mechanism of saccharopine reductase. *Molecules*, *16*(10), 8569–8589.

Barnard, G. F., & Akhtar, M. (1975). Stereochemistry of porphyrinogen carboxy-lyase reaction in heme biosynthesis. *Journal of the Chemical Society, Chemical Communications*, *13*, 494–496.

Barnard, G. F., & Akhtar, M. (1979). Stereochemical and mechanistic studies on the decarboxylation of uroporphyrinoen-III in heme biosynthesis. *Journal of the Chemical Society, Perkin Transactions*, *1*(10), 2354–2360.

Bearpark, M. J., Ogliaro, F., Vreven, T., Boggio-Pasqua, M., Frisch, M. J., Larkin, S. M., et al. (2007). CASSCF calculations for excited states of large molecules: Choosing when to use the RASSCF, ONIOM and MMVB approximations. In T. E. Simos & G. Maroulis (Eds.), *Computation in modern science and engineering vol 2, pts a and B: Vol. 2* (pp. 583–585). Melville, NY: American Institute of Physics.

Becke, A. D. (1993a). Density-functional thermochemistry. 3. The role of exact exchange. *The Journal of Chemical Physics*, *98*(7), 5648–5652.

Becke, A. D. (1993b). A mixing of Hartree–Fock and local density-functional theories. *The Journal of Chemical Physics*, *98*, 1372.

Blomberg, M. R. A., Borowski, T., Himo, F., Liao, R.-Z., & Siegbahn, P. E. M. (2014). Quantum chemical studies of mechanisms for metalloenzymes. *Chemical Reviews*, *114*(7), 3601–3658.

Borowski, T., & Broclawik, E. (2003). Catalytic reaction mechanism of lipoxygenase. A density functional theory study. *The Journal of Physical Chemistry. B*, *107*(19), 4639–4646.

Bruice, T. C. (1976). Some pertinent aspects of mechanism as determined with small molecules. *Annual Review of Biochemistry*, *45*, 331–373.

Bruice, T. C. (2002). A view at the millennium: The efficiency of enzymatic catalysis. *Accounts of Chemical Research*, *35*(3), 139–148.

Bushnell, E. A. C., Erdtman, E., Llano, J., Eriksson, L. A., & Gauld, J. W. (2011). The first branching point in porphyrin biosynthesis: A systematic docking, molecular dynamics and quantum mechanical/molecular mechanical study of substrate binding and mechanism of uroporphyrinogen-III decarboxylase. *Journal of Computational Chemistry*, *32*(5), 822–834.

Bushnell, E. A. C., & Gauld, J. W. (2013). An assessment of pure, hybrid, meta, and hybrid-meta GGA density functional theory methods for open-shell systems: The case of the nonheme iron enzyme 8R–LOX. *Journal of Computational Chemistry*, *34*(2), 141–148.

Careri, G., Fasella, P., & Gratton, E. (1979). Enzyme dynamics—Statistical physics approach. *Annual Review of Biophysics and Bioengineering*, *8*, 69–97.

Case, D. A., Cheatham, T. E., Darden, T., Gohlke, H., Luo, R., Merz, K. M., et al. (2005). The Amber biomolecular simulation programs. *Journal of Computational Chemistry*, *26*(16), 1668–1688.

Chaufan, G., de Molina, M. D. R., & de Viale, L. C. S. (2001). How does hexachlorobenzene treatment affect liver uroporphyrinogen decarboxylase? *The International Journal of Biochemistry & Cell Biology*, *33*(6), 621–630.

Cornell, W. D., Cieplak, P., Bayly, C. I., Gould, I. R., Merz, K. M., Ferguson, D. M., et al. (1995). A 2nd generation force-field for the simulation of proteins, nucleic acids, and organic molecules. *Journal of the American Chemical Society*, *117*(19), 5179–5197.

Cramer, C. J. (2002). *Essentials of computational chemistry: Theories and models*. New York: John Wiley & Sons Ltd.

Crosby, J., Stone, R., & Lienhard, G. E. (1970). Mechanisms of thiamine-catalyzed reactions—Decarboxylation of 2-(1-carboxy-1-hydroxyethyl)-3,4-dimethylthiazolium chloride. *Journal of the American Chemical Society*, *92*(9), 2891–2900.

Cruickshank, D. W. J. (1999). Remarks about protein structure precision. *Acta Crystallographica, Section D: Biological Crystallography*, *55*, 583–601.

Dal Peraro, M., Ruggerone, P., Raugei, S., Gervasio, F. L., & Carloni, P. (2007). Investigating biological systems using first principles Car–Parrinello molecular dynamics simulations. *Current Opinion in Structural Biology*, *17*(2), 149–156.

Dapprich, S., Komaromi, I., Byun, K. S., Morokuma, K., & Frisch, M. J. (1999). A new ONIOM implementation in Gaussian98. Part I. The calculation of energies, gradients, vibrational frequencies and electric field derivatives. *Journal of Molecular Structure: THEOCHEM*, *461*, 1–21.

de Verneuil, H., Sassa, S., & Kappas, A. (1983). Purification and properties of uroporphyrinogen decarboxylase from human erythrocytes. *The Journal of Biological Chemistry*, *258*(4), 2454–2460.

Devi-Kesavan, L. S., & Gao, J. L. (2003). Combined QM/MM study of the mechanism and kinetic isotope effect of the nucleophilic substitution reaction in haloalkane dehalogenase. *Journal of the American Chemical Society*, *125*(6), 1532–1540.

Erdtman, E., Bushnell, E. A. C., Gauld, J. W., & Eriksson, L. A. (2011). Computational studies on Schiff-base formation: Implications for the catalytic mechanism of porphobilinogen synthase. *Computational & Theoretical Chemistry*, *963*(2–3), 479–489.

Fan, J., Liu, Q., Hao, Q., Teng, M., & Niu, L. (2007). Crystal structure of uroporphyrinogen decarboxylase from *Bacillus subtilis*. *Journal of Bacteriology*, *189*, 3573–3580.

Field, M. J., Bash, P. A., & Karplus, M. (1990). A combined quantum-mechanical and molecular mechanical potential for molecular-dynamics simulations. *Journal of Computational Chemistry*, *11*(6), 700–733.

Fields, B. A., Bartsch, H. H., Bartunik, H. D., Cordes, F., Guss, J. M., & Freeman, H. C. (1994). Accuracy and precision in protein crystal-structure analysis—Two independent refinements of the structure of poplar plastocyanin at 173 K. *Acta Crystallographica, Section D: Biological Crystallography*, *50*, 709–730.

Frisch, M. J., Trucks, G. W., Schlegel, H. B., Scuseria, G. E., Robb, M. A., Cheeseman, J. R., et al. (2004). *Gaussian 03, revision C.02*. Wallingford, CT: Gaussian, Inc.

Frisch, M. J., Trucks, G. W., Schlegel, H. B., Scuseria, G. E., Robb, M. A., Cheeseman, J. R., et al. (2010). *Gaussian 09, revision B.01*. Wallingford, CT: Gaussian, Inc.

Gaffney, B. J. (1996). Lipoxygenases: Structural principles and spectroscopy. *Annual Review of Biophysics and Biomolecular Structure*, *25*, 431–459.

Gavish, B., & Werber, M. M. (1979). Viscosity-dependent structural fluctuations in enzyme catalysis. *Biochemistry*, *18*(7), 1269–1275.

Ghosh, A., Persson, B. J., & Taylor, P. R. (2003). Ab initio multiconfiguration reference perturbation theory calculations on the energetics of low-energy spin states of iron(III) porphyrins. *Journal of Biological Inorganic Chemistry*, *8*(5), 507–511.

Hagiwara, Y., Field, M. J., Nureki, O., & Tateno, M. (2010). Editing mechanism of aminoacyl-tRNA synthetases operates by a hybrid ribozyme/protein catalyst. *Journal of the American Chemical Society*, *132*(8), 2751–2758.

Hall, N. E., & Smith, B. J. (1998a). High-level ab initio molecular orbital calculations of imine formation. *The Journal of Physical Chemistry. A*, *102*(25), 4930–4938.

Hall, N. E., & Smith, B. J. (1998b). Solvation effects on zwitterion formation. *The Journal of Physical Chemistry. A*, *102*(22), 3985–3990.

Hammes, G. G., Benkovic, S. J., & Hammes-Schiffer, S. (2011). Flexibility, diversity, and cooperativity: Pillars of enzyme catalysis. *Biochemistry*, *50*(48), 10422–10430.

Hammes-Schiffer, S., & Benkovic, S. J. (2006). Relating protein motion to catalysis. *Annual Review of Biochemistry*, *75*, 519–541.

Heinemann, I. U., Jahn, M., & Jahn, D. (2008). The biochemistry of heme biosynthesis. *Archives of Biochemistry and Biophysics*, *474*, 238–251.

Henzler-Wildman, K. A., & Kern, D. (2007). Dynamic personalities of proteins. *Nature*, *450*(7172), 964–972. http://dx.doi.org/10.1038/nature06522.

Henzler-Wildman, K. A., Thai, V., Lei, M., Ott, M., Wolf-Watz, M., Fenn, T., et al. (2007). Intrinsic motions along an enzymatic reaction trajectory. *Nature*, *450*(7171), 838–844.

Himo, F. (2006). Quantum chemical modeling of enzyme active sites and reaction mechanisms. *Theoretical Chemistry Accounts*, *116*(1–3), 232–240.

Hu, L. H., Eliasson, J., Heimdal, J., & Ryde, U. (2009). Do quantum mechanical energies calculated for small models of protein-active sites converge? *The Journal of Physical Chemistry. A, 113*(43), 11793–11800.

Hu, L. H., Soderhjelm, P., & Ryde, U. (2011). On the convergence of QM/MM energies. *Journal of Chemical Theory and Computation*, 7(3), 761–777.

Huang, W. J., Llano, J., & Gauld, J. W. (2010). Redox mechanism of glycosidic bond hydrolysis catalyzed by 6-phospho-alpha-glucosidase: A DFT study. *The Journal of Physical Chemistry. B, 114*(34), 11196–11206.

Humbel, S., Sieber, S., & Morokuma, K. (1996). The IMOMO method: Integration of different levels of molecular orbital approximations for geometry optimization of large systems: Test for n-butane conformation and S(N)2 reaction: RCl+Cl. *The Journal of Chemical Physics, 105*(5), 1959–1967.

Hur, S., & Bruice, T. C. (2003). Enzymes do what is expected (chalcone isomerase versus chorismate mutase). *Journal of the American Chemical Society, 125*(6), 1472–1473.

Ivanov, I., Heydeck, D., Hofheinz, K., Roffeis, J., O'Donnell,, V. B., Kuhn, H., et al. (2010). Molecular enzymology of lipoxygenases. *Archives of Biochemistry and Biophysics, 503*(2), 161–174.

Jackson, A. H., Sancovich, H. A., Ferramola, A. M., Evans, N., Games, D. E., Matlin, S. A., et al. (1976). Macrocyclic intermediates in biosynthesis of porphyrins. *Philosophical Transactions of the Royal Society of London. Series B, Biological Sciences, 273*(924), 191–206.

Jencks, W. P. (1986). *Catalysis in chemistry and enzymology*. New York: Dover Publication.

Johansson, E., Steffens, J. J., Lindqvist, Y., & Schneider, G. (2000). Crystal structure of saccharopine reductase from *Magnaporthe grisea*, an enzyme of the alpha-aminoadipate pathway of lysine biosynthesis. *Structure, 8*(10), 1037–1047.

Juaréz, A. B., Aldonatti, C., Vigna, M. S., & Ríos de Molina, M. C. (2007). Studies on uroporphyrinogen decarboxylase from *Chlorella kessleri*. *Canadian Journal of Microbiology, 53*, 303–312.

Kamerlin, S. C. L., & Warshel, A. (2010). At the dawn of the 21st century: Is dynamics the missing link for understanding enzyme catalysis? *Proteins: Structure, Function, and Bioinformatics, 78*(6), 1339–1375.

Kepenekian, M., Calborean, A., Vetere, V., Le Guennic, B., Robert, V., & Maldivi, P. (2011). Toward reliable DFT investigations of Mn-porphyrins through CASPT2/DFT comparison. *Journal of Chemical Theory and Computation*, 7(11), 3532–3539.

Kollman, P. A., Kuhn, B., & Perakyla, M. (2002). Computational studies of enzyme-catalyzed reactions: Where are we in predicting mechanisms and in understanding the nature of enzyme catalysis? *The Journal of Physical Chemistry. B, 106*(7), 1537–1542.

Kosugi, T., & Hayashi, S. (2011). QM/MM reweighting free energy SCF for geometry optimization on extensive free energy surface of enzymatic reaction. *Journal of Chemical Theory and Computation, 8*(1), 322–334.

Kurplus, M., & McCammon, J. A. (1983). Dynamics of proteins: Elements and function. *Annual Review of Biochemistry, 52*(1), 263–300.

Lau, E. Y., & Bruice, T. C. (1998). Importance of correlated motions in forming highly reactive near attack conformations in catechol O-methyltransferase. *Journal of the American Chemical Society, 120*(48), 12387–12394.

Lee, J. K., & Houk, K. N. (1997). A proficient enzyme revisited: The predicted mechanism for orotidine monophosphate decarboxylase. *Science, 276*(5314), 942–945.

Lee, C., Yang, W., & Parr, R. G. (1988). Development of the Colle–Salvetti correlation-energy formula into a functional of the electron density. *Physical Review B: Condensed Matter, 37*(2), 785–789.

Lewis, C. A., & Wolfenden, R. (2008). Uroporphyrinogen decarboxylation as a benchmark for the catalytic proficiency of enzymes. *Proceedings of the National Academy of Sciences of the United States of America, 105*(45), 17328–17333.

Liu, H. N., & Gauld, J. W. (2008). Substrate-assisted catalysis in the aminoacyl transfer mechanism of histidyl-tRNA synthetase: A density functional theory study. *The Journal of Physical Chemistry. B, 112*(51), 16874–16882.

Liu, H. N., Llano, J., & Gauld, J. W. (2009). A DFT study of nucleobase dealkylation by the DNA repair enzyme AlkB. *The Journal of Physical Chemistry. B, 113*(14), 4887–4898.

Llano, J., & Gauld, J. W. (2010). Mechanistics of enzyme catalysis: From small to large active-site models. In C. F. Matta (Ed.), *Quantum biochemistry: Electronic structure and biological activity* (p. 920). Vol. 2 (p. 920). Weinheim: Wiley-VCH.

Lodola, A., Woods, C. J., & Mulholland, A. J. (2008). Chapter 9—Applications and advances of QM/MM methods in computational enzymology. In A. W. Ralph & C. S. David (Eds.), *Annual reports in computational chemistry: Vol. 4* (pp. 155–169). Amsterdam: Elsevier.

Lundberg, M., & Morokuma, K. (2007). Protein environment facilitates O-2 binding in non-heme iron enzyme. An insight from ONIOM calculations on isopenicillin N synthase (IPNS). *The Journal of Physical Chemistry. B, 111*(31), 9380–9389.

Martins, B. M., Grimm, B., Mock, H.-P., Huber, R., & Messerschmidt, A. (2001). Crystal structure and substrate binding modelling of the uroporphyrinogen-III decarboxylase from *Nicotianna tabacum. The Journal of Biological Chemistry, 276*(47), 44108–44116.

Maseras, F., & Morokuma, K. (1995). IMOMM—A new integrated ab-initio plus molecular mechanics geometry optimization scheme of equilibrium structures and transition-states. *Journal of Computational Chemistry, 16*(9), 1170–1179.

McCammon, J. A., Wolynes, P. G., & Karplus, M. (1979). Picosecond dynamics of tyrosine side-chains in proteins. *Biochemistry, 18*(6), 927–942.

Minor, W., Steczko, J., Bolin, J. T., Otwinowski, Z., & Axelrod, B. (1993). Crystallographic determination of the active-site iron and its ligands in soybean lipoxygenase L-1. *Biochemistry, 32*(25), 6320–6323.

Mock, H. P., Trainotti, L., Kruse, E., & Grimm, B. (1995). Isolation, sequencing and expression of cDNA sequences encoding uroporphyrinogen decarboxylase from tobacco and barley. *Plant Molecular Biology, 28*(2), 245–256.

Morokuma, K., Musaev, D. G., Vreven, T., Basch, H., Torrent, M., & Khoroshun, D. V. (2001). Model studies of the structures, reactivities, and reaction mechanisms of metalloenzymes. *IBM Journal of Research and Development, 45*(3–4), 367–395.

Nagel, Z. D., & Klinman, J. P. (2006). Tunneling and dynamics in enzymatic hydride transfer. *Chemical Reviews, 106*(8), 3095–3118.

Nagel, Z. D., & Klinman, J. P. (2010). Update 1 of: Tunneling and dynamics in enzymatic hydride transfer. *Chemical Reviews, 110*(12), PR41–PR67.

Neau, D. B., Gilbert, N. C., Bartlett, S. G., Boeglin, W., Brash, A. R., & Newcomer, M. E. (2009). The 1.85 angstrom structure of an 8R-lipoxygenase suggests a general model for lipoxygenase product specificity. *Biochemistry, 48*(33), 7906–7915.

Neese, F. (2006). A critical evaluation of DFT, including time-dependent DFT, applied to bioinorganic chemistry. *Journal of Biological Inorganic Chemistry, 11*(6), 702–711.

Nilsson, K., Lecerof, D., Sigfridsson, E., & Ryde, U. (2003). An automatic method to generate force-field parameters for hetero-compounds. *Acta Crystallographica, Section D: Biological Crystallography, 59*, 274–289.

Olsson, M. H. M., Mavri, J., & Warshel, A. (2006). Transition state theory can be used in studies of enzyme catalysis: Lessons from simulations of tunnelling and dynamical effects in lipoxygenase and other systems. *Philosophical Transactions of the Royal Society of London. Series B, Biological Sciences, 361*(1472), 1417–1432.

Olsson, M. H. M., & Ryde, U. (2001). Geometry, reduction potential, and reorganization energy of the binuclear Cu-A site, studied by density functional theory. *Journal of the American Chemical Society, 123*(32), 7866–7876.

Page, M. I., & Jencks, W. P. (1971). Entropic contributions to rate accelerations in enzymic and intramolecular reactions and chelate effect. *Proceedings of the National Academy of Sciences of the United States of America*, *68*(8), 1678–1683.

Pauling, L. (1948). Nature of forces between large molecules of biological interest. *Nature*, *161*(4097), 707–709.

Phillips, J. D., Whitby, F. G., Kushner, J. P., & Hill, C. P. (2003). Structural basis for tetrapyrrole coordination by uroporphyrinogen decarboxylase. *The EMBO Journal*, *22*(23), 6225–6233.

Reiher, M. (2002). Theoretical study of the Fe(phen)(2)(NCS)(2) spin-crossover complex with reparametrized density functionals. *Inorganic Chemistry*, *41*(25), 6928–6935.

Reiher, M., Salomon, O., & Hess, B. A. (2001). Reparameterization of hybrid functionals based on energy differences of states of different multiplicity. *Theoretical Chemistry Accounts*, *107*(1), 48–55.

Ribot, C., Hirsch, J., Batzergue, S., Tharreau, D., Notteghem, J. L., Lebrun, M. H., et al. (2008). Susceptibility of rice to the blast fungus, *Magnaporthe grisea*. *Journal of Plant Physiology*, *165*(1), 114–124.

Rosenberg, S., Silver, S. M., Sayer, J. M., & Jencks, W. P. (1974). Evidence for 2 concurrent mechanisms and a kinetically significant proton-transfer process in acid-catalyzed O-methyloxime formation. *Journal of the American Chemical Society*, *96*(26), 7986–7998.

Ryde, U. (2003). Combined quantum and molecular mechanics calculations on metalloproteins. *Current Opinion in Chemical Biology*, 7(1), 136–142.

Ryde, U. (2007). Accurate metal-site structures in proteins obtained by combining experimental data and quantum chemistry. *Dalton Transactions*, (6), 607–625.

Schenk, G., Pau, M. Y. M., & Solomon, E. I. (2004). Comparison between the geometric and electronic structures and reactivities of {FeNO}(7) and {FeO2}(8) complexes: A density functional theory study. *Journal of the American Chemical Society*, *126*(2), 505–515.

Schramm, V. L. (2012). Transition states, analogues, and drug development. *ACS Chemical Biology*, *8*(1), 71–81.

Schutz, C. N., & Warshel, A. (2001). What are the dielectric "constants" of proteins and how to validate electrostatic models? *Proteins*, *44*(4), 400–417.

Senn, H. M., & Thiel, W. (2009). QM/MM methods for biomolecular systems. *Angewandte Chemie International Edition*, *48*(7), 1198–1229.

Shaik, S., Cohen, S., Wang, Y., Chen, H., Kumar, D., & Thiel, W. (2010). P450 enzymes: Their structure, reactivity, and selectivity-modeled by QM/MM calculations. *Chemical Reviews*, *110*(2), 949–1017.

Shen, Y., & Ryde, U. (2004). The structure of sitting-atop complexes of metalloporphyrins studied by theoretical methods. *Journal of Inorganic Biochemistry*, *98*(5), 878–895.

Shurki, A., Strajbl, M., Villa, J., & Warshel, A. (2002). How much do enzymes really gain by restraining their reacting fragments? *Journal of the American Chemical Society*, *124*(15), 4097–4107.

Siegbahn, P. E. M. (2006). The performance of hybrid DFT for mechanisms involving transition metal complexes in enzymes. *Journal of Biological Inorganic Chemistry*, *11*(6), 695–701.

Siegbahn, P. E. M., & Blomberg, M. R. A. (1999). Density functional theory of biologically relevant metal centers. *Annual Review of Physical Chemistry*, *50*, 221–249.

Siegbahn, P. E. M., Blomberg, M. R. A., & Chen, S. L. (2010). Significant van der Waals effects in transition metal complexes. *Journal of Chemical Theory and Computation*, *6*(7), 2040–2044.

Siegbahn, P. E. M., & Himo, F. (2009). Recent developments of the quantum chemical cluster approach for modeling enzyme reactions. *Journal of Biological Inorganic Chemistry*, *14*(5), 643–651.

Sigfridsson, E., Olsson, M. H. M., & Ryde, U. (2001). Comparison of the inner-sphere reorganization energies of cytochromes, iron-sulfur clusters, and blue copper proteins. *The Journal of Physical Chemistry. B, 105*(23), 5546–5552.

Sigfridsson, E., & Ryde, U. (2002). Theoretical study of the discrimination between O-2 and CO by myoglobin. *Journal of Inorganic Biochemistry, 91*(1), 101–115.

Silva, P. J., & Ramos, M. J. (2005). Density-functional study of mechanisms for the cofactor-free decarboxylation performed by uroporphyrinogen III decarboxylase. *The Journal of Physical Chemistry. B, 109*, 18195–18200.

Singh, U. C., & Kollman, P. A. (1986). A combined ab initio quantum-mechanical and molecular mechanical method for carrying out simulations on complex molecular-systems—Applications to the $CH_3Cl + Cl^-$ exchange-reaction and gas-phase protonation of polyethers. *Journal of Computational Chemistry, 7*(6), 718–730.

Sumner, S., Söderhjelm, P., & Ryde, U. (2013). Effect of geometry optimizations on QM-cluster and QM/MM studies of reaction energies in proteins. *Journal of Chemical Theory and Computation, 9*, 4205–4214.

Svensson, M., Humbel, S., Froese, R. D. J., Matsubara, T., Sieber, S., & Morokuma, K. (1996). ONIOM: A multilayered integrated MO+MM method for geometry optimizations and single point energy predictions. A test for Diels–Alder reactions and Pt(P(t-Bu)(3))(2)+H-2 oxidative addition. *The Journal of Physical Chemistry, 100*(50), 19357–19363.

Talbot, N. J. (1995). Having a blast: Exploring the pathogenicity of *Magnaporthe grisea. Trends in Microbiology, 3*(1), 9–16.

Tian, B., Strid, Å., & Eriksson, L. A. (2011). Catalytic roles of active-site residues in 2-methyl-3-hydroxypyridine-5-carboxylic acid oxygenase: An ONIOM/DFT study. *The Journal of Physical Chemistry. B, 115*(8), 1918–1926.

Tuckerman, M. E., Marx, D., Klein, M. L., & Parrinello, M. (1997). On the quantum nature of the shared proton in hydrogen bonds. *Science, 275*(5301), 817–820.

Vancoillie, S., Zhao, H. L., Radon, M., & Pierloot, K. (2010). Performance of CASPT2 and DFT for relative spin-state energetics of heme models. *Journal of Chemical Theory and Computation, 6*(2), 576–582.

Vashishtha, A. K., West, A. H., & Cook, P. F. (2009). Chemical mechanism of saccharopine reductase from *Saccharomyces cerevisiae. Biochemistry, 48*(25), 5899–5907.

Villa, J., Strajbl, M., Glennon, T. M., Sham, Y. Y., Chu, Z. T., & Warshel, A. (2000). How important are entropic contributions to enzyme catalysis? *Proceedings of the National Academy of Sciences of the United States of America, 97*(22), 11899–11904.

Villa, J., & Warshel, A. (2001). Energetics and dynamics of enzymatic reactions. *The Journal of Physical Chemistry. B, 105*(33), 7887–7907.

Vreven, T., Byun, K. S., Komaromi, I., Dapprich, S., Montgomery, J. A., Morokuma, K., et al. (2006). Combining quantum mechanics methods with molecular mechanics methods in ONIOM. *Journal of Chemical Theory and Computation, 2*(3), 815–826.

Vreven, T., & Morokuma, K. (2000). On the application of the IMOMO (integrated molecular orbital plus molecular orbital) method. *Journal of Computational Chemistry, 21*(16), 1419–1432.

Vreven, T., & Morokuma, K. (2006). Chapter 3—Hybrid methods: ONIOM(QM:MM) and QM/MM. In C. S. David (Ed.), *Annual reports in computational chemistry: Vol. 2* (pp. 35–51). Amsterdam: Elsevier.

Vreven, T., Morokuma, K., Farkas, O., Schlegel, H. B., & Frisch, M. J. (2003). Geometry optimization with QM/MM, ONIOM, and other combined methods. I. Microiterations and constraints. *Journal of Computational Chemistry, 24*(6), 760–769.

Warshel, A. (1978). Energetics of enzyme catalysis. *Proceedings of the National Academy of Sciences of the United States of America, 75*(11), 5250–5254.

Warshel, A. (1987). Molecular biophysics—What about protein polarity. *Nature, 330*(6143), 15–16.
Warshel, A. (1998). Electrostatic origin of the catalytic power of enzymes and the role of preorganized active sites. *The Journal of Biological Chemistry, 273*(42), 27035–27038.
Warshel, A. (2002). Molecular dynamics simulations of biological reactions. *Accounts of Chemical Research, 35*(6), 385–395.
Warshel, A. (2003). Computer simulations of enzyme catalysis: Methods, progress, and insights. *Annual Review of Biophysics and Biomolecular Structure, 32*, 425–443.
Warshel, A., Aqvist, J., & Creighton, S. (1989). Enzymes work by solvation substitution rather than by desolvation. *Proceedings of the National Academy of Sciences of the United States of America, 86*(15), 5820–5824.
Warshel, A., & Levitt, M. (1976). Theoretical studies of enzymic reactions—Dielectric, electrostatic and steric stabilization of carbonium-ion in reaction of lysozyme. *Journal of Molecular Biology, 103*(2), 227–249.
Warshel, A., & Papazyan, A. (1998). Electrostatic effects in macromolecules: Fundamental concepts and practical modeling. *Current Opinion in Structural Biology, 8*(2), 211–217.
Warshel, A., Papazyan, A., & Kollman, P. A. (1995). On low-barrier hydrogen-bonds and enzyme catalysis. *Science, 269*(5220), 102–104.
Warshel, A., Sharma, P. K., Chu, Z. T., & Aqvist, J. (2007). Electrostatic contributions to binding of transition state analogues can be very different from the corresponding contributions to catalysis: Phenolates binding to the oxyanion hole of ketosteroid isomerase. *Biochemistry, 46*(6), 1466–1476.
Warshel, A., Sharma, P. K., Kato, M., & Parson, W. W. (2006). Modeling electrostatic effects in proteins. *Biochimica et Biophysica Acta, Proteins and Proteomics, 1764*(11), 1647–1676.
Warshel, A., Sharma, P. K., Kato, M., Xiang, Y., Liu, H., & Olsson, M. H. M. (2006). Electrostatic basis for enzyme catalysis. *Chemical Reviews, 106*(8), 3210–3235.
Williams, I. H. (1987). Theoretical modeling of specific solvation effects upon carbonyl addition. *Journal of the American Chemical Society, 109*(21), 6299–6307.
Xu, H. Y., Andi, B., Qian, J. H., West, A. H., & Cook, P. F. (2006). The alpha-aminoadipate pathway for lysine biosynthesis in fungi. *Cell Biochemistry and Biophysics, 46*(1), 43–64.
Young, D. (2001). *Computational chemistry: A practical guide for applying techniques to real world problems*. New York: Wiley-Interscience.
Zabriskie, T. M., & Jackson, M. D. (2000). Lysine biosynthesis and metabolism in fungi. *Natural Product Reports, 17*(1), 85–97.
Zhang, Y., Gan, Q. F., Pavel, E. G., Sigal, E., & Solomon, E. I. (1995). Epr definition of the nonheme ferric active-sites of mammalian 15-lipoxygenase—Major spectral difference relative to human 5-lipoxygenases and plant lipoxygenases and their ligand-field origin. *Journal of the American Chemical Society, 117*(28), 7422–7427.
Zhao, Y., & Truhlar, D. G. (2008a). Density functionals with broad applicability in chemistry. *Accounts of Chemical Research, 41*(2), 157–167.
Zhao, Y., & Truhlar, D. G. (2008b). The M06 suite of density functionals for main group thermochemistry, thermochemical kinetics, noncovalent interactions, excited states, and transition elements: Two new functionals and systematic testing of four M06-class functionals and 12 other functionals. *Theoretical Chemistry Accounts, 120*(1–3), 215–241.

CHAPTER SEVEN

QM and QM/MM Methods Compared: Case Studies on Reaction Mechanisms of Metalloenzymes

Tomasz Borowski*, Matthew Quesne*,†, Maciej Szaleniec*,[1]
*Jerzy Haber Institute of Catalysis and Surface Chemistry, Polish Academy of Sciences, Kraków, Poland
†Manchester Institute of Biotechnology and Faculty of Life Sciences, The University of Manchester, Manchester, United Kingdom
[1]Corresponding author e-mail address: ncszalen@cyfronet.pl

Contents

Abstract

The review focus is a comparison of QM and QM/MM modeling techniques applied to study of metalloenzymes. The chapter aim is to highlight many of the advantages and potential pitfalls of the exciting and revolutionary QM/MM techniques using both large QM/MM systems and QM-only modeling as references. The review is illustrated by case studies for isopenicillin *N* synthase, ethylbenzene dehydrogenase, cytochrome P450 enzyme, AlkB DNA repair enzyme as well as 4-hydroxyphenylpyruvate dioxygenase. We find many advantages in various QM/MM techniques, over the more traditional QM cluster approaches, while at the same time offering some advice about how to avoid potential complications arising from some of these approaches' most notable

Advances in Protein Chemistry and Structural Biology, Volume 100
ISSN 1876-1623
http://dx.doi.org/10.1016/bs.apcsb.2015.06.005

drawbacks. We conclude that while there will always be an important role for QM cluster models, in computational studies, the revolutionary developments in QM/MM techniques open a bright and exciting future of new research.

1. INTRODUCTION

Since its birth in the seventeenth century, science has owned its success to its ability to construct simple and mathematically manageable models of intricate real-life phenomena. The role of these models has always been to explain the major features of studied processes leaving the "small details" for more elaborate and complete theories/models. Once this first-generation model is mature and its strengths and weaknesses are well recognized, a second-generation model can be build up on top of the first one with the aim to provide a more complete description, covering (some of) the "details," of the phenomenon under investigation. We can encounter a similar ladder of models in computational chemistry, which is the subject of the present review into various computational studies on enzymatic reaction mechanisms. Here, the first-generation models could be equated to cluster models, where only active site residues plus substrate(s)/co-factor(s) are represented explicitly, whereas the remaining parts of the protein and solvent molecules are possibly taken into account only indirectly. Such models will be termed as "QM models" as they are typically used together with quantum mechanical (chemistry) methods, which are the methods of choice if one's aim is to provide accurate description of a chemical process (Blomberg, Borowski, Himo, Liao, & Siegbahn, 2014; Siegbahn & Borowski, 2006). The second-generation model tries to capture the effects that are due to residues not covered explicitly by the first one. As the computational cost grows rapidly with the number of atoms treated with a quantum chemistry method, some compromise must be met if one seeks a manageable computational model. One of the solutions of this problem, but certainly not the only one, rests on combining a QM method, which is applied to describe the key quantum-chemical process in the "active zone," with a molecular mechanical (MM) description of remaining parts of the system. The resulting methods are hence usually abbreviated as QM/MM.

Over the years two flavors of the methods combining QM with MM has emerged. The first one is a traditional additive QM/MM method, where the energy of the whole system combines energies of the QM- and MM-regions

(E^{QM}, E^{MM}) and a term describing interactions between the two (E^{QM-MM}) (Field, Bash, & Karplus, 1990; Warshel & Levitt, 1976):

$$E^{QM/MM} = E^{QM} + E^{MM} + E^{QM-MM} \tag{1}$$

The second is a particular (QM/MM) realization of a general extrapolation scheme which is known as ONIOM (Maseras & Morokuma, 1995). Here, one attempts to extrapolate the energy of the whole macromolecular system from more accurate energy of the key region—Model ($E^{Model,QM}$) and MM energy of the whole system ($E^{All,MM}$). Subtraction of a term $E^{Model,MM}$ guarantees we avoid double counting the contribution of the Model part:

$$E^{ONIOM} = E^{Model,QM} + E^{All,MM} - E^{Model,MM} \tag{2}$$

Despite differences in formulations, from a practical perspective the QM/MM and ONIOM(QM/MM) methods are in principle on a par with one another. A reader interested in an insightful dissection of the similarities and differences of the two approaches should refer to a paper by Vreven et al. (2006).

Dividing a macromolecular system into parts described with QM and MM methods usually means the QM–MM interface will go across some covalent bonds. Several approaches have been devised to deal with this issue (Senn & Thiel, 2007). One of them, probably the most popular, is the link-atom approach, where the dangling bond of the QM-part is saturated with some monovalent, usually hydrogen, atoms, which is not included in calculations employing the MM method. In the second approach, the dangling bond is represented with a frozen localized orbital.

As electrostatic interactions are critical for stability and function of biomolecules, an important issue is how they are treated in the QM/MM calculations. Within a mechanical-embedding scheme (ME), the electrostatic interactions between the QM- and MM-parts of the system are calculated classically, i.e., in exactly the same way as for the MM-part—usually using partial atomic charges. Hence, all atoms of the QM-part need to have atomic charges ascribed, and if a reaction coordinate is followed, they should be updated as the reaction moves forward. Moreover, as the QM-MM electrostatics are treated at the MM level, the electronic structure is not directly affected (i.e., polarized) by the charge distribution in the MM-part. Within the second scheme, which is called electronic embedding (EE), the QM-MM electrostatic interactions are calculated at the QM level by including electrostatic interactions between nuclei and electrons of the QM-part with atomic partial charges of the MM-part. Hence, in EE the electron cloud of the QM-region can be polarized by atomic charges from the MM-part, yet, the polarization of the MM atoms by the QM region is typically not accounted for. QM/MM methods including mutual polarization of the QM and MM parts of the system are currently (beginning in 2015) under development (Caprasecca, Jurinovich, Viani, Curutchet, & Mennucci, 2014).

2. ADVANTAGES OF QM/MM

The QM/MM formalism has a number of advantages which make this method especially suitable for studying enzymatic systems.

In its simplest implementation, the QM/MM operated on models that were as close to the real systems as it was possible to achieve, based on the diffraction studies of the crystallized proteins. Even this relatively small improvement provided a tremendous step forward in studies of the enzymes. It must be underlined that structural organization of the substrate(s) in the active site and their position with respect to the catalytically active residues or cofactors has a central place to most of the theories explaining enzyme catalytic power. Our understanding of enzyme catalysis is based on the knowledge on enzyme–substrate complex (ES) and the effect of the enzyme on the substrate conformation in the ES and in a transition state (TS). In addition to structural constraints acting on the active site constituents, another advantage of QM/MM techniques over QM-cluster models is that there is an explicit account taken of the QM surrounding. Using QM/MM techniques, there is no need to fix any atoms in the periphery of the QM model and in theory there is a more realistic system created. This approach also allowed computational chemists to address the very important question of reaction enantioselectivity, both introduced by appropriate substrate binding (like in the case of the short-chain alcohol dehydrogenases where appropriate binding of the alcohol/ketone with respect to NAD^+/NADH and Tyr H-bond donor in the during reactions, leading ES complex is responsible for reaction enantioselectivity (Dudzik et al., 2014)) or due to differences in the steric interactions during reactions, leading to different heights of the energy barriers observed for enantiomers or prochiral reagents (e.g., ethylbenzene dehydrogenase (EbDH) (Szaleniec et al., 2014) or cyclohexanone monooxygenase (Polyak, Reetz, & Thiel, 2012, 2013)).

The second big step forward was the introduction of enzyme electrostatics into the calculations. When working with QM-only models one has to decide which polar residues should be included into the model, and which not. In favorable cases, the active site is buried in a hydrophobic core of the protein and the choice is relatively simple, i.e., one includes in the QM-model only the polar residues making up the active site, whereas the next hydrophobic residues of the protein interior and polar residues from the protein surface are neglected. In less favorable cases, the active site is located close to the protein surface, which is usually more polar and more accessible for the solvent molecules, and hence, selection of a compact yet reliable active site model is more of a challenge. Energetic consequences of a particular choice of active site model had to be tested by modeling the reaction pathway with the use of different cluster models. As in the QM/MM model all protein (and solvent) residues are explicitly taken into account, their influence on the energetic profile of the catalyzed reaction, due to electrostatic and van der Waals interactions, has become directly computable. This improvement allows the calculation to take into account the electrostatic stabilization of TS originating both from the local residues (already well treated by QM-only models) and the long-range electrostatic interactions (like dipole moment of helixes).

Yet another criticism of the QM-only modeling steamed from a lack of the explicit treatment of the solvent. Although water molecules were frequently modeled in an explicit way if they were expected to be directly involved in the reaction mechanism (as proton acceptors or donors) in most other QM-only models, the solvent was usually treated with only an implicit approach. The application of QM/MM models allowed explicit treatment of waters that were either derived from the crystal models (so called structural water molecules) and/or water molecules that were introduced with MD simulations in the preparative steps preceding QM/MM modeling. Introduction of the solvent molecules into the model provides a more realistic representation of the hydrogen-bonding network in the studied model which certainly makes the QM/MM modeling more resembling a real enzyme system (at least on the conceptual level). However, introduction of the explicit solvent into the model delivers also a range of potential pitfalls into the modeling process, especially associated with mobility of the unconstrained solvent molecule in the model (see below).

Another advantage of the QM/MM models over the QM-only ones is an ease with which one can extend the QM-part of the former. In both approaches, one has to select which of the residues one will treat at the

QM-level of theory. However, in QM/MM studies all residues are represented in the model and hence moving a residue across the QM/MM border is a straightforward process requiring only a minor change in the input file, whereas adding a new residue to the cluster model usually requires the reaction path to be recomputed. This is because the cluster model has a considerable flexibility allowing it to adopt geometries, especially at its edge, where simple copying and pasting the geometry of the additional residue from the crystal structure is not possible. Although one has to reoptimize a new QM-2/MM model, the work already done is still a very good starting point for the updated QM/MM model.

Finally there is also a psychological factor involved especially when a theoretician has to confront his modeling results with the skepticism of his/her peers (especially those that specialize in more experimental approach to the study of biological systems). The small QM-only models are very often regarded as too oversimplified. Thus QM/MM approach may often seem more realistic as its complexity resembles the complexity of the biological systems.

The more realistic description offered by the QM/MM model comes, of course, at a price and several factors which help underline this are shortly presented in the following section.

3. DISADVANTAGES OF QM/MM

There are a number of factors that still make the QM-only cluster modeling more attractive than the QM/MM methodology. We are going to list them, without a particular order of importance in order to lay ground for examples provided further in the chapter.

The very first problem in starting the QM/MM calculation is the preparation of the model. QM/MM modeling has gained popularity relatively recently and as a consequence the development of the software used in model preparation is still lagging behind. Meanwhile the computational chemist has to change the scale of his/her models usually by 2 orders of magnitude as the QM-cluster models tend to be in the range of 100–300 atoms whereas 10,000–20,000 atom models are not very uncommon in QM/MM modeling. Another issue concerns finding atomic charges for the QM-part for ME calculations, as here the electrostatic interactions between QM and MM parts of the system are calculated classically. Similarly, nonstandard protein residues located in the MM fragment of the model need to be parametrized. Another decision to make, concerns which part, if not the whole of the model, should be allowed to relax in the modeling. Naturally, the more coordinates that are left for optimization the more iterations are required to optimize the model, yet on the other hand freezing too many atoms may yield a model not having necessary flexibility to accommodate geometry changes taking place along the studied reaction coordinate.

It is also important to point out that when calculating the energy profile of a reaction mechanism often freezing the coordinates of atoms far from the active region can actually give more reliable results. This is because for such a large system it can be impossible to ensure that each structure along the potential energy surface resides in the correct local minima. That is to say that all residues in the system are at the same conformation (except for the internal coordinate being modified); a change in one water molecule's hydrogen-bonding pattern can cause a total energy change <6 kcal/mol, and this molecule may be a very long way from the active site. Some of the solutions developed to solve this issue include: only allowing the movement of atoms within a selected distance of the active site, running the reaction in both directions until energy stability is achieved, or by calculating the reaction pathways of many different structures, taken from snapshots of the MD trajectory and averaging of the results (de Visser, Quesne, Martin, Comba, & Ryde, 2014; Lonsdale, Ranaghan, & Mulholland, 2010; Senn & Thiel, 2009; Warshel & Levitt, 1976).

Finally, the decision has to be made on which protein, and also possibly solvent, residues should be placed in the QM-part of the QM/MM model, yet a similar choice one makes when constructing a cluster model. In summary, the preparatory steps required to construct a usable QM/MM model are considerable more demanding compared to the QM cluster modeling. Convergence studies performed on QM-cluster models have shown that reaction energetic can change by 3–14 kcal/mol as the size of the QM system is increased from 27–83 to 161–220 atoms (Georgieva & Himo, 2010; Hu, Eliasson, Heimdal, & Ryde, 2009; Sevastik & Himo, 2007). The situation is often not much better in the QM/MM models, where electrostatic-embedding is included. Using these models, a similar absolute deviation was seen to those observed for the QM-only calculations. When the energetics of systems with over 40 different sized QM regions (40–446 atoms) were compared a mean absolute deviation of over 4 kcal/mol was observed (Hu, Söderhjelm, & Ryde, 2011; Kumar, Karamzadeh, Sastry, & de Visser, 2010; Quesne, Latifi, Gonzalez-Ovalle, Kumar, & de Visser, 2014). Therefore, both the number of atoms in the QM region as well as the type of embedding performed can have major effects on the potential energy surface obtained.

The immense size of the QM/MM models is also associated with another disadvantage. We are used to visual inspection of our theoretical models and association of the energy profile with molecular changes seen in the modeling. It is relatively easy to follow the structural changes of the few hundred atom model but rather more difficult if the part that is subjected to optimization encompasses more than 1000 atoms. To aid identification of structural changes, one often needs to resort to some means of highlighting the part of the model that changed the most, e.g., by appropriate color coding based on values of atom displacements calculated between the two structures being compared. Another issue connected with the size of the QM/MM model is a problem of multiple minima on the potential energy landscape. More specifically, it is not impossible that while following the reaction coordinate in the active site one or more residues in the MM region may change its geometry, i.e., forming a new H-bond, and this will of course affect the total QM/MM energy, which could be even misinterpreted, as demonstrated in one of our studies (Siegbahn & Borowski, 2011). Thus, it is strongly recommended that the optimized structures along the reaction coordinate are critically compared to one another and both the total QM/MM and the composite QM and MM energies are scrutinized along the full reaction pathway.

4. STERIC CONSTRAINS IN QM VERSUS QM/MM APPROACH

One of the clear advantages of the QM/MM approach over the ones based on a cluster models is the superior handling of the steric constraints imposed on the geometry of the active site residues by the global structure of the macromolecule. One study by Lundberg et al. offers a clear example of this, where QM-only and ONIOM models were compared for an O_2-binding step of the reaction catalyzed by isopenicillin *N* synthase (Lundberg & Morokuma, 2007). The study very clearly revealed that the QM cluster model, for the unbound state, has too much freedom to relax, which results in the energetics of dioxygen binding being affected by a 3–6 kcal/mol error, due to this effect. Another 3–4 kcal/mol discrepancy between the cluster and the ONIOM model comes from the protein–O_2 interactions, mostly of the van der Waals type, which are also more appropriately represented by the hybrid model. However, one should note here that this comparison could be much less unfavorable for the cluster model provided a much larger cluster was used, but then at the expense of the considerably larger computational burden. Notably, in this case polarization of the QM-part by the MM residues has a negligible effect on the energy of O_2 binding.

5. INFLUENCE OF THE EMBEDDING SCHEME ON THE REACTION CHEMISTRY: CASE OF EbDH

EbDH is a molybdoenzyme belonging to subfamily II of the DMSO reductase family (Hille, 1996; Kniemeyer & Heider, 2001) that catalyzes the oxygen-independent, stereospecific hydroxylation of ethylbenzene to (*S*)-1-phenylethanol. EbDH was the first known enzyme to be capable of the direct anaerobic hydroxylation of alkyl aromatic hydrocarbons (Johnson, Pelletier, & Spormann, 2001; Kniemeyer & Heider, 2001; Szaleniec et al., 2007), and it is involved in the mineralization of ethylbenzene by *Aromatoleum aromaticum*, a denitrifying *Betaproteobacterium* (Fuchs, Boll, & Heider, 2011; Heider, 2007; Rabus & Heider, 1998; Rabus et al., 2005). This enzyme becomes a prototype catalyst for a whole group of molybdenum-containing enzymes (so called EbDH-like enzymes), namely cholesterol C25 dehydrogenase (S25DH) (Chiang et al., 2008; Dermer & Fuchs, 2012), *p*-cymene dehydrogenase (CmdABC) (Strijkstra et al., 2014) as well as putative alkane hydroxylase from *Desulfococcus oleovorans*. EbDH-like enzymes are responsible for metabolic activation of hydrocarbons by introduction of hydroxyl group into benzylic secondary, primary, or aliphatic tertiary carbon atom (for EbDH, CmdABC, and S25DH, respectively). The molecular mechanism of these hydroxylations is very interesting as the whole process proceeds without molecular oxygen due to the anaerobic nature of the bacteria metabolism. These enzymes are also very interesting from the point of view of potential application as they exhibit high regioselectivity and EbDH is also highly enantioselective (+95% enantiomeric excess for more than 30 substrates) (Dudzik et al., 2013; Szaleniec et al., 2014). It was shown by isotope-labeling studies that oxygen is derived from a water molecule while the catalytic cofactor is oxidized by an external electron acceptor via enzyme's metal clusters including a high potential heme cofactor.

EbDH is an $\alpha\beta\gamma$ heterotrimer with subunits of 96, 43, and 23 kDa that contain a molybdenum cofactor (MoCo) and a heme b_{559} cofactor linked by a linear row of five iron–sulfur clusters (Kloer, Hagel, Heider, & Schulz, 2006). The enzyme active site is deeply buried in the α subunit and is composed of the MoCo and Asp_{223} and His_{192} side chains, which have been proposed to be involved in the catalytic process (Fig. 1). Based on analogy to other molybdenum enzymes within the DMSO reductase family that catalyze oxygen transfer reactions (George, Hilton, Temple, Prince, & Rajagopalan, 1999; Jormakka, Richardson, Byrne, & Iwata, 2004), it was suggested that an oxo ligand is coordinated at the molybdenum atom in the oxidized form of the catalytically active enzyme (Kloer et al., 2006; Szaleniec, Borowski, Schuhle, Witko, & Heider, 2010).

As the structure of EbDH has been solved in 2006 (Hoffken et al., 2006), the modeling of the reaction mechanism was possible. The initial studies focused on the formulation of the reaction hypothesis (Szaleniec et al., 2007; Szaleniec, Witko, & Heider, 2008) by means of QM-only modeling (Szaleniec et al., 2010; Szaleniec, Salwinski, Borowski, Heider, & Witko, 2012). All considered reaction hypotheses presumed C–H activation at the benzylic carbon atom by the oxidized MoCo in the first step, followed by formation of the intermediate product and an OH-rebound process yielding the alcohol product and a reduced MoCo (Fig. 2). However, various variants of this simple mechanism have been envisaged, depending on the type of C–H activation (homolytic vs. heterolytic) and the precise moment of the second electron transfer.

The extensive studies conducted with QM-only model showed that the pathway depicted on Fig. 2C is the most probable, i.e., the C–H activation proceeds in a homolytic manner which leads to formation of the radical intermediate and the second electron is transferred during the OH-rebound process (Szaleniec et al., 2010). This results with formation

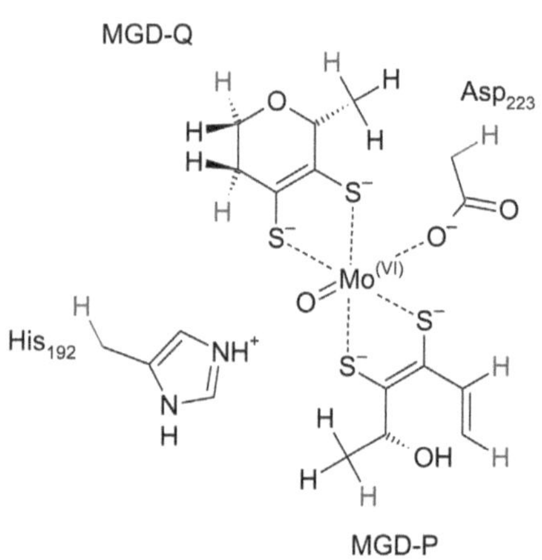

Figure 1 A small cluster model of EbDH active site used in QM-only studies. The coordinates of atoms in red (gray in the print version) were frozen.

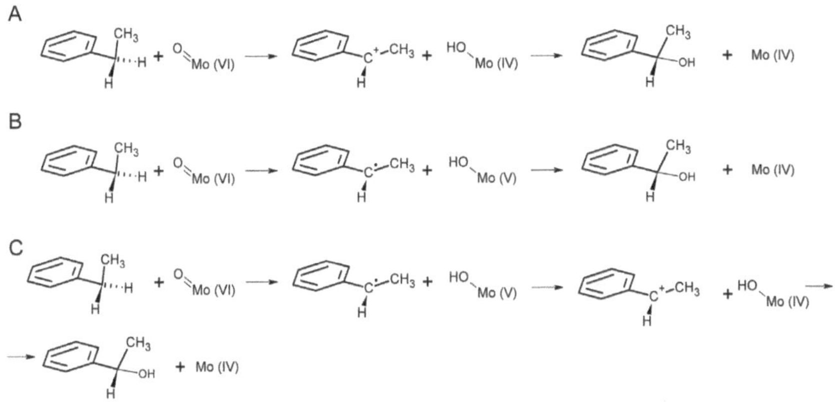

Figure 2 Hypothetical variants of reaction mechanism catalyzed by EbDH: (A) heterolytic C–H cleavage (TS1) by carbocation formation and the formation of the alcohol (TS2) by transfer of a hydroxide ion; (B) homolytic cleavage of the C—H bond leading to a radical intermediate and formation of the product by transfer of a hydroxyl radical equivalent; (C) "mixed" mechanism comprising of homolytic cleavage of the C—H bond and then electron transfer yielding carbocation species and Mo(IV) and finally resulting in formation of the product by transfer of hydroxide ion.

of a bond between OH^- ion and the hydrocarbon carbocation in TS2. The stage at which the second electron is being transferred from substrate to the MoCo was additionally correlated with electronic features of the reagents, which depended on electron withdrawing/donating effect introduced by different substituents (Szaleniec et al., 2012). Moreover, the computed energy profiles and kinetic isotope effects (KIEs), when combined with respective experimental KIE studies, showed that the C–H activation (TS1) is a process that controls the kinetics of the overall reaction for most of the substrates and it is also responsible for rate variation observed in the experiments done for various substrates.

The second important issue was how EbDH enforces such a high (*S*)-enantioselectivity of hydroxylation (Szaleniec, Dudzik, Pawul, & Kozik, 2009). From previous studies it was known that the enzyme has a relatively large active site which probably does not restrict rotation of the substrate. Therefore, the formation of the ES complex could not be the main factor responsible for such a high enantioselectivity. It was believed that the enzyme is imposing additional steric constrains that prevent the EbDH from activating pro(*R*) C—H bond and a subsequent hydroxylation leading to (*R*)-1-phenylethanol. However, to prove this one had to apply either a very big cluster model (more than 200 atoms) or use QM/MM methodology, in this case ONIOM(B3LYP:amber). The QM-part comprised only a minimum model, i.e., a first coordination sphere of molybdenum atom (oxo ligand, fragments of two pterin cofactors, and a fragment of Asp_{223} residue), fragment of His_{192} residue as well as the ethylbenzene (Fig. 3A). The (electronic structure) calculations were conducted on the same level of theory as in the case of QM-only studies, i.e., B3LYP/6-31g(d,p)/LANL2DZ effective core potential//triple-zeta basis set lacv3p** and AMBER force field for the MM region within the ONIOM approach implemented in Gaussian 09. The electronic energy was corrected with ZPE calculated for QM-part on the B3LYP/6-31g(d,p) level of theory. The whole model comprised of 15,907 atoms (total charge −3) and described the whole catalytic α subunit together with all crystal water molecules. Initially, all geometry optimizations were conducted using ME approach with the intention of calculating single-point corrections with EE for stationary points. However, it turned out that the choice of the embedding (mechanical vs. electronic) influenced the chemistry of the process, which in turn determined structure and energetics of the modeled system (see Table 1).

The choice of the embedding approach turned out to be critical to the determination of the preferential subtype of the pathway. Although both methods predicted homolytic C–H activation in TS1 (as depicted by the presence of spin density on the reactant—Table 1), the intermediate (I) differed dramatically both in geometry and electronic structure. In case of ME, a radical intermediate and Mo(V)-cofactor was observed, similarly to the QM-only model. On the other hand in the case of EE approach, a second electron transfer occurs after the TS1 is traversed, which leads to

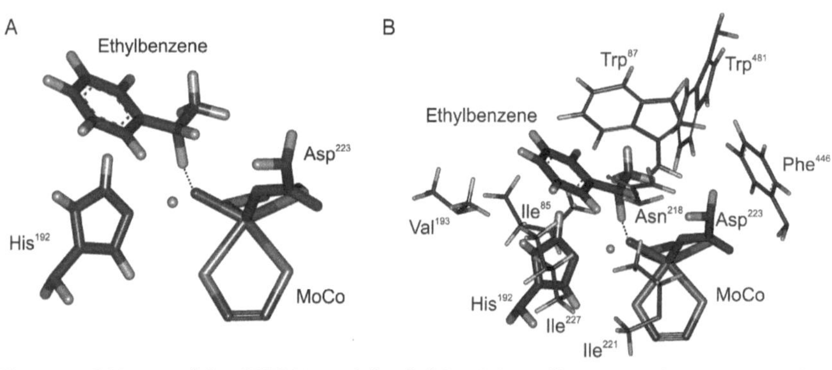

Figure 3 QM-part of the ONIOM models of EbDH (A) small QM1 used in ME/EE studies, (B) big QM2 used for single point EE corrections. Dotted line indicate pro(*S*) H-atom being activated by MoCo oxo ligand (Szaleniec et al., 2014).

Table 1 Geometric and Electronic Structure Differences Obtained in Mechanical-Embedding (ME) and Electronic-Embedding (EE) Studies on the EbDH Reaction Pathway with QM1/MM Approach

Geometric Parameter	*d* (C–H) (Å)	*d* (H–O) (Å)	$S_1S_2S_3S_4$ (°)	Total Spin Population/Charge of the Activated Substrate	Spin Population on C^{α}	Spin Population on Mo
TS1						
ME	1.30	1.26	−10.9	0.4/−0.06	0.29	−1.24
EE	1.30	1.36	−15.26	0.71/0.48	0.26	−1
Geometric Parameter	***d* (C–O)**	***d* (Mo–O)**	**$S_1S_2S_3S_4$ (°)**	**Spin Population of the Activated Substrate**	**Charge of the Activated Substrate**	**Spin Population on Mo**
I						
ME	2.94	2.23	−9	0.95	0.05	−1.05
EE	2.94	2.37	−10.56	0	0.96	0
Geometric Parameter	***d* (C–O)**	***d* (Mo–O)**	**$S_1S_2S_3S_4$ (°)**	**Spin Population of the Activated Substrate**	**Charge of the Activated Substrate**	**Spin Population on Mo**
TS2						
ME	**2.03**	**2.37**	−9.4	**0**	**0.38**	**0**
EE	1.92	2.53	−10.68	0	0.72	0
Geometric Parameter	**∡ benzyl ring–ethyl [°]**	***d* (Mo–O)**	**$S_1S_2S_3S_4$ (°)**	**Spin Population of Activated Substrate**	**Charge of the Activated Substrate**	**Spin Population on Mo**
P						
ME	20	4.41	−5.2	0	0.08	0
EE	35	3.23	−7.3	0	0.22	0

The *d* (C–H) and *d* (H–O) refer to distances between benzylic carbon and activated H atom and H atom and oxo ligand of molybdenum cofactor for the C–H activation process. The *d* (C–O) and *d* (Mo–O) represent distances between benzylic atom and O atom of the OH group as well as between O atom of the OH group and Mo atom in the OH-rebound process. $S_1S_2S_3S_4$ dihedral (see Fig. 1) describes the relative twist of pterin ligands coordinating Mo atom. Spin population on C^{α}—spin on benzylic carbon atom.
The values that were significantly different between ME and EE scheme were marked in bold.

a carbocation intermediate and to the reduction of the MoCo to Mo(IV). In the next step again models seemingly converged, as in TS2 both approaches predict carbocation intermediate reacting with OH^- yielding 1-phenylethanol bound to the cofactor by a H-bond. However, when the potential energy profiles are inspected (Fig. 4) one sees a completely different reaction profiles. The QM-only and ME to some extent are comparable (albeit there is a shift in a rate limit step from TS1 to TS2 when moving from the QM-only to the ME approach) and qualitatively yield the same mechanism. Surprisingly, a substantial stabilization occurs when protein electrostatic is introduced into the QM energy. This results with a system preference for the closed-shell carbocation versus radical open-shell state of the intermediate. Moreover, the bigger the charge separation the higher is the stabilization resulting in the biggest stabilization (~100 kJ/mol) for the carbocation species (I and TS2).

Of course not all systems will be this susceptible to change of the ME/EE approach, as in other instances it was shown that consistent results can be obtained with optimization at the ME-level followed by single-point correction done with the EE (Kawatsu, Lundberg, & Morokuma, 2011). However, one has to bear in mind that whenever there is a charge transfer step the corresponding part of the computed QM/MM PES can be susceptible to the level at which the electrostatic interactions are treated.

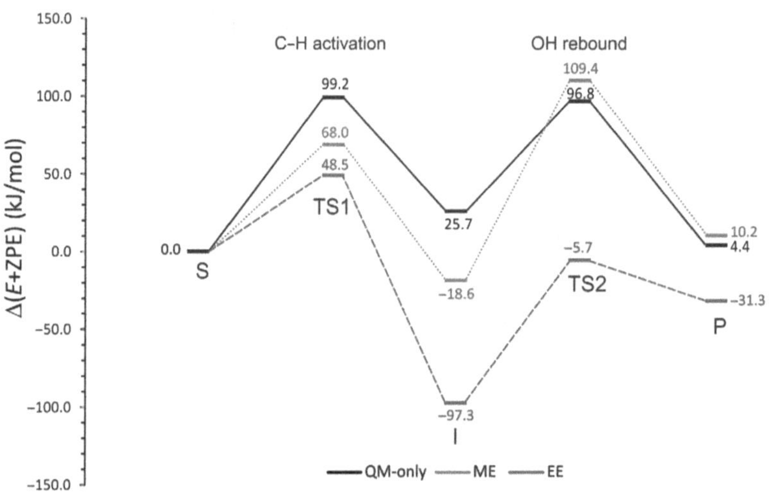

Figure 4 Comparison of the potential energy profiles obtained for QM-only (black solid line, $E+ZPE$ with CPCM solvent correction), QM/MM with ME (green dotted line (light gray in the print version)), and QM/MM with EE (red dashed line (dark gray in the print version)).

6. THE SIZE OF QM-PART AND THE OVER POLARIZATION EFFECT

As it was shown earlier the EE is prone to over stabilization of the charged species. In this specific case, the origin of such an artifact can be associated with the relatively small size of the QM-part. Ethylbenzene is directly exposed to MM-part of the active site, i.e., its electronic density is being polarized by point charges of the MM-part. Interestingly, in the close vicinity of ethylbenzene there are not so many charged residues which could do such a job and the most obvious candidate, protonated His_{192}, is treated with a QM method. In order to verify if the observed effect is true or artificial, one can expand the QM-part by including residues that could act as shields to polarization introduced by MM point charges (with their own electron density and by increasing a distance between the MM-part and ethylbenzene). In this particular case, the extension of QM1 part to QM2 part (shift from 52 atoms to 168 atoms treated with density functional theory (DFT)—Fig. 3) gave a model for which geometry optimization would be painstakingly slow. Therefore, only single-point corrections were calculated for geometries obtained with QM1:MM/EE model system.

The obtained energy correction was tremendous—an increase of TS1 energy by almost 40 kJ/mol, I energy by 100 kJ/mol but without any significant shift for TS2 (1 kJ/mol) and only slight increase of energy for the product (10 kJ/mol) (Fig. 5). The correction again yields a mechanistically different picture, as TS1 is again a rate-limiting step while TS2 proceeds apparently with a very small or no barrier at all. Interestingly, the introduction of the bigger QM-part did not change the system preference for the carbocation form of the intermediate. The approximate energy difference between radical and carbocation solution was estimated to be 70 kJ/mol (calculated as the difference between Mo(V) triplet and Mo(IV) singlet solution). As a result, the QM2:MM/EE calculations seem to provide the best available (up to date) description of the EbDH reactivity (Szaleniec et al., 2014).

This result is moreover consistent with experimental findings: (a) suggesting C–H activation is a rate-limiting step, (b) agreement of a calculated and predicted KIE, (c) agreement with experimentally established activation energy (~70 kJ/mol), and (d) with the overall tendency of molybdenum enzymes to catalyze a two-electron processes with, if possible, elimination of organic radical intermediates that could pose a danger for the enzyme. It should be also underlined that QM/MM approach allowed

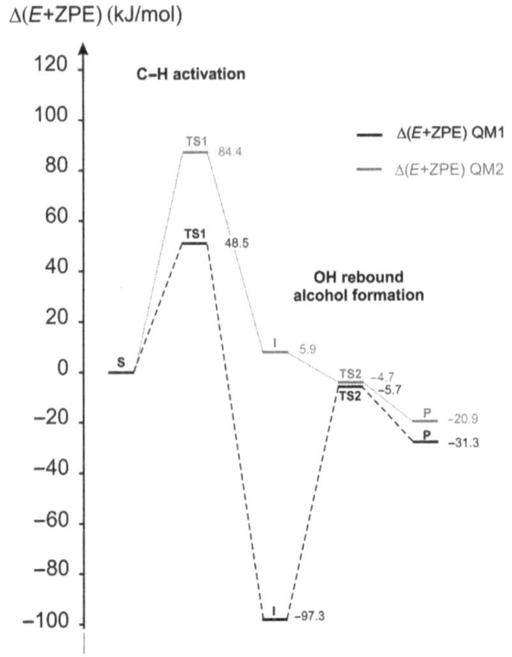

Figure 5 Energy profile obtained for different sizes of QM-part in QM/MM modeling with electronic embedding. QM1 (black dashed line; small model—Fig. 3A) and QM2 (blue dotted line (light gray in the print version); big model—Fig. 3B).

elucidation of the enzyme enantioselectivity pointing at C–H activation (TS1) as the stereo-controlling step. The activation of the pro(*S*) H atom turned out to proceed with a significantly lower barrier than the activation of the pro(*R*) H atom (Δ(*E*+ZPE) (pro(*R*)-pro(*S*)) equals 15.7 and 17 kJ/mol for QM1 and QM2, respectively) (Szaleniec et al., 2014).

Finally, it should be also mentioned that one needs to be careful when expanding the QM-part of the QM/MM or just the QM model. Charged residues, i.e., carboxyl groups, at the edge of the QM-treated region are very vulnerable to oxidation, especially if a highly reactive species is formed in the core of the active site. In such instances, one needs to include also H-bonding partners of such polar residues into the QM-region to prevent such, most likely artificial, electron transfer reaction.

7. HOW CAN A SPECIFIC ENZYMES ENVIRONMENT ALTER THE INTRINSIC NATURE OF A REACTION?

The cytochrome P450 family of enzyme are a diverse group of biocatalysts that are extremely well conserved throughout nature, with over

20,000 different sequences having already been characterized (de Montellano, 2004, 2010; Denisov, Makris, Sligar, & Schlichting, 2005; Groves, 2003; Kadish, Smith, & Guilard, 2010; Meunier, De Visser, & Shaik, 2004; Nelson, 2009; Sono, Roach, Coulter, & Dawson, 1996). Humans use a large number of P450 enzymes, concentrated in hepatocytes, to metabolize xenobiotics; with this class of enzyme also playing additional roles in processes like hormone synthesis (Guengerich, 2001; Munro, Girvan, & McLean, 2007; Posner & O'Neill, 2004). Many of the major adaptations of the P450's seem to be located in their binding pockets. For instance, the $P450_{BM3}$ enzymes have a relatively large binding pocket and hydroxylate lipids (Atkins & Sligar, 1987; Gelb, Heimbrook, Malkonen, & Sligar, 1982), of varying length, while $P450_{cam}$ chemoselectively hydroxylate their small camphor substrate and has a very tight binding pocket (Davydov, Hui Bon Hoa, & Peterson, 1999; Ruettinger, Wen, & Fulco, 1989). Many P450 isozymes have been shown to be very efficient in regioselective and/or stereospecific activation of their substrates, which could be very important in helping the pharmaceutical industry in future drug development (Grogan, 2011; O'Reilly, Köhler, Flitsch, & Turner, 2011).

Although cytochrome P450s are structurally and functionally diverse there are many properties that are conserved throughout this super family of enzymes. Their name is derived from a 450-nm absorption band seen once CO binds to the heme cofactor (Omura & Sato, 1962), which is linked to the protein via a thiolate linkage between the iron center and the sulfur of a cysteinate residue (Auclair, Moënne-Loccoz, & Ortiz de Montellano, 2001; Poulos, Finzel, Gunsalus, Wagner, & Kraut, 1985; Schlichting et al., 2000). Figure 6 illustrates the active site of $P450_{cam}$ with its heme cofactor covalently linked to cysteine 357 and the substrate positioned above its iron center. The active oxidant has been shown to be an iron(IV)-oxo heme cation radical species called Compound I (Cpd I) (Rittle & Green, 2010); in $P450_{cam}$ the activated oxygen would sit in the empty coordination position directly below the substrate (see Fig. 6). Computational modeling reveals that Cpd I contain close-lying spin states that can lead to a two-state reactivity profile (Kamachi & Yoshizawa, 2003; Shaik, Kumar, de Visser, Altun, & Thiel, 2005) with a bifurcation that could theoretically lead to different rearrangement patterns of species on the each reaction surface (de Visser, Ogliaro, & Shaik, 2001; Kumar, de Visser, Sharma, Cohen, & Shaik, 2004). It is known that a number of the P450 enzymes form both alcohol and olefin products (Forkert & Lee, 1997; Gunes et al., 2007; Lee, Parkinson, & Forkert, 1998; Loch, Potter, & Bachmann, 1995; Rettie, Boberg, Rettenmeier, & Baillie, 1988; Rettie, Rettenmeier, Howald, & Baillie, 1987; Rettie et al., 1995;

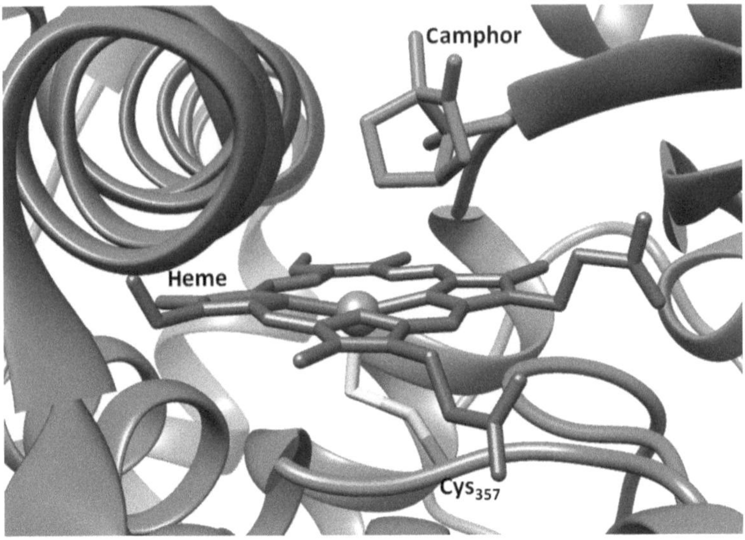

Figure 6 Extract of the active region of $P450_{cam}$ taken from the PDB: 2CPP structure.

Sadeque, Fisher, Korzekwa, Gonzalez, & Rettie, 1997; Wen, Wang, Kivisto, Neuvonen, & Backman, 2001) and the bifurcated reaction mechanisms of these biocatalysts is characterized in Fig. 7. Each possible potential energy surface starts with an initial hydrogen abstraction to form the radical intermediate. Subsequently, there is one possible product channel available for the OH rebound and a second leading to a further hydrogen abstraction, resulting in the formation of the alcohol and olefin products, respectively.

The large variation in the type of product produced by P450s must clearly be due to either adaptation of the catalyst or some innate properties of the substrate. Recently our group published a study that aimed to provide insight into the origin of the chemoselectivity of hydroxylation versus desaturation by using a combination of DFT, quantum mechanical/molecular mechanical (QM/MM), and valence bond (VB) modeling (Ji et al., 2015). Only by combining all of these methods were we able to quantify the effect that both the type of substrate and the active site of the enzyme played in the final determination of the product mix.

In our initial DFT calculations, we studied the possible reaction pathways highlighted in Fig. 7 using the substrates: valproic acid (VA), ethane (ET), and dihydroanthracene (DHA). We chose these substrates in order to hopefully provide evidence for a substrate-induced chemoselectivity. *In vitro* while VA can indeed lead to the formation of both hydroxylated

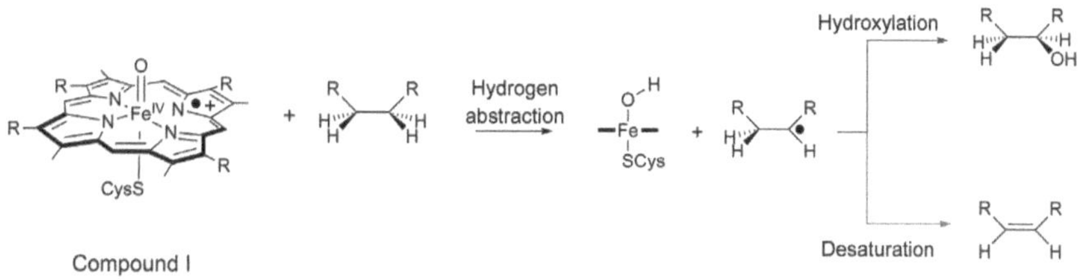

Figure 7 Possible hydroxylation and desaturation metabolism pathways catalyzed by P450 Cpd I.

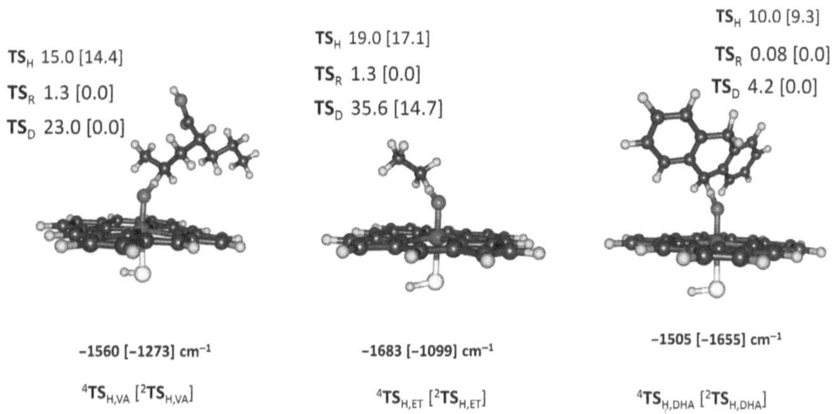

Figure 8 Hydrogen abstraction transition state geometries for VA, ET, and DHA by ^{4,2}Cpd I of P450 as calculated with DFT, both quartet and [doublet] values are shown. The imaginary frequencies (in wave numbers) are calculated using the UB3LYP hybrid functional. Free energies of activation of hydrogen atom abstraction TS_H, rebound TS_R, and desaturation TS_D barriers are also given.

and desaturated products, it is known that DHA is desaturated to form anthracene and that ET will only go through the alcohol product channel. Figure 8 depicts the initial TS complexes for each of the substrates and also gives ΔG values, at 298 K in kcal/mol, relative to each reactant complex. While these results do seem to provide a reasonable explanation for the exclusive hydroxylation of ET, things get a lot more complicated when we tried to assign a possible desaturation pathway to VA or DHA. For these substrates, on the doublet spin state surface, both the radical rebound and second hydrogen atom abstraction processes are barrierless, whereas high barriers for the desaturation pathways were found on the quartet surface. This seems to imply that while the quartet surface would lead in both cases to a purely alcohol product formation, there should be an even mix of products found on the low-spin surface. The fact that this is not the case *in vitro* means that product specificity is not completely kinetically controlled, since the inherent barriers calculated with the cluster models do not completely convey these specificities, but rather this is, at least in part, influenced by the protein environment. In order to take into account the shape and size of the substrate-binding pocket, we decided to model the $OleT_{JE}$ P450 isozyme, which converts two eicosanoic acid (EA) substrates to two olefins, $2CO_2$ and water, using compound 1 (Rude et al., 2011).

Using the crystal structure coordinates of the ES complex (Belcher et al., 2014), a well-established and -comprehensive QM/MM protocol (Kumar, Thiel, & de Visser, 2011; Quesne et al., 2014) was run along the whole reaction coordinate. The aim of these cluster model calculation was to establish the relative availability of the possible decarboxylation versus hydroxylation reaction pathways. Using DFT only methods, the two potential energy surfaces followed a familiar trend, to that seem with the desaturation versus hydroxylation pathways of the other substrates, with barrierless doublet reactivity after the formation of the radical intermediate and a dominant alcohol pathway on the quartet surface. However, when modeling the whole ES complex we find a very different story. Now the quartet surface is the dominant pathway and the $^4\mathbf{TS_D}$ is reduced from 17.8 kcal/mol, in the DFT model, to 5.1 kcal/mol, in the QM/MM. The $^4\mathbf{TS_R}$ barrier is increased from 4.9 to 6.6 kcal/mol. This means that now the barrier for the hydroxyl rebound is above that of the decarboxylaion transition, making olefin production more likely. We can explain this phenomenon, at least in part, by looking at the optimized structures of $^4\mathbf{I_R}$ and $^4\mathbf{TS_R}$, in Fig. 3.

Creating a DFT model that is solvated correctly, to replicate the environment present inside an enzyme, is one of the largest drawbacks of the cluster method. Usually an implicit solvent model is added to correct gas phase energies using a dielectric constant that estimate of a typical inner protein environment. However, clearly the hydrophilic nature of a proteins active region is very much protein dependant. Figure 9 indicates that a network of water molecules bridge the iron(IV)-hydroxo to the amine groups of Arg_{245}, in the radical intermediate $^4\mathbf{I_{EA}}$. This may be a mechanism that makes the rotation of the hydroxyl group harder and in effect destabilizing the OH-rebound pathway. It seems to take relatively little hydroxyl rotation to reach the decarboxylation TS $^4\mathbf{TS_{D,EA}}$, which could suggest one reason for the swapping of chemoselectivity. In this context, it is also interesting to remember that the active site of $P450_{cam}$, as shown in Fig. 6, has a very tight binding character. It may be interesting to speculate that one factor affecting the purely hydroxylated product, of this enzyme, may be its relatively

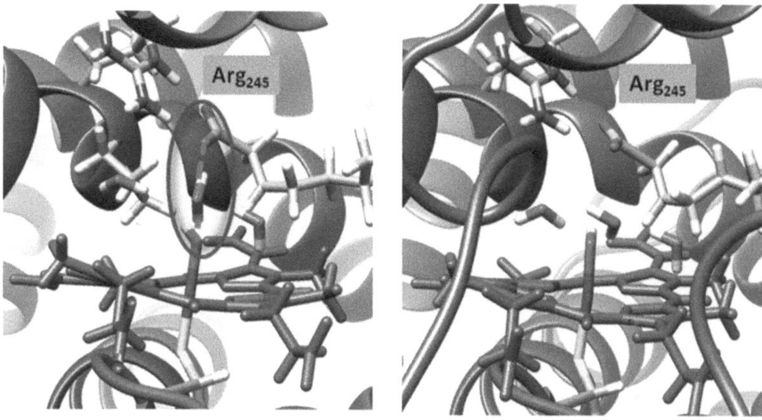

$^{4}I_{R,EA}$ $^{4}TS_{D,EA}$

Figure 9 QM/MM optimized geometries (CHARMM/B3LYP) of the radical intermediate ($^{4}I_{R,EA}$) and decarboxylation transition ($^{4}TS_{D,EA}$).

hydrophobic pocket, which might prevent forming analogous H-bonded networks engaging the OH ligand.

In order to assign a value to the innate favorability of each pathway, we subsequently set up DFT calculation, to enable VB analysis. This analysis has proved very useful in previous reaction analysis of substrate oxidation by both heme and nonheme metal–oxo/nitrido complexes (Kumar et al., 2014; Sainna et al., 2015). The innate reaction energy of the hydroxyl rebound G_{reb} can be represented by VB analysis, as shown in Fig. 10. The height of the activation barrier depends on the strength of the CO bond formed $[\Psi_P{}^* - \Psi_P]$ (BDE_{CO}) as well as on the excitation energy $[\Psi_I - \Psi_I{}^*]$ ($E_{ex,FeOH}$) whereby the iron is reduced by moving an electron from the hydroxyl, Eq. (3). The activation energy of the desaturation pathway, by contrast, is described as the difference in strength of the C—H bond that is broken (BDE_{CH}) and the O—H bond that is formed (BDE_{OH}). In this case as well as the excitation energy ($E_{ex,FeOH}$), from iron reduction, there is an additional energy associated with the two radicals, on adjacent carbon atoms, that have now formed a new π-bond with energy E_π, Eq. (4).

$$G_{reb} = BDE_{CO} + E_{ex,\,FeOH} \quad (3)$$

$$G_D = BDE_{OH} - BDE_{CH} + E_\pi + E_{ex,\,FeOH} \quad (4)$$

The desaturation pathway is thermodynamically favored if $\Delta E_{reb}{}^{\ddagger} > \Delta E_D{}^{\ddagger}$, and the barrier heights are expected to follow the trends in reaction energies,

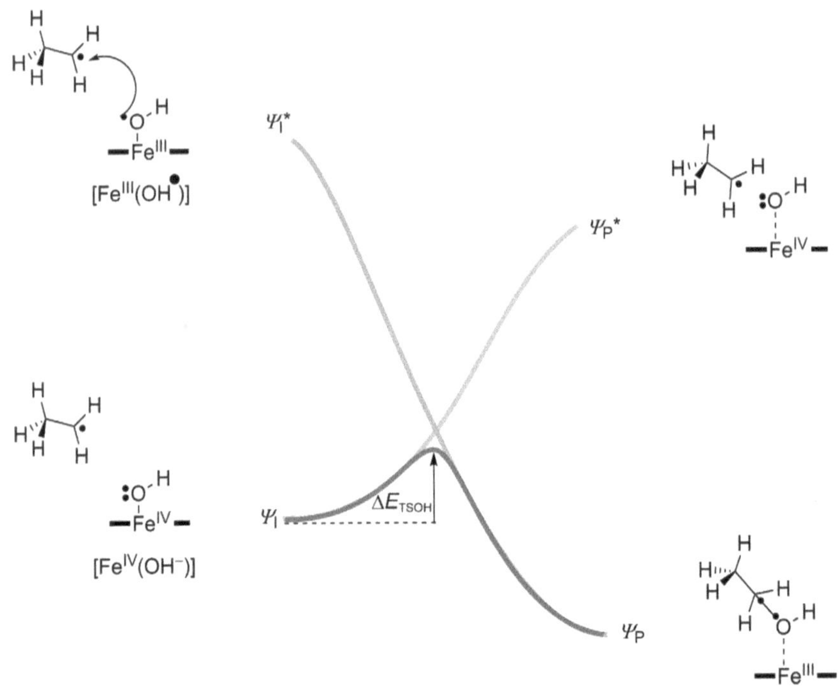

Figure 10 VB analysis for the hydroxylation of ethane. Ψ_I is the ground state intermediate with the oxidized cofactor, Ψ_I* is the excited intermediate with hydroxyl radical able to form a covalent bond with the substrate Ψ_P. The hydroxyl anion in Ψ_I is connected to a high-energy excited product Ψ_P*, where no covalent bonding is allowed. The point at which these two paths cross can give us the innate hydroxylation barrier of ethane ΔE_{TSOH}.

i.e., the more exergonic the reaction is the lower the expected barrier. The BDE_{OH} values for the iron(III)-hydroxo/iron(IV)–water complexes has been previously calculated for this catalyst as 70.8 kcal/mol (Kumar et al., 2009). The corresponding C—H bond strength of the substrate that is broken ranges from 36.7 for $C_2H_5^{\bullet}$ to 21.8 kcal/mol for cyclo-$C_6H_7^{\bullet}$. By contrast, the C—O bond strength of the alcohol product ranges from BDE_{CO} = 85.8 kcal/mol for ET to 59.7 kcal/mol for cyclohexadiene. This gives a difference in $G_{reb} - G_D$ of +16 kcal/mol for ET and −25 kcal/mol for cyclohexadiene. These values implicate a strong effect of the substrate on the regioselectivity of hydroxylation over desaturation. Ethanol forms a much stronger C—O bond (by 26 kcal/mol) than the corresponding cyclohexadiene which in turn forms a weaker C—H bond (by another 15 kcal/mol) during a second hydrogen atom abstraction, in the desaturation process. Since ET has a $G_{reb} - G_D$ of +16 kcal/mol, the perturbations effect of substrate binding in the enzymes active site would have to overcome 16 kcal/mol to swap the regioselectivity of hydroxylation to that of desaturation. We also calculated the $G_{reb} - G_D$ difference for EC, VA, and DHA to find values of 10, 13, and −4 kcal/mol, respectively. This implies that only the DHA substrate is innately driven toward desaturation and the formation of anthracene product. However, we know that there is some desaturation occurring in both EC and VA metabolism; and therefore, the P450 isozymes involved have a specific substrate-binding pocket that will change the balance in favor of desaturation rather than hydroxylation.

In summary, in this study the QM/MM method turned out to be very useful as we can clearly see that a protein environment can have a dramatic effect on the energy profile catalyzed by an enzyme system. One of the most important things to underline is that often no one method can tell the whole story and that only by removing variables, in other words by simplifying ones system, can we quantify the effects of the innate versus environments components of our energetic. Although of course in this case we could not have even started to understand these factors without first including the protein environment into our QM/MM model.

8. NOVEL MODIFICATIONS IN ENZYME STRUCTURES MAY PRODUCE REACTIVITY PATTERNS ONLY OBSERVED USING QM/MM

Another highly conserved group of the iron oxygenase is the AlkB DNA repair biocatalysts, which demethylate DNA/RNA bases damaged by intra- or extracellular chemicals (Falnes, Johansen, & Seeberg, 2002; O'Brien, 2006; Trewick, Henshaw, Hausinger, Lindahl, & Sedgwick, 2002; Yi, Yang, & He, 2009). Such covalent modifications of nucleic acids can have a dramatic effect on cellular function by preventing the formation of Watson–Crick base-pairs, eventually leading to a block on DNA replication (Ladwein & Jung, 2011; Mishina, Duguid, & He, 2006; Mishina & He, 2006). AlkB repair enzymes demethylate alkylated bases by catalyzing a reaction, which uses one molecule of alpha ketoglutarate (αKG) and another of molecular oxygen (Fu et al., 2010; Giri, Sun, Chen, Costa, & Maroney, 2011; Li et al., 2012; Yu & Hunt, 2009). A catalytic mechanism, based on the function of related enzymes, has been proposed. In this hypothesis, molecular oxygen is first activated by the iron center before a superoxo

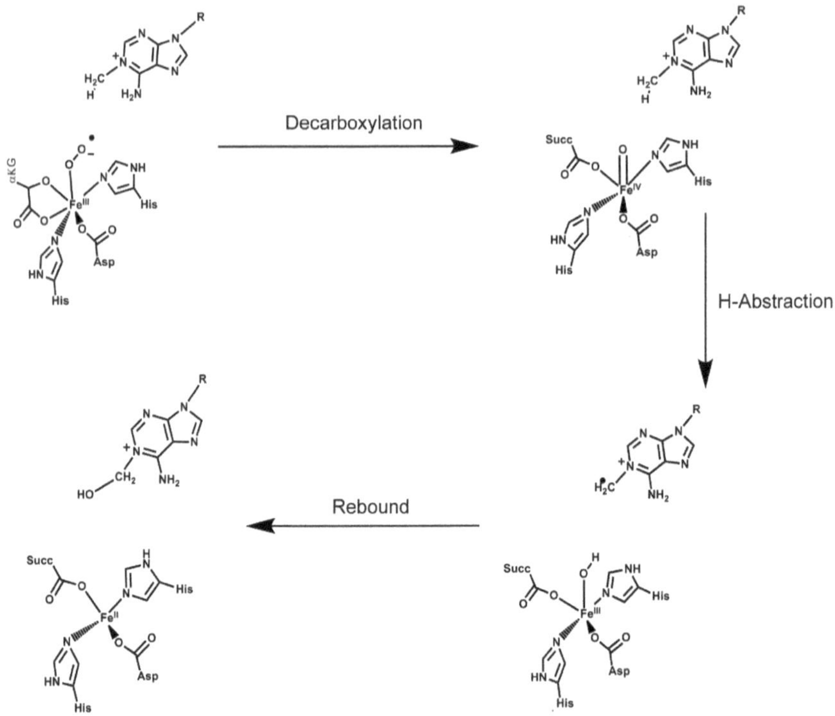

Figure 11 General mechanism of substrate hydroxylation catalyzed by nonheme iron(IV)-oxo enzymes.

attack on the beta carbon of αKG leads to decarboxylation and the formation of a ferryl-oxo species capable of abstracting hydrogen from the substrate before a hydroxyl-rebound step ends in the formation of the alcohol product (Fig. 11; Falnes et al., 2002; O'Brien, 2006; Trewick et al., 2002; Yi et al., 2009).

However, this mechanism seems to be complicated by crystallographic data, which at the same time as confirming a 2-His/1-Asp iron ligand system, common to many nonheme dioxygenases (Yu et al., 2006), also indicates that the oxygen-binding site is located not at the coordination site most proximal to the substrate (Fig. 12). Our group decided to investigate the reaction mechanism catalyzed by this enzyme using a combination of QM and QM/MM techniques (Quesne et al., 2014). When molecular oxygen was inserted into this sixth coordination position *trans* to His_{131} and the structure was minimized there was no sterochemical clashes of note observed. However, when molecular oxygen was positioned *trans* to His_{187}, in location with the distal oxygen atom in line with a hydrogen

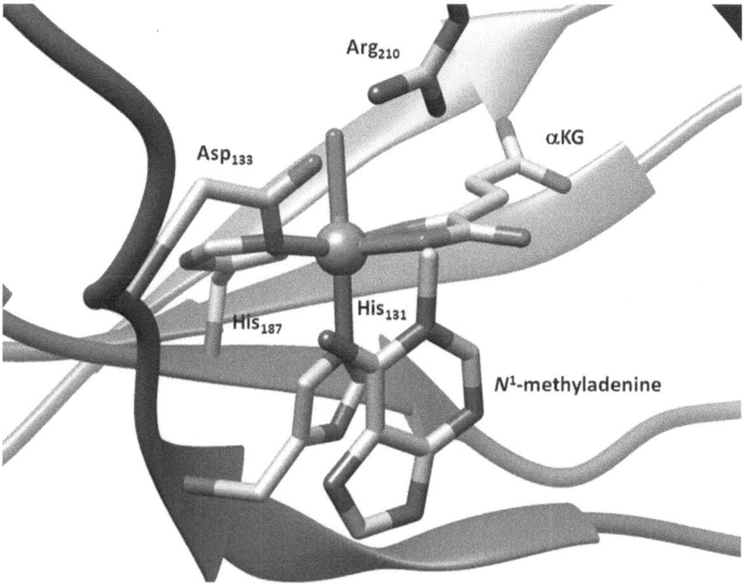

Figure 12 Extract taken from the PDB: 3I2O structure, showing the active site of the AlkB repair enzyme. An oxygen atom taken the place of a water molecule *trans* to His_{131}.

of the substrate, considerable stereochemical constraints were noted. In this position, the proximal oxygen atom of the iron(III)-superoxo is <1.7 Å away from atoms on the substrate and Asp_{133} and the distal oxygen atom is positions within 1.75 Å of the αKG molecule. We therefore decided to start our calculations with an iron(IV)-oxo structure located *trans* to His_{131} (**R**) before rotating the oxo trans to His_{187} (**R′**) as the first step of our reaction mechanism.

Computational studies on the reaction mechanism of the AlkB enzyme by the Gauld group found a rate determining hydrogen atom abstraction barrier of 20.9 kcal/mol for methylated adenine, using a small cluster model of the active region (Liu, Llano, & Gauld, 2009). This barrier is high for a nonheme iron(IV)-oxo catalyzed reaction and a barrier with a magnitude over 20 kcal/mol may indeed be indicative of a turnover rate that is too low to happen in nature. One common problem with QM-cluster models, as stated above, is that the unrealistic movement of the substrate during a reaction mechanism can lead to unphysically large barriers. Working with the hypothesis that the protein's local environment may act to lower the barrier found in the models of Gauld et al., we decided to use a combined QM/MM approach. Our methods have been carefully benchmarked and calibrated (Godfrey, Porro, & de Visser, 2008) and start from the 3I20 crystal structure coordinates, i.e., these that were also used to prepare Fig. 12 (Yu et al., 2006). The QM region contained the iron(IV)-oxo group, truncated His_{131} His_{187} and Asp_{133} residues as well as an acetate for the succinate (decarboxylated coenzyme) and the N^1-methyladenine part of the substrate. Hydrogen atoms were added using PROPKA and a molecular dynamics simulation was run for 800 ps on the solvated protein, using the Charmm software package (Brooks, Bruccoleri, Olafson, Swaminathan, & Karplus, 1983). Finally three snapshots were selected from the MD trajectory, at 300, 400, and 500 ps. The reaction energy profiles were calculated for all snapshots and all possible spin states using B3LYP, B3LYP* BP86, and M06 (Becke, 1993; Borowski, Noack, Radoń, Zych, & Siegbahn, 2010; Lee, Yang, & Parr, 1988; Reiher, Salomon, & Artur Hess, 2001) but in all cases gave the spin state ordering of quintet < triplet < singlet.

The isomerization barrier ($^5\mathbf{TS_I}$) from $^5\mathbf{R}$ to $^5\mathbf{R}'$ in the snapshot taken at 500 ps (Sn_{500}) was $\Delta E + ZPE = 9.0$ kcal/mol with an isomerization energy of $\Delta E + ZPE = -6.0$ kcal/mol. In $^5\mathbf{R}$ the σ^*_{x2-y2} is singly occupied and the σ^*_{z2} orbital is virtual, whereas the ordering is reversed for $^5\mathbf{R}'$ and the σ^*_{z2} orbital is singly occupied instead. In this orientation, the σ^*_{z2} orbital in $^5\mathbf{R}'$ has lesser antibonding interactions than the HOMO, i.e., σ^*_{x2-y2}, in $^5\mathbf{R}$ and consequently $^5\mathbf{R}'$ is more stable than $^5\mathbf{R}$. Previous work by our group (Borowski et al., 2010) on the nonheme iron halogenase SyrB2 showed an isomerization barrier of 13.4 kcal/mol, which is slightly higher in energy because in SyrB2 there is a halide as well as an oxygen rotation included in the barrier height. As $^5\mathbf{R}$ rotates to $^5\mathbf{R}'$, the spin density on the oxygen atom decreases from 0.75 to 0.69 resulting in a slightly less efficient oxidant. The LUMO orbital of $^5\mathbf{R}$ shows antibonding character along the Fe—O bond and very little involvement of other ligands, while the LUMO orbital in $^5\mathbf{R}'$ displays interactions with atoms on Asp_{133}, His_{187}, and succinate. One possible reason why the enzyme uses the less powerful oxidant for the rate-limiting hydrogen abstraction is to separate the dioxygen and the substrate-binding processes to avoid harmful side reactions. This was tested using DFT models where we found a hydrogen atom abstraction barrier of $\Delta E + ZPE = 32.8$ kcal/mol for the iron(III)-superoxo, compared to a value of $\Delta E + ZPE = 23.4$ kcal/mol for the iron(IV)-oxo. This seems to indicate that the iron(III)-superoxo is not an efficient oxidant of the substrate. However, even a small percentage of iron(III)-peroxo could be devastating for the enzyme.

Although there is still an argument to be had about why the enzyme needs to rotate the oxo group, we were more clear about how it stabilized the rotation. The oxo group forms a hydrogen bond with Arg_{210} that assists oxo rotation by shortening from 2.414 Å in ^{5}**R** to 1.888 Å in ^{5}**TS**$_I$ and then elongates to 2.470 Å in ^{5}**R**′. We then found two competing hydrogen abstraction pathways where ^{5}**R**′, either transferred an electron (from abstracted hydrogen atom) into the virtual σ^*_{x2-y2} orbital ($^5\sigma$-pathway) or transferred a beta electron to double occupy the π^*_{xz} orbital ($^5\pi$-pathway). We would usually expect that the $^5\sigma$-pathway would be dominant in the quintet PES of an nonheme iron(IV)-oxo catalyst, since the angle of substrate approach moves from approximately 180° in the σ-pathway to closer to 120° in the π-pathway and the latter is often stereochemically unfavorable (Borowski, Bassan, & Siegbahn, 2004; de Visser, 2006; Sahu et al., 2013; Wong et al., 2011; Ye & Neese, 2011). In the σ-pathway, an alpha electron is transferred to the σ^*_{x2-y2} orbital forming (^{5}**I**$_\sigma$) with $\pi^*_{xz}{}^{\uparrow}\ \pi^*_{yz}{}^{\uparrow}\ \pi^*_{xy}{}^{\uparrow}\ \sigma^*_{z2}{}^{\uparrow}\ \sigma^*_{x2-y2}{}^{\uparrow}\ \phi_{Sub}{}^{\downarrow}$ configuration, whereas the π-pathway transfers a beta electron into a singly occupied π orbital to give the intermediate (^{5}**I**$_\pi$) with configuration $\pi^*_{xz}{}^{2}\ \pi^*_{yz}{}^{\uparrow}\ \pi^*_{xy}{}^{\uparrow}\ \sigma^*_{z2}{}^{\uparrow}\ \sigma^*_{x2-y2}{}^{0}\ \phi_{Sub}{}^{\uparrow}$. In the ^{5}**TS**$_{H,\sigma}$ structure, a Fe–O–C angle of 141.4° is found compared to the smaller angle of 131.4° in ^{5}**TS**$_{H,\pi}$, making the π hydrogen abstraction channel more close to optimum then the σ. Figure 13 depicts the PES from ^{5}**R** to ^{5}**I** using Sn_{500} and UB3LYP theory where both ^{5}**TS**$_{H,\sigma}$ and ^{5}**TS**$_{H,\pi}$ have comparable energies: 18.2 versus 18.6 kcal/mol, which is well within the margin of error for this level of theory. We got numbers of

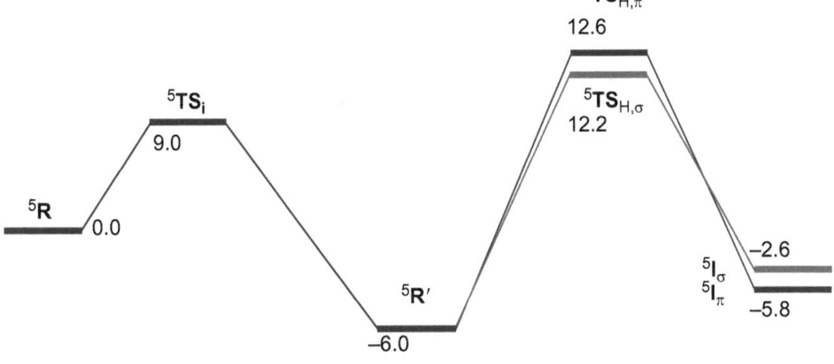

Figure 13 Potential energy surface from reactant to intermediate for Sn_{500}, at UB3LYP level of theory.

15.0 versus 15.2 kcal/mol at UB3LYP* level of theory and found $^{5}\mathbf{TS}_{H,\pi}$ of 13.0 and 15.1 kcal/mol for Sn_{300} and Sn_{400}, respectively. Since DFT models indicated barriers of over 20 kcal/mol, it is apparent that the QM/MM approach gave much more favorable barrier heights.

In both $^{5}\mathbf{I}_{\sigma}$ and $^{5}\mathbf{I}_{\pi}$, the hydroxyl group forms hydrogen bonds with the carboxylate group of Asp_{133} leading to a significant rebound barrier of >13.1 kcal/mol. This large rebound barrier contrasts with 0.6 kcal/mol found with DFT model calculations (Liu et al., 2009). This is due in part to the strong hydrogen-bonding interactions seen in the QM/MM model stabilize the radical intermediates and raise the OH transfer barrier. We also calculated hydrogen abstraction barriers for N^1-methylguanine (27.9 kcal/mol), N^1-methyladenine (23.4 kcal/mol), N^3-methylcytosine (6.5 kcal/mol), and N^3-methylthymine (10.9 kcal/mol) using DFT cluster model complexes. Finally, we attempted to correlates these barriers with the relative strengths of the C—H bond that is broken for each substrate. Each BDE_{CH} was calculated from the energy difference between the full substrate and the sum of the substrate without a hydrogen atom and the energy of a hydrogen atom and found that the BDE_{CH} values fall within 5 kcal/mol of each other. Inspection of the optimized geometries seems to suggest that using DFT models the substrates are able to reorient unrealistically and form multiple hydrogen-bonding interactions with oxygen atoms on the succinate, Asp_{133} and the iron-oxo group. Since we can assume that these hydrogen-bonding interactions, which are artifacts of the cluster model endowed with too much freedom to optimize, can reduce catalytic efficiency, by stabilizing the reactant, these model complexes may not be a good representative of the actual enzyme. Hence, in this particular case study, the QM/MM approach offered a far more realistic description of the ES interactions than the cluster QM model. Taking into account protein residues surrounding the active site and forming the substrate pocket prevented the system from falling into spurious minima featuring H-bonding interactions that are not feasible in the real system.

9. RING HYDROXYLATION AND REARRANGEMENT BY 4-HYDROXYPHENYLPYRUVATE DIOXYGENASE

Another enzyme utilizing ferryl-oxo species to elicit substrate oxidation is 4-hydroxyphenylpyruvate dioxygenase (HPPD), which is mechanistically related to α-ketoglutarate-dependent enzymes as it also uses an α-ketoacid moiety (of the substrate) to generate (with the evolvement of O_2) the key

$Fe^{4+}=O$ intermediate. The reaction mechanism of the ring oxidation and the coupled rearrangement has been elusive as, until very recently, experimental and computational studies have not been able to work out a consensus reaction scheme. A recent ONIOM study has delivered some new insights into the course of the reaction allowing us to put forth a mechanistic hypothesis consistent with both experimental and computational results (Wojcik et al., 2014).

More specifically, experimentally determined secondary KIE, measured with normal and ring-deuterated substrates, is markedly inverse, which was originally interpreted in favor of the synchronous epoxidation of the ring (**a** → **c** in Fig. 14), as such a reaction would involve a hybridization change on one of the ring's carbon atoms and this, in turn, would explain the size of KIE. On the other hand, results of a previous cluster-based QM study argued against such a synchronous oxygen atom insertion and suggested a step-wise radical mechanism for side chain rearrangement (**a** → **b** → **e** → **d**). Since the time of the original QM-only study, a crystal structure has been solved for an E-product complex of an enzyme closely related to HPPD—hydroxymandelate synthase and this enabled us to model, with the help of classical molecular dynamics simulations, the structure of HPPD-Fe^{4+} $=$O-HPA—the key reaction intermediate. Notably, as of the now no structural information is available for HPPD-(its substrate or product)

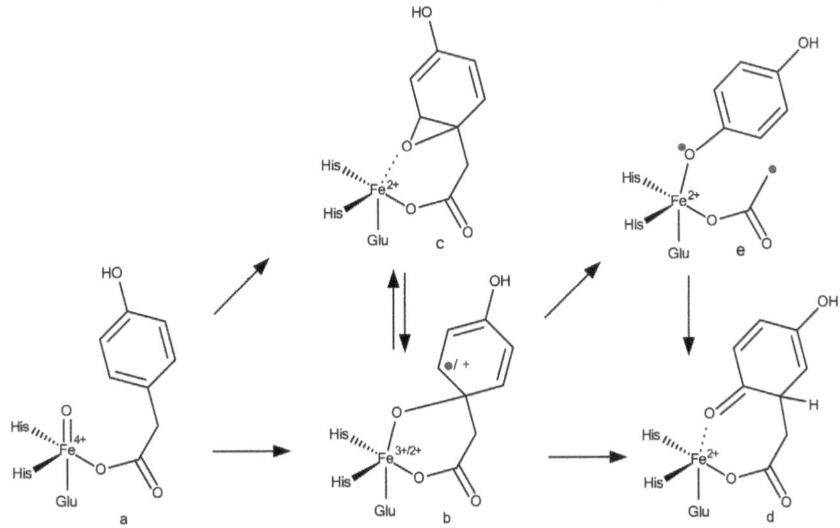

Figure 14 Reaction mechanisms suggested for ring hydroxylation and side chain migration steps of the catalytic reaction of HPPD.

complexes. With such a structure at hand, an ONIOM(B3LYP:amber) model was constructed where the first-shell ligands of Fe and second-shell residues H-bonding to the first shell were included in the QM-part of the model, whereas all residues with atoms within 15 Å sphere from Fe were allowed to relax while the others were kept in their positions, derived from the MD simulation (see Fig. 15).

An important feature of the recent ONIOM model with respect to the previous QM-one is that it accounts for the second-shell residues forming H-bonds with the first-shell ligands (Admittedly, an appropriate QM model including these second-shell residues would also do well in this respect). These H-bonding contacts turned out to be pivotal for realistic description of the reaction mechanism as they tune the redox potential of the iron. In consequence, the preferred mechanism goes through the radical **b**, in which an electron transfer from the ring to iron is facilitated by the presence of the aforementioned H-bonds and this process takes place before the TS for the side chain migration is reached. Hence, the migration is a single-step

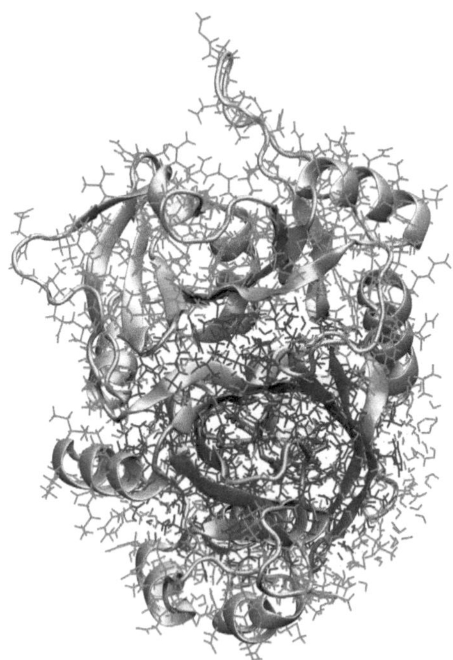

Figure 15 ONIOM(B3LYP:amber) model of HPPD. In violet (dark gray in the print version) sticks Fe and its ligands, in blue (black in the print version) the 15 Å sphere allowed to optimize, and in red (gray in the print version) atoms with fixed coordinates.

concerted process. The computed reaction energy profile, obtained at the ONIOM(B3LYP:amber)-EE level, reveals the barriers for electrophilic attack on the ring (TS1) and the subsequent side chain migration (TS2) amount to 18.0 and 14.7 kcal/mol, respectively (corresponding values computed with ONIOM-ME are: 19.2 and 15.3 kcal/mol). Importantly, KIE values computed with the ONIOM-ME for TS1 (**a**→**b**) and TS2 (**b**→**c**), which are 0.94 and 0.83, respectively, offered an explanation for the experimentally observed value (0.84), provided one takes into account the inherent energy uncertainty of the B3LYP method (±5 kcal/mol). Within this margin of error, TS2 may be actually the barrier limiting the decay of the $Fe^{4+}{=}O$ species, which was the process that was followed experimentally with isotopic substitution. Hence, a mechanism that is in agreement with the computational results and explains the experimental KIE data has been proposed. The direct oxygen atom insertion into the ring, that was originally invoked to explain the size of KIE, was found, similarly as in the previous QM-cluster study, very unlikely. Importantly, the ONIOM model turned out to be superior to the QM-only cluster one which consisted of the same residues as the QM-part of the ONIOM model. Even though the same mechanism has been found with the cluster model, KIE values turned out to be fairly sensitive to the details how the geometrical constraints on the QM-region are imposed. More specifically, KIE computed for TS2 with the cluster model amounts to 1.01 while the value obtained with ONIOM-ME is 0.83, which is evidently a better match for the experimental value of 0.84. Apparently, the QM/MM model, where the constrained atoms are much further away of the reaction center, is more robust in KIE computations than the QM-cluster model, where atoms at the edge of the model needs to be constrained to prevent unrealistic geometry changes during optimization.

10. CONCLUSIONS

In this short review, we have surveyed several case studies that highlight many of the advantages and potential pitfalls of the exciting and revolutionary QM/MM techniques, which are now available to researchers. We have examined how these techniques use explicit solvent molecules and a flexible protein environment to more accurately replicate the hydrogen bonding networks and substrate orientations of a protein *in vitro*. We have also considered the possible drawbacks of using various electronic-embedding protocols and looked at the possible future use of cluster models to complement QM/MM. There is no doubt that as these techniques develop ever more exciting frontiers of computational study there will be a bright future for QM/MM.

ACKNOWLEDGMENTS

M.S. acknowledges support from a computational grants MNiSW/SGI4700/PAN/038/2007, MNiSW/IBM_BC_HS21/PAN/038/2007, and "Biotransformations for pharmaceutical and cosmetics industry" project No. POIG.01.03.01-00-158/09-00, partially financed by the European Union within the European Regional Development Fund. T.B. acknowledges support from the National Science Centre, Poland (2011/03/B/NZ1/04999). M.Q. acknowledges support from the Marian Smoluchowski Krakow Research Consortium—a Leading National Research Centre KNOW grant. This research was also supported in part by PL-Grid Infrastructure.

REFERENCES

Atkins, W. M., & Sligar, S. G. (1987). Metabolic switching in cyctochrome P-450cam: Deuterium isotope effects on regiospecificity and the monooxygenase/oxidase ratio. *Journal of the American Chemical Society, 109*(12), 3754–3760.

Auclair, K., Moënne-Loccoz, P., & Ortiz de Montellano, P. R. (2001). Roles of the proximal heme thiolate ligand in cytochrome P450cam. *Journal of the American Chemical Society, 123*(21), 4877–4885.

Becke, A. D. (1993). Density-functional thermochemistry. III. The role of exact exchange. *Chemical Physics, 98*, 5648–5652.

Belcher, J., McLean, K. J., Matthews, S., Woodward, L. S., Fisher, K., Rigby, S. E. J., et al. (2014). Structure and biochemical properties of the alkene producing cytochrome P450 OleTJE (CYP152L1) from the Jeotgalicoccus sp. 8456 bacterium. *Journal of Biological Chemistry, 289*(10), 6535–6550.

Blomberg, M. R. A., Borowski, T., Himo, F., Liao, R. Z., & Siegbahn, P. E. M. (2014). Quantum chemical studies of mechanisms for metalloenzymes. *Chemical Reviews, 114*(7), 3601–3658.

Borowski, T., Bassan, A., & Siegbahn, P. E. (2004). Mechanism of dioxygen activation in 2-oxoglutarate-dependent enzymes: A hybrid DFT study. *Chemistry, 10*(4), 1031–1041.

Borowski, T., Noack, H., Radoń, M., Zych, K., & Siegbahn, P. E. (2010). Mechanism of selective halogenation by SyrB2: A computational study. *Journal of the American Chemical Society, 132*(37), 12887–12898.

Brooks, B. R., Bruccoleri, R. E., Olafson, B. D., Swaminathan, S., & Karplus, M. (1983). CHARMM: A program for macromolecular energy, minimization, and dynamics calculations. *Journal of Computational Chemistry, 4*(2), 187–217.

Caprasecca, S., Jurinovich, S., Viani, L., Curutchet, C., & Mennucci, B. (2014). Geometry optimization in polarizable QM/MM models: The induced dipole formulation. *Journal of Chemical Theory and Computation, 10*(4), 1588–1598.

Chiang, Y. R., Ismail, W., Heintz, D., Schaeffer, C., Van Dorsselaer, A., & Fuchs, G. (2008). Study of anoxic and oxic cholesterol metabolism by sterolibacterium denitrificans. *Journal of Bacteriology, 190*(3), 905–914.

Davydov, D. R., Hui Bon Hoa, G., & Peterson, J. A. (1999). Dynamics of protein-bound water in the heme domain of P450BM3 studied by high-pressure spectroscopy: Comparison with P450cam and P450 2B4. *Biochemistry, 38*(2), 751–761.

de Montellano, P. R. O. (2004). *Cytochrome P450: Structure, mechanism, and biochemistry.* New York, Boston, Dordrecht, London, Moscow: Kluwer Acadennic/Plenum Publishers.

de Montellano, P. R. O. (2010). Hydrocarbon hydroxylation by cytochrome P450 enzymes. *Chemical Reviews*, *110*(2), 932.

Denisov, I. G., Makris, T. M., Sligar, S. G., & Schlichting, I. (2005). Structure and chemistry of cytochrome P450. *Chemical Reviews*, *105*(6), 2253–2277.

Dermer, J., & Fuchs, G. (2012). Molybdoenzyme that catalyzes the anaerobic hydroxylation of a tertiary carbon atom in the side chain of cholesterol. *Journal of Biological Chemistry*, *287*(44), 36905–36916.

de Visser, S. P. (2006). Propene activation by the oxo-iron active species of taurine/α-ketoglutarate dioxygenase (TauD) enzyme. How does the catalysis compare to heme-enzymes? *Journal of the American Chemical Society*, *128*, 9813–9824.

de Visser, S. P., Ogliaro, F., & Shaik, S. (2001). How does ethene inactivate cytochrome P450 En route to its epoxidation? A density functional study the research is supported in part by the ISF and in part by the ministry of science, culture, and sport. F.O. Acknowledges the European union for a Marie curie fellowship. *Angewandte Chemie*, *40*(15), 2871–2874.

de Visser, S. P., Quesne, M. G., Martin, B., Comba, P., & Ryde, U. (2014). Computational modelling of oxygenation processes in enzymes and biomimetic model complexes. *Chemical Communications*, *50*(3), 262–282.

Dudzik, A., Kozik, B., Tataruch, M., Wójcik, A., Knack, D., Borowski, T., et al. (2013). The reaction mechanism of chiral hydroxylation of p-OH and p-NH2 substituted compounds by ethylbenzene dehydrogenase. *Canadian Journal of Chemistry*, *91*, 775–786.

Dudzik, A., Snoch, W., Borowiecki, P., Opalinska-Piskorz, J., Witko, M., Heider, J., et al. (2014). Asymmetric reduction of ketones and beta-keto esters by (S)-1-phenylethanol dehydrogenase from denitrifying bacterium Aromatoleum aromaticum. *Applied Microbiology and Biotechnology*, *99*, 5055–5069.

Falnes, P. O., Johansen, R. F., & Seeberg, E. (2002). AlkB-mediated oxidative demethylation reverses DNA damage in Escherichia coli. *Nature*, *419*(6903), 178–182.

Field, M. J., Bash, P. A., & Karplus, M. (1990). A combined quantum-mechanical and molecular mechanical potential for molecular-dynamics simulations. *Journal of Computational Chemistry*, *11*(6), 700–733.

Forkert, P.-G., & Lee, R. P. (1997). Metabolism of ethyl carbamate by pulmonary cytochrome P450 and carboxylesterase isozymes: Involvement of CYP2E1 and hydrolase a. *Toxicology and Applied Pharmacology*, *146*(2), 245–254.

Fu, Y., Dai, Q., Zhang, W., Ren, J., Pan, T., & He, C. (2010). The AlkB domain of mammalian ABH8 catalyzes hydroxylation of 5-methoxycarbonylmethyluridine at the wobble position of tRNA. *Angewandte Chemie*, *122*(47), 9069–9072.

Fuchs, G., Boll, M., & Heider, J. (2011). Microbial degradation of aromatic compounds—From one strategy to four. *Nature Reviews. Microbiology*, *9*(11), 803–816.

Gelb, M. H., Heimbrook, D. C., Malkonen, P., & Sligar, S. G. (1982). Stereochemistry and deuterium isotope effects in camphor hydroxylation by the cytochrome P450cam monooxygenase system. *Biochemistry*, *21*(2), 370–377.

George, G. N., Hilton, J., Temple, C., Prince, R. C., & Rajagopalan, K. V. (1999). Structure of the molybdenum site of dimethyl sulfoxide reductase. *Journal of the American Chemical Society*, *121*(6), 1256–1266.

Georgieva, P., & Himo, F. (2010). Quantum chemical modeling of enzymatic reactions: The case of histone lysine methyltransferase. *Journal of Computational Chemistry*, *31*(8), 1707–1714.

Giri, N. C., Sun, H., Chen, H., Costa, M., & Maroney, M. J. (2011). X-ray absorption spectroscopy structural investigation of early intermediates in the mechanism of DNA repair by human ABH2. *Biochemistry*, *50*(22), 5067–5076.

Godfrey, E., Porro, C. S., & de Visser, S. P. (2008). Comparative quantum mechanics/molecular mechanics (QM/MM) and density functional theory calculations on the oxo-iron species of taurine/alpha-ketoglutarate dioxygenase. *Journal of Physical Chemistry A, 112*(11), 2464–2468.

Grogan, G. (2011). Cytochromes P450: Exploiting diversity and enabling application as biocatalysts. *Current Opinion in Chemical Biology, 15*(2), 241–248.

Groves, J. T. (2003). The bioinorganic chemistry of iron in oxygenases and supramolecular assemblies. *Proceedings of the National Academy of Sciences, 100*(7), 3569–3574.

Guengerich, F. P. (2001). Common and uncommon cytochrome P450 reactions related to metabolism and chemical toxicity. *Chemical Research in Toxicology, 14*(6), 611–650.

Gunes, A., Bilir, E., Zengil, H., Babaoglu, M. O., Bozkurt, A., & Yasar, U. (2007). Inhibitory effect of valproic acid on cytochrome P450 2C9 activity in epilepsy patients. *Basic & Clinical Pharmacology & Toxicology, 100*(6), 383–386.

Heider, J. (2007). Adding handles to unhandy substrates: Anaerobic hydrocarbon activation mechanisms. *Current Opinion in Chemical Biology, 11*(2), 188–194.

Hille, R. (1996). The mononuclear molybdenum enzymes. *Chemical Reviews, 96*(7), 2757–2816.

Hoffken, H. W., Duong, M., Friedrich, T., Breuer, M., Hauer, B., Reinhardt, R., et al. (2006). Crystal structure and enzyme kinetics of the (S)-specific 1-phenylethanol dehydrogenase of the denitrifying bacterium strain EbN1. *Biochemistry, 45*(1), 82–93.

Hu, L. H., Eliasson, J., Heimdal, J., & Ryde, U. (2009). Do quantum mechanical energies calculated for small models of protein-active sites converge? *Journal of Physical Chemistry A, 113*(43), 11793–11800.

Hu, L., Söderhjelm, P. R., & Ryde, U. (2011). On the convergence of QM/MM energies. *Journal of Chemical Theory and Computation,* 7(3), 761–777.

Ji, L., Faponle, A. S., Quesne, M. G., Sainna, M. A., Zhang, J., Franke, A., et al. (2015). Drug metabolism by cytochrome P450 enzymes: What distinguishes the pathways leading to substrate hydroxylation over desaturation? *Chemistry—A European Journal, 21*, 9083–9092.

Johnson, H. A., Pelletier, D. A., & Spormann, A. M. (2001). Isolation and characterization of anaerobic ethylbenzene dehydrogenase, a novel Mo-Fe-S enzyme. *Journal of Bacteriology, 183*(15), 4536–4542.

Jormakka, M., Richardson, D., Byrne, B., & Iwata, S. (2004). Architecture of NarGH reveals a structural classification of Mo-bisMGD enzymes. *Structure, 12*(1), 95–104.

Kadish, K. M., Smith, K. M., & Guilard, R. (2010). *Handbook of porphyrin science.* World Scientific.

Kamachi, T., & Yoshizawa, K. (2003). A theoretical study on the mechanism of camphor hydroxylation by compound I of cytochrome P450. *Journal of the American Chemical Society, 125*(15), 4652–4661.

Kawatsu, T., Lundberg, M., & Morokuma, K. (2011). Protein free energy corrections in ONIOM QM:MM modeling: A case study for isopenicillin N synthase (IPNS). *Journal of Chemical Theory and Computation,* 7(2), 390–401.

Kloer, D. P., Hagel, C., Heider, J., & Schulz, G. E. (2006). Crystal structure of ethylbenzene dehydrogenase from aromatoleum aromaticum. *Structure, 14*(9), 1377–1388.

Kniemeyer, O., & Heider, J. (2001). Ethylbenzene dehydrogenase, a novel hydrocarbon-oxidizing molybdenum/iron-sulfur/heme enzyme. *The Journal of Biological Chemistry, 276*(24), 21381–21386.

Kumar, D., de Visser, S. P., Sharma, P. K., Cohen, S., & Shaik, S. (2004). Radical clock substrates, their C-H hydroxylation mechanism by cytochrome P450, and other reactivity patterns: What does theory reveal about the clocks' behavior? *Journal of the American Chemical Society, 126*(6), 1907–1920.

Kumar, S., Faponle, A. S., Barman, P., Vardhaman, A. K., Sastri, C. V., Kumar, D., et al. (2014). Long-range electron transfer triggers mechanistic differences between iron

(IV)-Oxo and iron (IV)-imido oxidants. *Journal of the American Chemical Society*, *136*(49), 17102–17115.

Kumar, D., Karamzadeh, B., Sastry, G. N., & de Visser, S. P. (2010). What factors influence the rate constant of substrate epoxidation by compound I of cytochrome P450 and analogous iron(IV)-Oxo oxidants? *Journal of the American Chemical Society*, *132*(22), 7656–7667.

Kumar, D., Tahsini, L., Visser, S. P. D., Kang, H. Y., Kim, S. J., & Nam, W. (2009). Effect of porphyrin ligands on the regioselective dehydrogenation versus epoxidation of olefins by oxoiron (IV) mimics of cytochrome P450. *The Journal of Physical Chemistry A*, *113*(43), 11713–11722.

Kumar, D., Thiel, W., & de Visser, S. P. (2011). Theoretical study on the mechanism of the oxygen activation process in cysteine dioxygenase enzymes. *Journal of the American Chemical Society*, *133*(11), 3869–3882.

Ladwein, K. I., & Jung, M. (2011). Oxidized cytosine metabolites offer a fresh perspective for active DNA demethylation. *Angewandte Chemie, International Edition*, *50*(51), 12143–12145.

Lee, R. P., Parkinson, A., & Forkert, P. G. (1998). Isozyme-selective metabolism of ethyl carbamate by cytochrome P450 (CYP2E1) and carboxylesterase (hydrolase a) enzymes in murine liver microsomes. *Drug Metabolism and Disposition: The Biological Fate of Chemicals*, *26*(1), 60–65.

Lee, C., Yang, W., & Parr, R. G. (1988). Development of the colle-salvetti correlation-energy formula into a functional of the electron density. *Physical Review B*, *37*, 785–789.

Li, D., Delaney, J. C., Page, C. M., Yang, X., Chen, A. S., Wong, C., et al. (2012). Exocyclic carbons adjacent to the N 6 of adenine are targets for oxidation by the Escherichia coli adaptive response protein AlkB. *Journal of the American Chemical Society*, *134*(21), 8896–8901.

Liu, H., Llano, J., & Gauld, J. W. (2009). A DFT study of nucleobase dealkylation by the DNA repair enzyme AlkB. *The Journal of Physical Chemistry B*, *113*(14), 4887–4898.

Loch, J. M., Potter, J., & Bachmann, K. A. (1995). The influence of anesthetic agents on rat hepatic cytochromes P450 in vivo. *Pharmacology*, *50*(3), 146–153.

Lonsdale, R., Ranaghan, K. E., & Mulholland, A. J. (2010). Computational enzymology. *Chemical Communications*, *46*(14), 2354–2372.

Lundberg, M., & Morokuma, K. (2007). Protein environment facilitates O-2 binding in non-heme iron enzyme. An insight from ONIOM calculations on isopenicillin N synthase (IPNS). *Journal of Physical Chemistry B*, *111*(31), 9380–9389.

Maseras, F., & Morokuma, K. (1995). Imomm—A new integrated Ab-initio plus molecular mechanics geometry optimization scheme of equilibrium structures and transition-states. *Journal of Computational Chemistry*, *16*(9), 1170–1179.

Meunier, B., De Visser, S. P., & Shaik, S. (2004). Mechanism of oxidation reactions catalyzed by cytochrome P450 enzymes. *Chemical Reviews-Columbus*, *104*(9), 3947–3980.

Mishina, Y., Duguid, E. M., & He, C. (2006). Direct reversal of DNA alkylation damage. *Chemical Reviews*, *106*(2), 215–232.

Mishina, Y., & He, C. (2006). Oxidative dealkylation DNA repair mediated by the mononuclear non-heme iron AlkB proteins. *Journal of Inorganic Biochemistry*, *100*(4), 670–678.

Munro, A. W., Girvan, H. M., & McLean, K. J. (2007). Variations on a (t) heme—Novel mechanisms, redox partners and catalytic functions in the cytochrome P450 superfamily. *Natural Product Reports*, *24*(3), 585–609.

Nelson, D. R. (2009). The cytochrome p450 homepage. *Human Genomics*, *4*(1), 59.

O'Brien, P. J. (2006). Catalytic promiscuity and the divergent evolution of DNA repair enzymes. *Chemical Reviews*, *106*(2), 720–752.

Omura, T., & Sato, R. (1962). A new cytochrome in liver microsomes. *Journal of Biological Chemistry*, *237*(4), C1375–PC1376.

O'Reilly, E., Köhler, V., Flitsch, S. L., & Turner, N. J. (2011). Cytochromes P450 as useful biocatalysts: Addressing the limitations. *Chemical Communications, 47*(9), 2490–2501.
Polyak, I., Reetz, M. T., & Thiel, W. (2012). Quantum mechanical/molecular mechanical study on the mechanism of the enzymatic Baeyer–Villiger reaction. *Journal of the American Chemical Society, 134*(5), 2732–2741.
Polyak, I., Reetz, M. T., & Thiel, W. (2013). Quantum mechanical/molecular mechanical study on the enantioselectivity of the enzymatic Baeyer–Villiger reaction of 4-hydroxycyclohexanone. *The Journal of Physical Chemistry B, 117*(17), 4993–5001.
Posner, G. H., & O'Neill, P. M. (2004). Knowledge of the proposed chemical mechanism of action and cytochrome P450 metabolism of antimalarial trioxanes like artemisinin allows rational design of new antimalarial peroxides. *Accounts of Chemical Research, 37*(6), 397–404.
Poulos, T. L., Finzel, B. C., Gunsalus, I. C., Wagner, G. C., & Kraut, J. (1985). The 2.6-a crystal structure of pseudomonas putida cytochrome P-450. *Journal of Biological Chemistry, 260*(30), 16122–16130.
Quesne, M. G., Latifi, R., Gonzalez-Ovalle, L. E., Kumar, D., & de Visser, S. P. (2014). Quantum mechanics/molecular mechanics study on the oxygen binding and substrate hydroxylation step in AlkB repair enzymes. *Chemistry-a European Journal, 20*(2), 435–446.
Rabus, R., & Heider, J. (1998). Initial reactions of anaerobic metabolism of alkylbenzenes in denitrifying and sulfate-reducing bacteria. *Archives of Microbiology, 170*(5), 377–384.
Rabus, R., Kube, M., Heider, J., Beck, A., Heitmann, K., Widdel, F., et al. (2005). The genome sequence of an anaerobic aromatic-degrading denitrifying bacterium, strain EbN1. *Archives of Microbiology, 183*(1), 27–36.
Reiher, M., Salomon, O., & Artur Hess, B. (2001). Reparameterization of hybrid functionals based on energy differences of states of different multiplicity. *Theoretical Chemistry Accounts: Theory, Computation, and Modeling (Theoretica Chimica Acta), 107*(1), 48–55.
Rettie, A. E., Boberg, M., Rettenmeier, A. W., & Baillie, T. A. (1988). Cytochrome P-450-catalyzed desaturation of valproic acid *in vitro*. Species differences, induction effects, and mechanistic studies. *Journal of Biological Chemistry, 263*(27), 13733–13738.
Rettie, A. E., Rettenmeier, A. W., Howald, W. N., & Baillie, T. A. (1987). Cytochrome P-450—Catalyzed formation of delta 4-VPA, a toxic metabolite of valproic acid. *Science, 235*(4791), 890–893.
Rettie, A. E., Sheffels, P. R., Korzekwa, K. R., Gonzalez, F. J., Philpot, R. M., & Baillie, T. A. (1995). CYP4 isoenzyme specificity and the relationship between. Omega.-hydroxylation and terminal desaturation of valproic acid. *Biochemistry, 34*(24), 7889–7895.
Rittle, J., & Green, M. T. (2010). Cytochrome P450 compound I: Capture, characterization, and CH bond activation kinetics. *Science, 330*(6006), 933–937.
Rude, M. A., Baron, T. S., Brubaker, S., Alibhai, M., Del Cardayre, S. B., & Schirmer, A. (2011). Terminal olefin (1-alkene) biosynthesis by a novel P450 fatty acid decarboxylase from jeotgalicoccus species. *Applied and Environmental Microbiology, 77*(5), 1718–1727.
Ruettinger, R. T., Wen, L.-P., & Fulco, A. J. (1989). Coding nucleotide, 5' regulatory, and deduced amino acid sequences of P-450BM-3, a single peptide cytochrome P-450: NADPH-P-450 reductase from bacillus megaterium. *Journal of Biological Chemistry, 264*(19), 10987–10995.
Sadeque, A. J. M., Fisher, M. B., Korzekwa, K. R., Gonzalez, F. J., & Rettie, A. E. (1997). Human CYP2C9 and CYP2A6 mediate formation of the hepatotoxin 4-ene-valproic acid. *Journal of Pharmacology and Experimental Therapeutics, 283*(2), 698–703.
Sahu, S., Widger, L. R., Quesne, M. G., de Visser, S. P., Matsumura, H., Moënne-Loccoz, P., et al. (2013). Secondary coordination sphere influence on the reactivity of nonheme iron (II) complexes: An experimental and DFT approach. *Journal of the American Chemical Society, 135*(29), 10590–10593.

Sainna, M. A., Kumar, S., Kumar, D., Fornarini, S., Crestoni, M. E., & de Visser, S. P. (2015). A comprehensive test set of epoxidation rate constants for iron (iv)–oxo porphyrin cation radical complexes. *Chemical Science, 6*(2), 1516–1529.

Schlichting, I., Berendzen, J., Chu, K., Stock, A. M., Maves, S. A., Benson, D. E., et al. (2000). The catalytic pathway of cytochrome P450cam at atomic resolution. *Science, 287*(5458), 1615–1622.

Senn, H. M., & Thiel, W. (2007). QM/MM studies of enzymes. *Current Opinion in Chemical Biology, 11*(2), 182–187.

Senn, H. M., & Thiel, W. (2009). QM/MM methods for biomolecular systems. *Angewandte Chemie, International Edition, 48*(7), 1198–1229.

Sevastik, R., & Himo, F. (2007). Quantum chemical modeling of enzymatic reactions: The case of 4-oxalocrotonate tautomerase. *Bioorganic Chemistry, 35*(6), 444–457.

Shaik, S., Kumar, D., de Visser, S. P., Altun, A., & Thiel, W. (2005). Theoretical perspective on the structure and mechanism of cytochrome P450 enzymes. *Chemical Reviews, 105*(6), 2279–2328.

Siegbahn, P. E. M., & Borowski, T. (2006). Modeling enzymatic reactions involving transition metals. *Accounts of Chemical Research, 39*(10), 729–738.

Siegbahn, P. E. M., & Borowski, T. (2011). Comparison of QM-only and QM/MM models for the mechanism of tyrosinase. *Faraday Discussions, 148*, 109–117.

Sono, M., Roach, M. P., Coulter, E. D., & Dawson, J. H. (1996). Heme-containing oxygenases. *Chemical Reviews, 96*(7), 2841–2888.

Strijkstra, A., Trautwein, K., Jarling, R., Wohlbrand, L., Dorries, M., Reinhardt, R., et al. (2014). Anaerobic activation of p-cymene in denitrifying betaproteobacteria: Methyl group hydroxylation versus addition to fumarate. *Applied and Environmental Microbiology, 80*(24), 7592–7603.

Szaleniec, M., Borowski, T., Schuhle, K., Witko, M., & Heider, J. (2010). Ab inito modeling of ethylbenzene dehydrogenase reaction mechanism. *Journal of the American Chemical Society, 132*(17), 6014–6024.

Szaleniec, M., Dudzik, A., Kozik, B., Borowski, T., Heider, J., & Witko, M. (2014). Mechanistic basis for the enantioselectivity of the anaerobic hydroxylation of alkylaromatic compounds by ethylbenzene dehydrogenase. *Journal of Inorganic Biochemistry, 139*, 9–20.

Szaleniec, M., Dudzik, A., Pawul, M., & Kozik, B. (2009). Quantitative structure enantioselective retention relationship for high-performance liquid chromatography chiral separation of 1-phenylethanol derivatives. *Journal of Chromatography A, 1216*(34), 6224–6235.

Szaleniec, M., Hagel, C., Menke, M., Nowak, P., Witko, M., & Heider, J. (2007). Kinetics and mechanism of oxygen-independent hydrocarbon hydroxylation by ethylbenzene dehydrogenase. *Biochemistry, 46*(25), 7637–7646.

Szaleniec, M., Salwinski, A., Borowski, T., Heider, J., & Witko, M. (2012). Quantum chemical modeling studies of ethylbenzene dehydrogenase activity. *International Journal of Quantum Chemistry, 112*(8), 1990–1999.

Szaleniec, M., Witko, M., & Heider, J. (2008). Quantum chemical modeling of the C–H cleavage mechanism in oxidation of ethylbenzene and its derivates by ethylbenzene dehydrogenase. *Journal of Molecular Catalysis A: Chemical, 286*(1–2), 128–136.

Trewick, S. C., Henshaw, T. F., Hausinger, R. P., Lindahl, T., & Sedgwick, B. (2002). Oxidative demethylation by Escherichia coli AlkB directly reverts DNA base damage. *Nature, 419*(6903), 174–178.

Vreven, T., Byun, K. S., Komaromi, I., Dapprich, S., Montgomery, J. A., Morokuma, K., et al. (2006). Combining quantum mechanics methods with molecular mechanics methods in ONIOM. *Journal of Chemical Theory and Computation, 2*(3), 815–826.

Warshel, A., & Levitt, M. (1976). Theoretical studies of enzymic reactions: Dielectric, electrostatic and steric stabilization of the carbonium ion in the reaction of lysozyme. *Journal of Molecular Biology, 103*(2), 227–249.

Wen, X., Wang, J. S., Kivisto, K. T., Neuvonen, P. J., & Backman, J. T. (2001). *In vitro* evaluation of valproic acid as an inhibitor of human cytochrome P450 isoforms: Preferential inhibition of cytochrome P450 2C9 (CYP2C9). *British Journal of Clinical Pharmacology*, *52*, 547–553.

Wojcik, A., Broclawik, E., Siegbahn, P. E. M., Lundberg, M., Moran, G., & Borowski, T. (2014). Role of substrate positioning in the catalytic reaction of 4-hydroxyphenylpyruvate dioxygenase-a QM/MM study. *Journal of the American Chemical Society*, *136*(41), 14472–14485.

Wong, S. D., Bell, C. B., Liu, L. V., Kwak, Y., England, J., Alp, E. E., et al. (2011). Nuclear resonance vibrational spectroscopy on the FeIV = OS = 2 non-heme site in TMG3tren: Experimentally calibrated insights into reactivity. *Angewandte Chemie, International Edition*, *50*(14), 3215–3218.

Ye, S., & Neese, F. (2011). Nonheme oxo-iron(IV) intermediates form an oxyl radical upon approaching the C-H bond activation transition state. *Proceedings of the National Academy of Sciences of the United States of America*, *108*(4), 1228–1233.

Yi, C., Yang, C. G., & He, C. (2009). A non-heme iron-mediated chemical demethylation in DNA and RNA. *Accounts of Chemical Research*, *42*(4), 519–529.

Yu, B., Edstrom, W. C., Benach, J., Hamuro, Y., Weber, P. C., Gibney, B. R., et al. (2006). Crystal structures of catalytic complexes of the oxidative DNA/RNA repair enzyme AlkB. *Nature*, *439*(7078), 879–884.

Yu, B., & Hunt, J. F. (2009). Enzymological and structural studies of the mechanism of promiscuous substrate recognition by the oxidative DNA repair enzyme AlkB. *Proceedings of the National Academy of Sciences of the United States of America*, *106*(34), 14315–14320.

CHAPTER EIGHT

QM/MM Studies Reveal How Substrate–Substrate and Enzyme–Substrate Interactions Modulate Retaining Glycosyltransferases Catalysis and Mechanism

Hansel Gómez[*,†,1], **Fernanda Mendoza**[*,†], **José M. Lluch**[*,†], **Laura Masgrau**[*,2]

[*]Institut de Biotecnologia i de Biomedicina (IBB), Universitat Autònoma de Barcelona, Bellaterra (Barcelona), Spain

[†]Departament de Química, Universitat Autònoma de Barcelona, Bellaterra (Barcelona), Spain

[2]Corresponding author: e-mail address: Laura.Masgrau@uab.cat

Contents

Abstract

Glycosyltransferases (GTs) catalyze the biosynthesis of glycosidic linkages by transferring a monosaccharide from a nucleotide sugar donor to an acceptor substrate, and they do that with exquisite regio- and stereospecificity. Retaining GTs act with retention of the configuration at the anomeric carbon of the transferred sugar. Their chemical mechanism has been under debate for long as conclusive experimental data to confirm the

[1] Present address: Joint BSC-CRG-IRB Program in Computational Biology, Molecular Modeling & Bioinformatics Group, Institute for Research in Biomedicine (IRB Barcelona), Barcelona, Spain.

Advances in Protein Chemistry and Structural Biology, Volume 100
ISSN 1876-1623
http://dx.doi.org/10.1016/bs.apcsb.2015.06.004

mechanism have been elusive. In the past years, quantum mechanical/molecular mechanical (QM/MM) calculations have shed light on the mechanistic discussion. Here, we review the work carried out in our group investigating three of these retaining enzymes (LgtC, α3GalT, and GalNAc-T2). Our results support the controversial front-side attack mechanism as the general mechanism for most retaining GTs. The latest structural data are in agreement with these findings. QM/MM calculations have revealed how enzyme–substrate and substrate–substrate interactions modulate the transfer reaction catalyzed by these enzymes. Moreover, they provide an explanation on why in some cases a strong nucleophilic residue is found on the β-face of the sugar, opening the door to a shift toward a double-displacement mechanism.

1. INTRODUCTION

Glycosyltransferases (GTs) catalyze the biosynthesis of a new glycosidic linkage by transferring a monosaccharide from an activated sugar donor to an acceptor substrate which can be another sugar molecule, a lipid, a peptide, DNA, or an organic molecule (Breton, Fournel-Gigleux, & Palcic, 2012). Glycosylation of proteins is the most abundant type of post-translational modification and plays critical roles in molecular recognition processes such as cell–cell interactions, cell development, or cell-pathogen recognition (Bennett et al., 2012; Hart & Copeland, 2010; Solá, Rodríguez-Martínez, & Griebenow, 2007). Misregulation of GTs activity has been related to several diseases (inflammation, cancer) and to infection (Ohtsubo & Marth, 2006; Zhang et al., 2006). Glycosylation is also crucial for many natural products' properties (Křen, 2008). However, and despite the important role that GTs have in many biological functions and processes, some fundamental aspects regarding their activity remain unanswered. This is the case for the chemical mechanism followed by retaining GTs, which is still under strong debate (Lairson, Henrissat, Davies, & Withers, 2008).

Retaining GTs form a new glycosidic bond with net retention of the configuration at the anomeric carbon of the transferred sugar. By analogy with retaining glycosyl hydrolases, and despite the lack of any relatedness, a double-displacement mechanism was first proposed (Fig. 1A). In this mechanism, a nucleophilic residue from the enzyme, located on the β-face of the sugar ring and close to the anomeric carbon, would attack the donor substrate to form a covalent glycosyl–enzyme complex (CGE). In a second step, the acceptor substrate would attack the CGE leading to the final product with net retention of the configuration. Some experimental results were in agreement with this mechanism; but not conclusive data were obtained

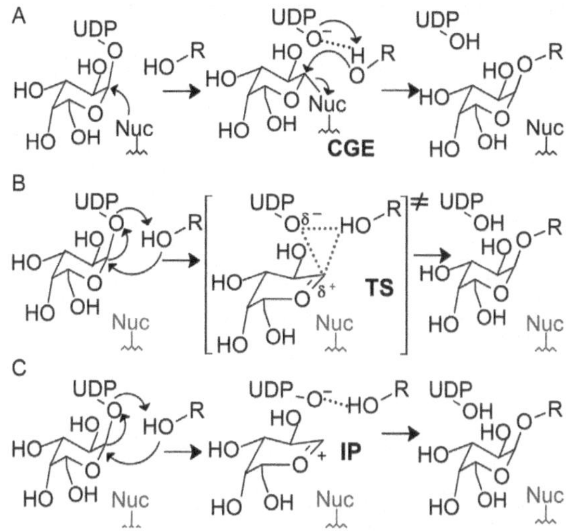

Figure 1 Main reaction mechanisms proposed in the literature for enzymatic glycosyl transfer with retention of the configuration: (A) double-displacement mechanism with formation of a covalent glycosyl–enzyme complex (CGE), and front-side attack mechanism through an oxocarbenium; (B) transition state; or (C) a short-lived ion pair intermediate (IP).

for wild-type enzymes (i.e., the experiments were carried out on enzymes with the putative nucleophile mutated). For example, the CGE was trapped in the active site of the E303C mutant of blood-type glycosyltransferase GTB (a GT6 family member) in the absence of the acceptor substrate (Soya, Fang, Palcic, & Klassen, 2011). In another experiment, the activity of the E317A mutant of bovine α-1,3-galactosyltransferase (α3GalT, family GT6) was partially recovered by the addition of a chemical nucleophile (azide) (Monegal & Planas, 2006). Nevertheless, it has become apparent that most retaining GTs do not have a suitably positioned nucleophile to carry out this role and alternative mechanisms had to be proposed (Lairson et al., 2008; Persson et al., 2001). The main one was a front-side attack mechanism in which the acceptor nucleophile would attack the anomeric carbon by the same face of the sugar where the UDP leaving group is with formation of an oxocarbenium ion transition state (Fig. 1B) or oxocarbenium–phosphate ion pair intermediate (Fig. 1C). In principle, this mechanism does not require the presence of a nucleophilic residue, although if present, it could provide electrostatic stabilization of the oxocarbenium species (Lairson et al., 2008). Such a mechanism, that received the $S_{N}i$ (intramolecular nucleophilic change) or "internal return" name, was first invoked to rationalize the unusual stereochemical outcomes of simple alkyl halides and has also been inferred for the solvolysis reaction of α-glycosyl fluorides (Chan, Tang, & Bennet, 2012). Against this mechanism goes an expected steric hindrance between UDP and the acceptor substrate (especially for proposal 1B). However, structural data of enzyme complexes with sugar analogues (Lira-Navarrete et al., 2014; Persson et al., 2001) or product mimetics (Errey et al., 2010), together with kinetic data (Lee et al., 2011), are in agreement with it.

In the past years, hybrid quantum mechanics/molecular mechanics (QM/MM) calculations have proved to be very valuable in the study of retaining GTs mechanism (Ardèvol & Rovira, 2011; Bobovská, Tvaroška, & Kóňa, 2014, 2015; Gómez, Lluch, & Masgrau, 2012, 2013; Gómez, Polyak, Thiel, Lluch, & Masgrau, 2012; Gómez et al., 2014; Lira-Navarrete et al., 2014; Rojas-Cervellera, Ardèvol, Boero, Planas, & Rovira, 2013; Trnka, Kozmon, Tsvaroska, & Koca, 2015). Most of the systems for which a crystallographic structure was available have been studied theoretically and much insight has been gained in the description and understanding of how these enzymes catalyze the formation of a new glycosidic bond. Here, we review our work on three of these enzymes, namely *Neisseria meningitidis* lipopolysaccharyl-α-1,4-galactosyltransferase C (LgtC) (Gómez, Polyak, et al., 2012), bovine α3GalT (Gómez, Lluch, et al., 2012; Gómez et al., 2013), and human polypeptide-*N*-acetylgalactosamine transferase 2 (GalNAc-T2) (Gómez et al., 2014). Special attention has been given in showing how biomolecular interactions between the enzyme and the substrates and, very importantly, between donor and acceptor substrates modulate the chemical reaction mechanism of these enzymes.

2. METHODOLOGICAL OVERVIEW

In this section, a general overview of the methodological protocol used in our studies of retaining GTs is presented. The particular details vary from one another and can be found in the original works (Gómez, Lluch, et al., 2012; Gómez, Polyak, et al., 2012; Gómez et al., 2013, 2014).

2.1 Model Preparation from the Crystallographic Data

The theoretical study of reactions catalyzed by enzymes with models that include the full solvated enzyme require the knowledge of the three-dimensional coordinates of the enzyme:substrate complex (Michaelis complex). The coordinates of the enzyme are usually taken from crystallographic data available at the Protein Data Bank (PDB), but they rarely contain those of the substrate. Crystal structures of enzymes with substrate analogues bound that do not react or of mutated enzymes that are inactive may be available. Thus, some of the Michaelis complex coordinates need to be modeled either deriving the coordinates from other ligands/residues found in the PDB files or by performing ligand:enzyme docking calculations. In the case of retaining GTs, the Michaelis complex is actually a ternary complex, as two substrates are involved in the reaction (which usually follow a biordered mechanism). The donor substrate (UDP-monosaccharide for the systems reviewed here) binds first, sometimes accompanied by a conformational change linked to UDP binding; this is followed by the binding of the acceptor substrate, reaction, and product release (Boix et al., 2001; Jamaluddin, Tumbale, Withers, Acharya, & Brew, 2007; Lairson et al., 2008). Crystal structures of the enzyme with the hydrolyzed donor (i.e., enzyme:UDP) or inactive substrate analogues, the structure of the enzyme:UDP:acceptor or that of a mutated enzyme with the sugar donor substrate, are common for those systems. Combination of these data allowed us to prepare an initial ternary complex model (enzyme:donor:acceptor) for each of the systems studied. All the retaining GTs reviewed here present a GTA fold, a conserved DXD motif, and a divalent cation to which the pyrophosphate group of UDP is coordinated. The coordinates of the hydrogen atoms (not present in the crystal structures) were then built and the protonation state of the titratable residues (His, Glu, Asp, Arg, and Lys) were assigned based on the pK_a values given by the empirical PROPKA procedure (Li, Robertson, & Jensen, 2005) and verified through visual inspection.

The ternary complex was then solvated with explicit water molecules and counterions were added to neutralize the full system. Molecular dynamics (MD) simulations of the solvent were carried out to equilibrate it. MD simulations of the full system are also usually recommended in order to fully equilibrate the Michaelis complex, although in some cases small restrains to the acceptor heavy atoms had to be applied to keep it in a suitable position for catalysis (see the original papers for more details). For some of the systems, QM(SCC-DFTB)/MM MD (see below) of the Michaelis complex was also carried out. In the molecular mechanics (MM) calculations, we have employed the CHARMM22 force field (MacKerell, Feig, & Brooks, 2004; MacKerell et al., 1998) for the metal ion, protein atoms, and TIP3P water molecules; the CHARMM36 all-atom carbohydrates force field (Guvench, Hatcher, Venable, Pastor, & MacKerell,

2009; Guvench et al., 2008) for the sugar moieties and the disaccharide linkages (when present); and have derived the parameters for the UDP-monosaccharide linkage from the CHARMM existing ones. The CHARMM (Brooks et al., 2009) or the NAMD (Phillips et al., 2005) programs were used to run the MD simulations.

2.2 Study of the Glycosyl Transfer Reaction

One or several representative structures from the MD simulations of the Michaelis complex were used as starting points in the reactivity studies. These were carried out by hybrid QM/MM calculations (Senn & Thiel, 2009). For the QM/MM calculations, we have typically reduced the system to a 24–30 Å radius sphere around the anomeric center C1 (Fig. 2, left side) and have defined an active zone (atoms that are allowed to move in the optimizations) of 12–15 Å around C1 (insert in Fig. 2).

2.2.1 QM/MM Partition

QM/MM calculations are a very convenient approach to study enzyme catalysis as they allow the modeling of the chemical reaction (breakage and formation of new covalent bonds, which need a QM representation) taking place in the presence of the solvated enzyme (Lonsdale, Harvey, & Mulholland, 2012). The system is partitioned into the QM zone, which will be described at the QM level and that contains, at least, the atoms directly

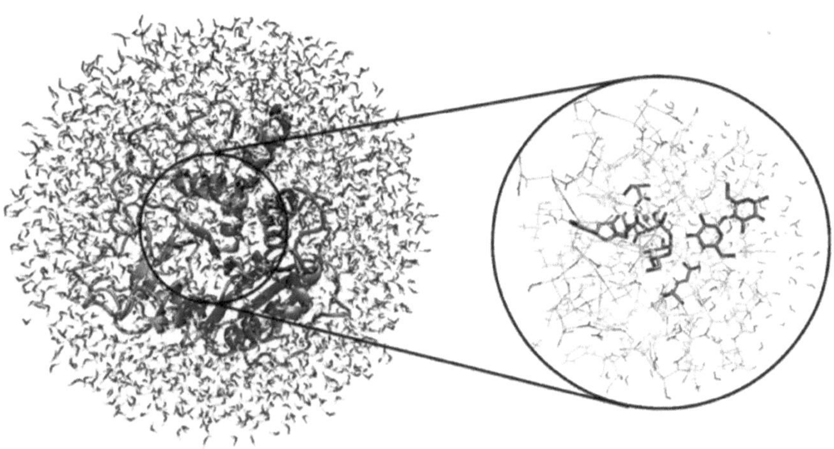

Figure 2 Molecular representation of a model system used in our QM/MM calculations. The insert highlights the active zone, with the QM atoms shown in licorise. *From Gómez et al. (2013).*

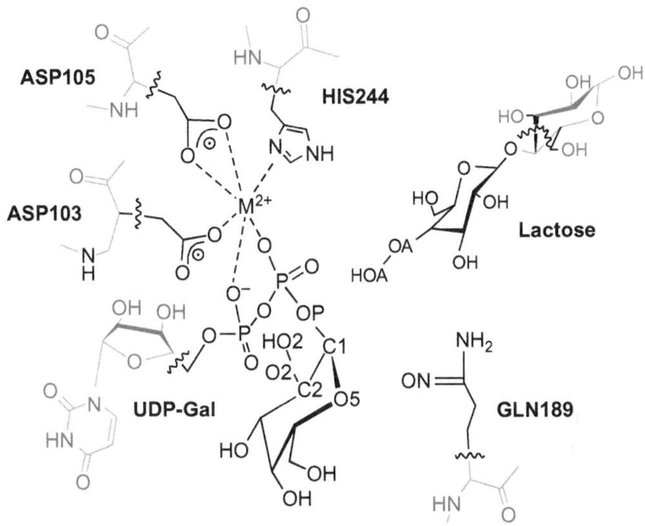

Figure 3 Scheme representing one of the QM/MM partitions used in our work. The atoms treated at the QM level of theory are depicted in black, while the gray ones belong to the MM zone. The wavy lines indicate the boundary between the QM and MM atoms. *Adapted from Gómez, Polyak, et al. (2012).*

involved in the reaction; and the MM zone, which includes the rest of the system and is described with a force field (MM level). Figure 3 shows an example of QM/MM partition that we have used in our work (Gómez, Polyak, et al., 2012). Typically, the QM zone was formed by 80–100 atoms and the MM zone by 7000–12,000 atoms. Link atoms within the charge-shift model were used in the boundaries between the QM and MM zones that involved covalent bonds (De Vries et al., 1999; Sherwood et al., 1997). For labeling consistency through this review, we will refer to the anomeric carbon as C1, to the donor substrate glycosidic oxygen as OP, to the attacking hydroxyl atoms as OA for the oxygen atom and HOA for the corresponding hydrogen atom, and to the oxygen atom of the putative nucleophile (if any) as ON. The QM zone included the monosaccharide to be transferred, part of the acceptor substrate, the divalent cation and its first coordination sphere (which includes the phosphate groups of UDP, some residues' side chains and some water molecules), and the putative nucleophile in the systems presenting one (LgtC and α3GalT). In the LgtC system, a larger QM zone was tested that included two protein residues that interact with the substrates; the potential energy profiles were not significantly affected, although Bobovská et al. (2015) have recently reported that the inclusion of residues interacting with the phosphate oxygen atoms of UDP was important to characterize the transition states in their study on the blood group GTB glycosyltransferase. In some cases, we have not been able to validate the stationary points found by frequency calculations, but we have attributed this to the flatness of the potential energy surface (PES) around these points.

2.2.2 QM/MM Methods

The reactivity studies started with QM/MM minimization of the structures selected from the MD simulations of the Michaelis complex. The system was then driven from reactants to products in adiabatic mapping calculations following a defined reaction coordinate that suitably described the mechanism(s) under study. The reaction coordinate was usually scanned in steps of 0.2 Å. For the front-side attack, we have found that the reaction coordinate is generally well defined by the lineal combination of the breaking/forming distances: $RC_i = [d(OP-C1) - d(OA-C1) - d(HOA-OP)]$. For the first step of the double-displacement mechanism, we used $RC_{dd1} = [d(OP-C1) - d(ON-C1) - d(HOA-OP)]$, and $RC_{dd2} = [d(ON-C1) - d(OA-C1) - d(HOA-OP)]$ for the second step. The information of these potential energy profiles was used to search for transition state structures and possible intermediates in the reaction, which were verified by frequency calculations of the QM zone.

Geometry optimizations were generally carried out at the QM(BP86/SVP)/CHARMM level of theory, using the BP86 functional (Becke, 1988; Perdew, 1986; Perdew & Yue, 1986; Slater, 1951; Vosko, Wilk, & Nusair, 1980) and the SVP basis set (Schäfer, Horn, & Ahlrichs, 1992) in combination with the resolution-of-the-identity approximation (Eichkorn, Treutler, Ohm, Marco, & Ahlrichs, 1995; Eichkorn, Weigend, Treutler, & Ahlrichs, 1997). The energies were improved with QM/MM single-point energy calculations using the M05-2X functional (Zhao, Schultz, & Truhlar, 2006), which has been shown to properly describe carbohydrate chemistry (Csonka, French, Johnson, & Stortz, 2009) and the UDP-Gal breaking process (Kóňa & Tvaroška, 2009), and a larger basis set (TZVP) (Schäfer, Huber, & Ahlrichs, 1994). Other functionals and basis sets have also been tested (Gómez, Polyak, et al., 2012; Gómez et al., 2013).

The metal cofactor found in the crystal structures is usually Mn^{2+}. However, these enzymes are usually also active with Mg^{2+}, which is computationally more convenient. Thus, we have used Mg^{2+} in our models. The cation cofactor is proposed to act as a Lewis acid in retaining GTs catalysis and it is known that both Mn^{2+} and Mg^{2+} can adopt this role in a similar manner (Bock, Katz, Markham, & Glusker, 1999). Moreover, both divalent cations have given almost the same results in a validation study on dissociation of α-glycosidic bond in sugar phosphates (Kóňa & Tvaroška, 2009). For example, the M05-2X energies and bond lengths differ by less than 0.2 kcal/mol and 0.01 Å, respectively, from the reference method. Nevertheless, the effect of having a Mn^{2+} cation instead of the modeled Mg^{2+} on the chemical step itself was tested for the LgtC system (Gómez, Polyak, et al., 2012). The calculated reaction mechanism is the same for both metal ions, with the potential energy barriers being slightly different (~1.5 kcal/mol lower for Mn^{2+}, thus, with Mn^{2+} the reaction is somewhat faster), which is in agreement with recent studies on the retaining α-1,2-mannosyltransferase Krep2p/Mnt1p (Bobovská et al., 2014) and GalNAc-T2 (Trnka et al., 2015).

Although we have seen that the QM/MM calculations using QM=SCC-DFTB (Elstner et al., 1998; Frauenheim, Seifert, Elstner, & Sternberg, 2002) highly overestimate the potential energy barriers of these glycosyl transfer reactions, its semiempirical nature allows a computationally efficient way to estimate free energy profiles. Thus, comparison of the QM(SCC-DFTB)/MM potential and free energy profiles was used in some of the systems to qualitatively evaluate the effect of thermal and entropic contributions to the reaction. Umbrella sampling QM/MM MD simulations at the QM(SCC-DFTB)/CHARMM level were performed to compute the potential of mean force and the free energy profile at 27 °C. The reaction coordinate was scanned in steps of 0.1 Å. In the MD simulations, a force constant was applied on the reaction coordinate. The same active region as in QM/MM optimizations was subjected to NVT dynamics with the Nosé–Hoover thermostat (Hoover, 1985; Nosé, 1984). The SHAKE procedure was applied at every step for the O–H bonds in the water molecules (Ryckaert, Ciccotti, & Berendsen, 1977). The umbrella integration analysis (Kästner & Thiel, 2005, 2006) and the weighted histogram analysis method (Kumar, Rosenberg, Bouzida, Swendsen, & Kollman, 1992; Souaille & Roux, 2001) were used to compute the free energy profile.

The QM/MM calculations were carried out with the modular program package ChemShell (Sherwood et al., 2003), using DL_POLY (Smith & Forester, 1996) to compute the MM energies and gradients (with the CHARMM force field); and the QM energies and gradients were calculated with Gaussian09 (Frisch et al., 2009) for QM=M05-2X, MNDO (Thiel, 2005) for QM=SCC-DFTB, and TURBOMOLE (Ahlrichs, Bär,

Häser, Horn, & Kölme, 1989) for QM=BP86. No cutoffs were applied for the MM and QM/MM nonbonding interactions, the latter being treated within an electronic embedding scheme. Energy minimizations were carried out with the limited-memory Broyden–Fletcher–Goldfarb–Shanno (L-BFGS) algorithm (Liu & Nocedal, 1989; Nocedal, 1980); transition state searches were done with the microiterative TS optimizer that combines L-BFGS and the partitioned rational function optimizer P-RFO (Baker, 1986; Banerjee, Adams, Simons, & Shepard, 1985), all of them implemented in the HDLCopt module (Billeter, Turner, & Thiel, 2000) of ChemShell. The potential of mean force calculations was carried out using the dynamics module within ChemShell.

2.3 Further Analysis

Besides describing the structural and energetic features of the reaction, we also analyzed other aspects in order to get a better understanding of the glycosyl transfer mechanism and catalysis. Natural population analysis (Reed, Weinstock, & Weinhold, 1985) charges were determined from the QM/MM calculations. A natural bond orbital (NBO) analysis (Foster & Weinhold, 1980; Reed, Curtiss, & Weinhold, 1988; Reed & Weinhold, 1983; Reed et al., 1985) was also performed for some of the stationary points in order to find and visualize the most significant molecular orbital interactions. This was done with the NBO program v3.1 (Glendening, Reed, Carpenter, & Weinhold, 2009) included in Gaussian09. An estimation of the electrostatic stabilization/destabilization of the substrates by their interaction with enzyme's residues was carried out by setting the MM charges of these residues to zero in additional single-point energy calculations. The effect of mutating a residue or changing the substrate was also examined with new *in silico* models of the mutant enzymes or substrate modifications and assuming no major structural changes. These results were compared with experimental data (already available or newly generated). Some of these results will not be reviewed here but can be found in the original works.

3. RETAINING GTs MECHANISM

3.1 *Neisseria meningitidis* LgtC

The first QM/MM works that included the full enzyme in the study of the reaction mechanism of retaining GTs were those of OtsA (Ardèvol &

Rovira, 2011) and LgtC (Gómez, Polyak, et al., 2012). Ardèvol et al. used QM(DFT)/MM metadynamics simulations to study trehalose-6-phosphate synthase (OtsA), a retaining GT of the GT20 family with a GTB fold, and obtained a front-side attack mechanism via formation of a very short-lived ion pair intermediate (Fig. 1C). We applied QM(DFT)/MM calculations to study the mechanism of LgtC (family GT8, GTA fold) and concluded an asynchronous front-side attack mechanism through a single oxocarbenium transition state for this enzyme (Fig. 1B). The main results that we obtained for LgtC (Gómez, Polyak, et al., 2012) are summarized in this section.

LgtC (EC 2.4.1.44) catalyzes galactosyl transfer from the UDP-Gal donor substrate to a lactose (LAT) terminated acceptor with an α-1,4 regiospecificity. The enzyme is involved in the biosynthesis of lipo-oligosaccharides that cover the cell surface of the bacteria and that have been related with its great virulence (Tzeng & Stephens, 2000). Crystal structures of the enzyme with substrate analogues revealed that a Gln residue (Gln189) is positioned on the β-face of Gal, with its side-chain oxygen atom at 3.5 Å from the anomeric carbon (Persson et al., 2001). As can be seen in the structure of the optimized reactant complex (R) depicted in Fig. 4, the two substrates seem ready for reaction through a front-side mechanism: the attacking hydroxyl has its oxygen atom (OA) at a distance of 3.11 Å from the anomeric carbon (C1) and its hydrogen atom (HOA) makes an hydrogen bond with the UDP β-phosphate (d(OP–HOA) = 1.67 Å). UDP has been proposed to act as the catalytic base to deprotonate (i.e., activate) the attacking hydroxyl group. The oxygen atom (ON) of the putative nucleophile Gln189 in a double-displacement mechanism is at 3.38 Å from C1, in agreement with the crystallographic structure. Despite this relatively short distance glutamine is expected to be a poor nucleophile. This is reflected in the fact that no CGE could be characterized in

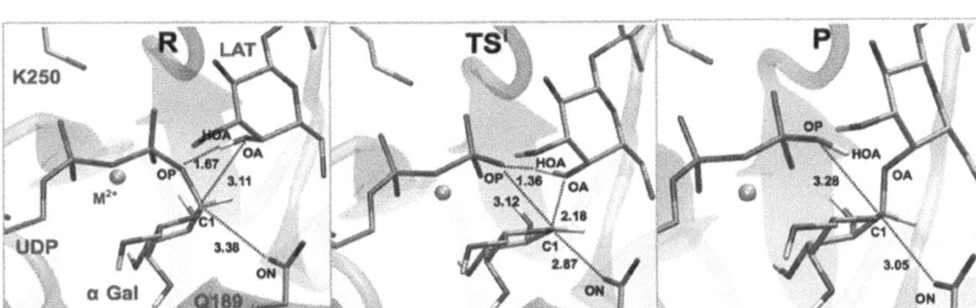

Figure 4 Molecular representation of the optimized reactants (R), transition state (TS^i), and products (P) for the front-side attack mechanism in LgtC. Distances are given in Å.

our work on this system. The front-side attack mechanism was then considered. By performing geometry optimizations (at QM(BP86/SVP)/MM level) for restrained values of the reaction coordinate (defined as $RC = [d(OP - C1) - d(OA - C1)]$), the system was driven from reactants to products through a front-side attack path. Notice that for this particular enzyme, the inclusion of the *d*(HOA–OP) distance in the definition of the reaction coordinate was not necessary as the interaction already existed in R (i.e., its inclusion did not change significantly the energy profile (unpublished results)). The highest energy structure along the calculated reaction pathway was used as a starting structure to initiate a saddle point search (TS^i in Fig. 4), which was characterized as having a single imaginary frequency corresponding to the chemical process under study. The bond distances of these two structures (TS^i guess and optimized TS^i) were very similar, reflecting the suitability of the geometrical reaction coordinate used to describe the transfer reaction. The energies were recalculated at several levels of theory, and our best estimation corresponded to single-point energies at the QM(M05-2X/TZVP)/MM level (see Section 2.2 for discussion). The calculated potential energy barrier for this TS^i was 14.6 kcal/mol, in good agreement with the experimentally derived phenomenological free energy barrier of ~16 kcal/mol (experimental $k_{cat}=14$–$34\ s^{-1}$ at 303 K), which strongly suggests that the debated front-side mechanism is possible for LgtC.

Inspection of the structures depicted in Fig. 4 shows that the TS^i is highly dissociative with regard to the UDP–Gal bond: at TS^i, the bond is already broken (3.12 Å) and the attacking hydroxyl (OA) has enough room to start its attack (*d*(OA–C1) = 2.18 Å). Transfer of HOA to UDP, which is confirmed as the catalytic base that activates the attacking hydroxyl group, has also started (*d*(HOA–OP) = 1.36 Å). TS^i presents some oxocarbenium character, with a positive charge increase of 0.17 a.u. at the anomeric center ($q_{C1} + q_{O5}$) and of 0.30 a.u. when the whole α-Gal ring is considered. This is concomitant with the shortening of the $C1–O5_{\alpha\text{-Gal}}$ distance (by 0.07 Å to be 1.29 Å in TS^i). In going from reactants to the transition state, the α-Gal ring conformation changes from an initial distorted chair $^4C_1$ ($\theta=236°$, $\varphi=9°$) to a conformation between an ^4E envelope and a $^4H_5$ half-chair ($\theta=250°$, $\varphi=41°$). This change in the puckering of the ring brings the anomeric carbon C1 closer to the oxygen atom of Gln189 ($d(C1–ON_{Gln189})=3.38$ and 2.87 Å, in R and TS^i, respectively), which provides electrostatic stabilization to the positive charge developed at the anomeric center. Gln189 (via its side-chain amide nitrogen) also forms an hydrogen bond with the $O6_{LAT}$ atom of the acceptor substrate ($d(O6_{LAT}–N_{Gln189})$ = 2.00 and 1.89 Å, in R and TS^i, respectively). Thus, Gln189 is involved in acceptor binding and oxocarbenium stabilization. An estimation of the electrostatic stabilization provided by Gln189 along the reaction path is given in Fig. 5: its maximum effect is of 2.5 kcal/mol on the route to TS^i (so when the glycosidic bond is being broken), and at TS^i it is of 2 kcal/mol. This is in agreement with the experimental k_{cat} for the Q189A mutant, which is 3% of that for the wild-type enzyme and which corresponds to an increase of ~2 kcal/mol in the phenomenological free energy barrier.

The electrostatic contribution of other significant residues that interact with the substrates is also presented in Fig. 5. Besides Gln189, Asp188 (hydrogen bonded to $O4_{\alpha\text{-Gal}}$ and $O6_{\alpha\text{-Gal}}$) also stabilizes the α-Gal oxocarbenium structures. The highest stabilization along the transfer reaction, though, comes from Lys250, which interacts with the phosphates of the UDP leaving group. This residue is highly conserved in the GT8 family to which LgtC belongs to, and it is part of one of the loops that fold upon the binding of the donor substrate. His87 and Gly247, which form hydrogen bonds with UDP, also provide some stabilization to the leaving group. Finally, Asp130 (hydrogen bonded to $O6_{LAT}$) increases its contribution as the reaction proceeds.

On the other hand, and as can be seen in Fig. 4, there are several substrate–substrate interactions that are already present in R and that are prevalent (presumably stronger) at the TS. Thus, these interactions may help

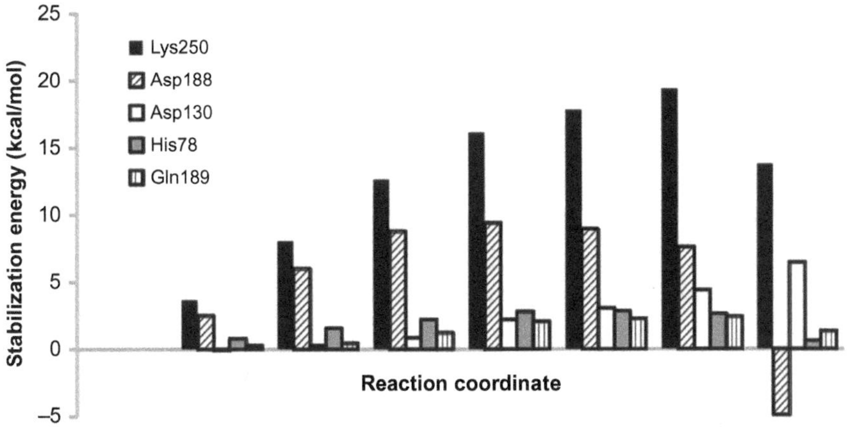

Figure 5 Electrostatic stabilization of the QM zone along the front-side mechanism by surrounding residues of LgtC. Calculations were performed at the QM(BP86/SVP)/MM level of theory. *Adapted from Gómez, Polyak, et al. (2012).*

orienting the substrates to achieve the desired reaction specificity, facilitate UDP–Gal bond breakage, and provide stabilization of the TS. There are two intra-substrate interactions present in UDP-Gal: (1) an hydrogen bond between the donor substrate atoms $O2_{\alpha\text{-Gal}}$ and the β-phosphate oxygen coordinating the metal cation ($d(HO2_{\alpha\text{-Gal}}–O_{UDP})=1.68$ and 1.64 Å, in R and TS^{i}, respectively); and (2) the hydrogen bond between one oxygen of the α-phosphate of UDP and the O3 of the ribose of UDP. Moreover, there are two inter-substrate interactions between the donor and acceptor: (3) the hydrogen bond between the attacking hydroxyl (OA) and the glycosidic oxygen (OP), with $d(HOA_{LAT}–OP)=1.67$ and 1.36 Å, in R and TS^{i}, respectively; and (4) the interaction of OP with the neighboring hydroxyl of the attacking hydroxyl, that is, with $O3_{LAT}$ ($d(HO3_{LAT}–OP)=1.94$ and 1.65 Å, in R and TS^{i}, respectively). In order to further analyse these interactions, we calculated the transfer reaction energy barriers (reaction paths and optimized stationary points) for *in silico* variants of the substrates (UDP-2-deoxygalactose and 3-deoxylactose). The computed barrier heights were 2.3 and 3.3 kcal/mol higher than those obtained for the unmodified substrates, confirming that substituting one of these hydroxyls by a hydrogen atom impedes the transfer reaction significantly.

Finally, for the wild-type enzyme with the unmodified substrates, we computed an estimation of the free energy profile. The theoretical calculation of the free energy barrier based on umbrella sampling MD simulations at the QM(SCC-DFTB)/MM level of theory showed that enthalpic and entropic effects were not significant for this system (Figure S3 in Gómez, Polyak, et al., 2012).

Thus, all together, these results conclude that the front-side attack is the most plausible mechanism for catalysis by LgtC. Moreover, they show that such a mechanism is possible because it starts with UDP-Gal dissociation, which is stabilized by enzyme–substrate and substrate–substrate interactions.

3.2 Bovine α3GalT

α3GalT (EC 2.4.1.87) catalyzes galactosyl transfer from the UDP-Gal donor substrate to a lactose (LAT) terminated acceptor (β-galactosyl groups in glycoconjugates) with an α-1,3 regiospecificity. The enzyme and its products (α-epitope) are present in most mammals, but old world primates, including humans, do not have active forms of the enzyme due to inactivating mutations (Galili & Swanson, 1991). As a result, antibodies against this epitope are produced in humans, which serve as a defense against infection by pathogens but are also a barrier for xenotransplantation of organs from species that produce active forms of α3GalT. α3GalT presents a GTA fold and is a member of the GT6 family of GTs (Campbell, Davies, Bulone, & Henrissat, 1997), the only family for which a nucleophile suitably positioned to act in a double-displacement mechanism has been reported. GTA and GTB, GTs implicated in blood-type biosynthesis, also belong to this family. Thus, α3GalT is a very good system to test the proposal of the double-displacement mechanism.

After building and preparing the initial model, the QM(BP86/SVP)/MM optimized reactant (R) shown in Fig. 6 was obtained. Notice that in the original work, four different snapshots from the MD simulations of the Michaelis complex were considered (Gómez et al., 2013). For simplicity, here we will only present the structural results corresponding to one of these snapshots together with the averaged energy barriers. The full details can be found in the original paper. As can be seen in Fig. 6, donor and acceptor substrates are close (d(OA–C1) = 3.04 Å) but, contrary to what was observed in LgtC, the HOA–OP hydrogen bond is not formed yet as HOA is interacting with the neighboring $O4_{LAT}$. The ON of Glu317, the putative nucleophile, is located at 4.25 Å from the C1 atom, which is further away than OA but at a distance from which a nucleophilic attack could take place. Both mechanisms were therefore investigated for α3GalT.

For driving the system from reactants to products through a front-side attack mechanism, the geometrical reaction coordinate used was $RC_i = [d(\mathrm{OP}-\mathrm{C1}) - d(\mathrm{OA}-\mathrm{C1}) - d(\mathrm{HOA}-\mathrm{OP})]$. Notice that the HOA–OP distance had to be explicitly included in the definition of RC for the present system. Structures from this path were used as initial guess for transition state search and ion pair minimization. TS_1^i, IP, and TS_2^i shown in Fig. 6 were characterized. The averaged potential energies of TS_1^i and TS_2^i are very similar (16.2 ± 2.0 and 16.3 ± 3.7 kcal/mol, respectively, at the QM(M05-2X/TZVP//BP86/SVP) level) and close to the experimentally derived phenomenological free energy barrier of ~17 kcal/mol ($k_{cat} = 6.4\ s^{-1}$ at 310 K), indicating that the front-side attack is plausible for α3GalT. The IP formed has a potential energy between 0.6 and 3.2 kcal/mol lower than the transition states, indicating its short-lived nature. The reaction begins with reorientation of HOA toward OP. Once the HOA–OP interaction is formed, the UDP–Gal bond can break, which leads to IP formation through TS_1^i. In the IP, the attacking hydroxyl (OA) is at 2.79 Å from C1, and the ring has adopted a flat conformation around the anomeric center (d(C1–O5) = 1.28 Å) that brings C1 at only 2.51 Å from

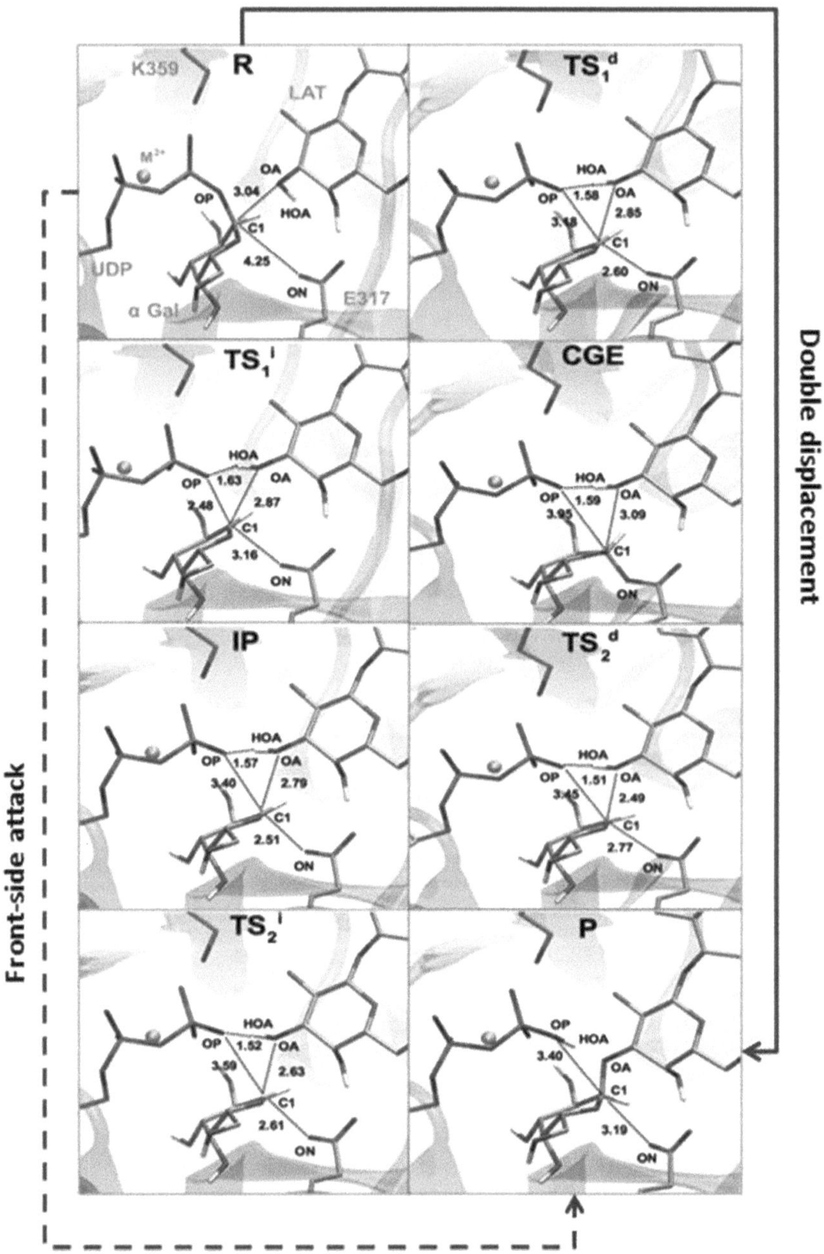

Figure 6 Molecular representations of the stationary points localized for going from reactants (R) to products (P) in α3GalT catalysis through a front-side attack (TS_1^i, IP, TS_2^i) or a double-displacement mechanism (TS_1^d, CGE, TS_2^d). Distances are given in Å.

the ON atom of Glu317. By this point, a positive charge increase of 0.32 a.u. has occurred at the anomeric center (q_{C1} +q_{O5}). This is strongly favored by the closer presence of Glu317. Glu317 is structurally equivalent to Gln189 in LgtC, and it is negatively charged. Therefore, its electrostatic stabilization of the oxocarbenium species is much more significant. We studied the front-side mechanism for the E317Q and E317A mutants and found increases of up to ~15 kcal/mol in barrier heights (Figure 4 of Gómez et al., 2013). Thus, we concluded that the presence of Glu317 on the β-face of α-Gal in α3GalT, which is also involved in acceptor binding, is crucial for reaction even in the front-side attack mechanism. We called this *nucleophilically assisted* catalysis and warned it could mislead in the interpretation of the experimental mutagenesis data (i.e., inactivation of α3GalT in the E317A mutant is also expected in this mechanism). From the IP, as the attacking hydroxyl (OA) is being activated by its interaction with the leaving group (UDP), OA attack takes place readily and HOA is finally transferred to UDP. The final product (P) is formed, with net retention of the configuration.

The possibility of a double-displacement mechanism in α3GalT was also investigated. CGE formation was modeled considering the reaction coordinate $RC_{dd1} = [d(OP-C1) - d(ON-C1) - d(HOA-OP)]$ and the attack of the acceptor substrate to the CGE using $RC_{dd2} = [d(ON-C1) - d(OA-C1) - d(HOA-OP)]$. The optimized and characterized TSs and CGE intermediate are also depicted in Fig. 6. The averaged potential energy barriers corresponding to $TS_1{}^d$ and $TS_2{}^d$ are again very similar (15.1±3.5 and 15.9±3.1 kcal/mol, respectively) and in agreement with the experimentally derived ones. CGE formation is quite endoergic (11.4±4.2 kcal/mol above reactants) and the barriers for going back to reactants or evolving toward the products are relatively small. These two facts may explain why it is so difficult to experimentally isolate this complex for the wild-type enzyme, as has also been recently highlighted for another GT6 family member (Bobovská et al., 2015). From a structural point of view, the TS for the first step ($TS_1{}^d$) is very dissociative, with d(C1–OP) = 3.18 Å, and the attacking ON atom at 2.60 Å from C1. As $TS_1{}^i$ and IP in the front-side attack, $TS_1{}^d$ is an oxocarbenium species ($\Delta q_{R\rightarrow TS}$(C1 + O5) = 0.33 a.u., d(C1–O5$_{\alpha\text{-Gal}}$) = 1.28 Å). It is important to note that the reaction has also started with the reorientation of HOA, which provides stabilization to the leaving group (OP–HOA hydrogen bond). This interaction seems to be very important for UDP-sugar bond breakage despite of the mechanism, and we have referred to it as *substrate-assisted catalysis*. The lactose OA is at 2.85 Å from C1 in TS_1^d, a similar distance to the one obtained for TS_1^i. But in this mechanism, it is ON the nucleophile that attacks, leading to the formation of the CGE (d(ON–C1) = 1.56 Å). The attack of LAT on the α-face of the ring leads to the final product with net retention of the configuration and passing through TS_2^d, which resembles very much TS_2^i but exhibits slightly shorter d(C1–OA) and longer d(C1–ON).

The approximate location of the optimized TSs and intermediates is indicated in the two-dimensional PES shown in Fig. 7 (calculated at the QM(BP86/SVP)/MM level of theory). The PES was calculated by taking the CGE formation path and, at each point, calculating the front-side attack path at fixed ON–C1 distances. For clarity, the y- and x-axes represent the

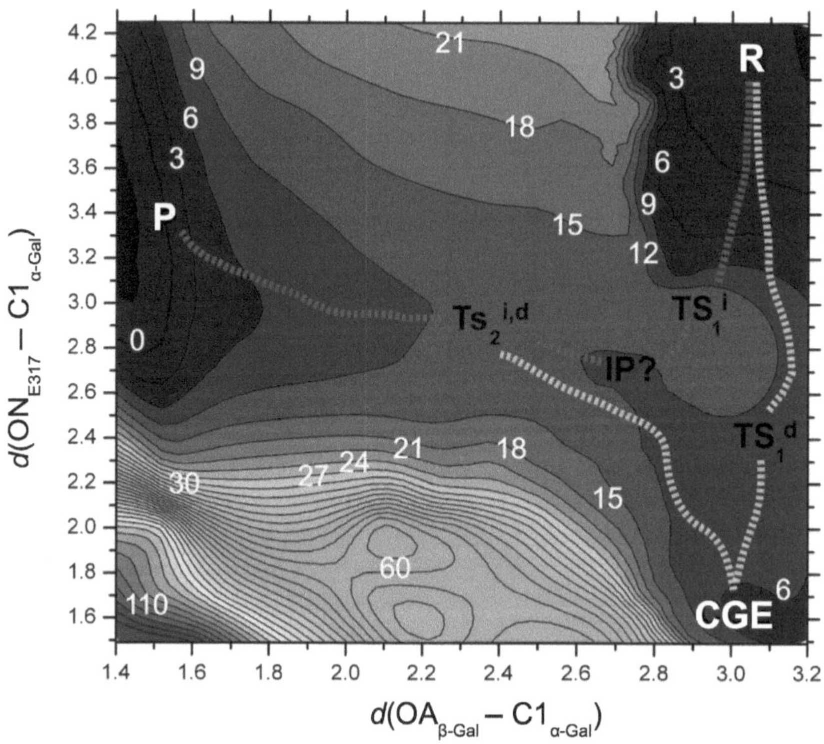

Figure 7 Two-dimensional potential energy surface for galactosyl transfer catalyzed by α3GalT. Both the front-side attack and the double-displacement mechanisms are indicated. Energies are given in kcal/mol and distances in Å. Contour lines at intervals of 3 kcal/mol. *Adapted from Gómez et al. (2013).*

d(ON–C1) and d(OP–C1) distances, respectively. It can be observed that the different oxocarbenium species share a similar energy, resulting in a flat zone around the IP. When comparing both mechanisms, the front-side attack barriers are ~0.4 kcal/mol higher than double-displacement ones (QM = (M05-2X/TZVP//BP86/SVP)), suggesting that the latter could be the preferred mechanism as it would be faster. However, the differences found are within the error of the QM methods so that no clear conclusion can be drawn except that they share relatively similar potential energy barriers. In the study of GTB (GT6 family member), Bobovská et al. (2015) found that front-side attack and double displacement would share the initial formation of an IP (called OCII 2A in their work) and the final step from a slightly different IP (OCII 2B) to product formation, being this last one the rate-determining step for both mechanisms. According to this, they concluded that both mechanisms could not be distinguished. Ardèvol et al. studied bovine α3GalT but could only find the double-displacement mechanism (Rojas-Cervellera et al., 2013). Given the similar energies and structures between the two mechanisms reported by us and by Bobovská et al., this result is somehow surprising. Even though they are calculating free energy surfaces and that we are reporting PESs, entropic contributions are not expected to be so big to completely vanish the front-side attack mechanism. Intriguingly, during their simulations, the OA atom gets as close as ~2.5 Å from C1 but it is not able to attack. In fact, the important HOA–OP interaction (substrate-assisted catalysis) has not been formed at this point, so that the OA hydroxyl is not activated. In agreement with this, there is little dissociation of the C1–OP bond (~2.2 Å), so that there is not enough room for the acceptor to attack and nucleophilic attack by Glu317 is required in order to completely break the UDP–Gal bond. The results obtained with our model indicate that even in the case of the double-displacement mechanism, the *substrate-assisted catalysis* is determinant for CGE formation at suitable rates. Differences in the structure of the initial model used may be the origin of the different results obtained.

As for LgtC, our analysis reveals that enzyme–substrate and substrate–substrate interactions are very important in facilitating the reaction (for both mechanisms). These interactions provide stabilization to the UDP leaving group by interacting with its phosphate oxygen atoms (through Arg365, Tyr139, Lys359, $O2_{\alpha\text{-Gal}}$, HOA, and a water molecule coordinated to M^{2+}) or with the oxocarbenium species (through Asp316 and Glu317).

3.3 Human Polypeptide-*N*-Acetylgalactosamine Transferase 2

GalNAc-T2 (EC 2.4.1.41) is a GT27 family member with a GTA fold that catalyzes the transfer of GalNAc from the UDP-GalNAc donor to Ser/Thr residues with net retention of the configuration in what is the first step of mucin-type O-glycosylation (Ten Hagen, Fritz, & Tabak, 2003). Mucins are highly O-glycosylated glycoproteins ubiquitous in mucous secretions on cell surfaces and in body fluids. Disruptions in their glycans biosynthesis are linked to various diseases, including cancer. Crystallographic structures of GalNAc-T2 show that the only possible nucleophilic residues that could act in a double-displacement mechanism are positioned ~7 Å away from C1, suggesting that this is not the reaction mechanism followed by the enzyme (Fritz, Hurley, Trinh, Shiloach, & Tabak, 2004; Fritz, Raman, & Tabak, 2006; Kubota et al., 2006; Lira-Navarrete et al., 2014). Long time-scale MD simulations performed by Milac et al. did not change this observation (Milac, Buchete, Fritz, Hummer, & Tabak, 2007). Their simulations are more consistent with a front-side attack, as the acceptor hydroxyl (OA) is found at ~3 Å all along the simulations. The structural and simulations data also show that there is not any residue's side chain on the β-face of GalNAc that could provide stabilization to the oxocarbenium. However, the carbonyl backbone of Ala307 could play such a role. We thus performed QM(DFT)/MM calculations to explore and analyze this mechanism, in what is the first computational study of a retaining GT using a peptide (EA2; sequence PTTDSTTPAPTTK) as the acceptor substrate (Gómez et al., 2014). Two different snapshots from MD simulations were considered in the reactivity studies but, for simplicity, only the results for one of them will be summarized here.

The glycosyl transfer reaction was modeled using the reaction coordinate $RC_i = [d(OP - C1) - d(OA - C1) - d(HOA - OP)]$ (the same results were obtained with an RC that did not include the HOA–OP distance). The potential energy profile obtained presents a quite planar zone in the high energy structures corresponding to oxocarbenium species (see Figure 2 in Gómez et al., 2014). Taking this into account, we cannot discard the existence of a short-lived ion pair intermediate even though we were not able to localize an IP by unrestrained geometry optimization at the QM(BP86/SVP)/MM level. In fact, only one transition state could be found by unrestrained TS search optimization, but it could not be fully characterized by frequency calculations. Thus, the topology of the PES in this region, together with the limitations of the density functional method used to describe it (Trnka et al., 2015), did not allow us to completely describe the type of front-side mechanism used by GalNAc-T2 (Fig. 1B vs. C). What is clear, though, is that an asynchronous front-side mechanism is active in this enzyme (with bond dissociation and formation taking place one after the other). Later studies of this enzyme seem to indicate that a short-lived ion pair intermediate may be actually formed (Lira-Navarrete et al., 2014; Trnka et al., 2015). The calculated stationary points reported in our work are depicted in Fig. 8.

It can be seen that at the optimized reactants, the HOA–OP interaction is already present (d(HOA–OP) = 1.79 Å) and that the acceptor OA is at a quite short distance from C1 (2.86 Å). The carbonyl backbone oxygen atom (ON) is at 4.09 Å from C1. The reaction starts with the breakage of the UDP–GalNAc bond with an energy cost that accounts for most of the potential energy barrier for the overall transfer reaction. An estimation of the free energy profile at the QM(SCC-DFTB)/MM level was also done for this system (Figure S3 in Gómez et al., 2014). Comparison with the corresponding potential energy profile suggests that entropic effects may be relatively small for this system, indicating that the calculated potential energy barrier can be used as a good estimate of the free energy barrier. At the QM = (M05-2X/TZVP//BP86/SVP) level of theory, the calculated potential energy barrier, which is given by the energy of TS^i, is of 19.8 kcal/mol, in quite good agreement with the experimental phenomenological free energy barrier of 17.3 kcal/mol (derived from a k_{cat} value of 3.7 s^{-1} at 310 K).

As can be seen in Fig. 8, in TS^i the OA atom is attacking C1 (d(OA–C1) = 2.20 Å) and HOA is being transferred to OP (d(HOA–OP) = 1.24 Å). As observed for the other systems, the breakage of the glycosidic

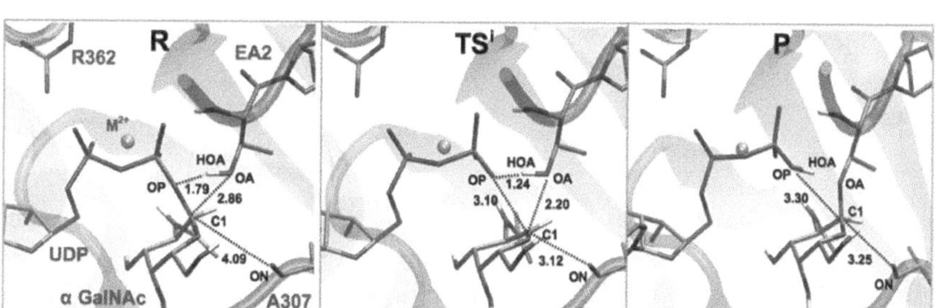

Figure 8 Molecular representations of the optimized reactants (R), transition state (TS^i), and products (P) for a front-side mechanism in GalNAc-T2.

bond provokes a positive charge development ($\Delta q(C1+O5_{GalNAc})=$ 0.25 a.u.) and a planar geometry at the anomeric center. This brings C1 at a distance of 3.12 Å from ON_{Ala307}. The analysis of electrostatic contributions estimates that this results in a ~2 kcal/mol stabilization of the oxocarbenium ion by the Ala307 carbonyl group, which is a moderate contribution similar to the one found for Gln189 in LgtC. Another enzyme residue that stabilizes the oxocarbenium formation is Glu334, located on the β-face of the GalNAc ring (hydrogen bonded to O4) and involved in donor binding. Glu334 is equivalent to Asp188 and Asp316 in LgtC and α3GalT, respectively, and they all provide a similar degree of stabilization. On the other side of the sugar donor, Arg362 comes out as the most stabilizing residue and it greatly contributes in making UDP a good leaving group.

Very interestingly, and as has been observed in the previous reviewed retaining GTs, substrate–substrate interactions appear again as having a crucial role in reaction. The most significant one is the HOA–OP interaction (*substrate-assisted catalysis*), which contributes to make UDP a good leaving group while it activates the attacking hydroxyl group (OA). In GalNAc-T2, another donor-to-acceptor interaction is found to stabilize the negative charge developing on UDP. This is the hydrogen bond between OP and the backbone amine of the acceptor substrate (Thr7 of the EA2 peptide, Fig. 8). Such an interaction could be recurrent in retaining GTs acting on peptide acceptor substrates. Finally, as the glycosidic bond is broken, the negative charge increase in UDP is stabilized by the internal interaction between the group at position 2 of the sugar ring (in this case the NAc group) and the oxygen atom of the β-phosphate coordinated to the metal cofactor.

Some of these interactions were tested *in silico* with substrate derivatives or mutated enzymes and the results were compared to experimental data. The readers can find these results in the original paper (Gómez et al., 2014).

3.4 Implications for Retaining GTs Mechanism

The QM/MM calculations carried out by our group (and by others) on retaining GTs have had an important role in characterizing the front-side attack mechanism as the general one followed by these enzymes (at least for the studied systems, and only with the possible exception of family GT6 members) (Ardèvol & Rovira, 2011; Bobovská et al., 2014, 2015; Gómez, Lluch, et al., 2012; Gómez et al., 2013; Gómez, Polyak, et al., 2012; Gómez et al., 2014; Lira-Navarrete et al., 2014; Rojas-Cervellera et al., 2013; Trnka et al., 2015). Crystallographic structural data with substrate analogues or product mimics also support this as the preferred mechanism. The clue to understand this same-face mechanism is the realization that the reaction starts with the breakage of the UDP-sugar bond, so that enough room is provided for the acceptor substrate to attack by the same side where the UDP leaving group is. This glycosidic bond breakage is accompanied by the development of a negative charge on UDP and of a positive one at the anomeric center of the sugar (oxocarbenium ion–phosphate ion structure).

A key aspect for UDP-sugar bond breakage to happen at suitable rates is the stabilization of the UDP leaving group. Different interactions with enzyme residues and that with the metal cofactor are observed for each system. Notably, substrate–substrate interactions are also involved in UDP stabilization. Among them, the most important one is that of the hydrogen atom from the attacking hydroxyl (HOA) with the glycosidic oxygen of UDP (OP) (*substrate-assisted catalysis*). This interaction is formed, if not present in the Michaelis complex, early in the reaction path and will lead to the final proton transfer to OP at the ending part of the reaction. Thus, UDP is confirmed as the catalytic base that activates the acceptor substrate in retaining GTs catalysis. Figure 9 depicts the molecular orbitals that participate in this interaction at transition state structures of the enzymes reviewed in this chapter (Gómez et al., 2013, 2014). Another recurrent substrate–substrate interaction that stabilizes UDP departure, although with a much smaller contribution, is the one between the substituent (—OH or —NAc) at the C2 position of the sugar in the donor substrate with the β-phosphate oxygen coordinated to M^{2+}. Finally, depending on the chemical nature of the acceptor substrate and on its binding mode

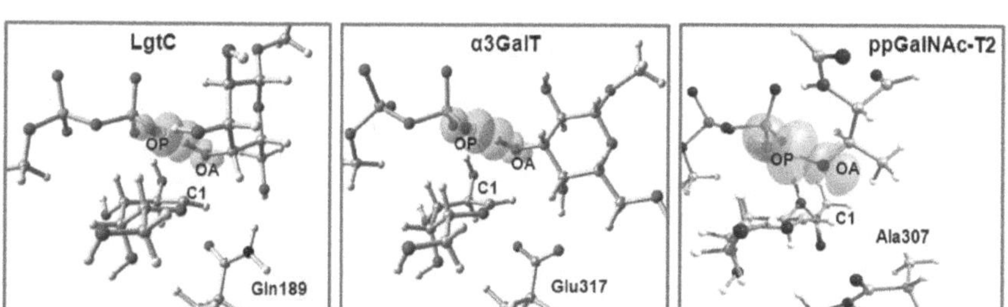

Figure 9 Interaction between molecular orbitals of the substrates corresponding to the crucial OP–HOA interaction at the transition state that stabilize UDP departure in LgtC, α3GalT, and GalNAc-T2. For clarity, just a fraction of the QM atoms is shown. *Figures adapted from Gómez et al. (2013, 2014).*

relative to the sugar donor, other donor-to-acceptor interactions may be established: $O3_{LAT}$–OP_{UDP} in LgtC, HN_{Thr7}–OP_{UDP} in GalNAc-T2, and none in α3GalT.

Another way to facilitate UDP-sugar bond breakage is by stabilizing the oxocarbenium species that are formed (TSs and IP, if present). The conserved $Asp188_{LgtC}$/$Asp316_{\alpha3GalT}$/$Glu334_{GalNAc\text{-}T2}$ residue, involved in sugar binding by interaction with the O4 (and sometimes O6) hydroxyl groups, contributes to it. Most importantly, stabilization by residues located on the β-face of the sugar and close to the C1 anomeric carbon is also encountered: $Gln189_{LgtC}$/$Glu317_{\alpha3GalT}$/$Ala307_{GalNAc\text{-}T2}$. In some systems, like the GT6 family member α3GalT, the UDP leaving group stabilization is not strong enough and oxocarbenium stabilization needs to be particularly enhanced. Thus, a strong nucleophile like Glu is found at this position (*nucleophilically assisted catalysis*). Therefore, Glu317 plays a key role in α3GalT catalysis via a front-side mechanism. This situation can be seen as a limit case of the front-side attack mechanism and, as a strong nucleophile is found in the proximity of C1, it can actually lead to CGE formation and, therefore, a switch to the double-displacement mechanism. Notice that for the reaction to proceed through the second step of the double-displacement mechanism, this CGE cannot be very stable. This may explain why it has not been possible to observe it experimentally for wild-type enzymes. It is also important to notice that, according to our calculations, the presence of a Glu is not enough to make bond breakage easy in α3GalT and that *substrate-assisted catalysis* is a requirement both for the front-side and the double-displacement mechanisms (see Figure 5 in Gómez et al., 2013). Thus, UDP stabilization related altogether with acceptor activation seems to be a priority in catalysis by retaining GTs. The front-side mechanism represents a shortcut for sugar transfer that requires less atomic reorganization than the double-displacement mechanism and, by using the OP–HOA interaction as a key element, it may prevent fast hydrolysis of the donor substrate in the absence of the acceptor substrate.

4. CONCLUSIONS

The key role of glycans in molecular recognition processes related to cell and life development, including disease and infection, is now recognized and an expanding field of research. GTs are responsible for the formation of new glycosidic bonds in glycan biosynthesis, but still many aspects of their activity remain unknown. This is the case for the chemical reaction mechanism used by the retaining enzymes, a question that is under strong debate. A controversial front-side mechanism in which the acceptor substrate nucleophilically attacks by the same side where the leaving group is has been proposed, together with a double-displacement mechanism involving a nucleophilic residue from the enzyme and a covalent glycosyl–enzyme intermediate. QM/MM calculations have brought a detailed description of these mechanisms and have allowed us to propose the front-side mechanism as the general mechanism followed by these enzymes. On the other hand, the double displacement can be seen as a limit case of the front-side attack that may operate in particular (yet few) cases where a strong nucleophile is present in the β-face of the transferred sugar (so far, family GT6 members). The presence of such nucleophile seems to be related to an insufficient UDP stabilization as a leaving group, as well as with acceptor binding. Our analyses reveal how biomolecular interactions modulate the particular mechanistic strategy used by each retaining GT. In brief, if the enzyme–substrate and substrate–substrate interactions are strong enough to make UDP a good leaving group (and in consequence giving to OA the necessary space to attack), the front-side mechanism will be the most efficient mechanism and the preferred one for glycosyl transfer with retention of the configuration. Thus, UDP stabilization related to acceptor activation seems to be a priority in catalysis by retaining GTs. However, if UDP stabilization as a leaving group is not good enough, strong stabilization needs to be provided to the oxocarbenium ion and, if a strong nucleophile is used for this purpose (like in GT6 family members), it may lead to a shift toward a double-displacement mechanism. As a general mechanism, though, the front-side mechanism represents an efficient shortcut for sugar transfer which, by using the OP–HOA interaction as a key element, would also ensure that the donor substrate is not easily hydrolyzed in the absence of the acceptor substrate.

ACKNOWLEDGMENTS

The authors want to thank all the collaborators in the original works reviewed here: W. Thiel, I. Polyak, L.A. Tabak, R. Rojas and D. Patel. We are grateful for the financial support given by the Spanish "Ministerio de Economia y Competitividad" through projects CTQ2011-24292 and CTQ2014-53144-P, and the "Ayudas para la formación postdoctoral 2013" Program (to H.G.); the "UAB-Banco Santander" Program (to L.M.) and the "Becas de Doctorado en el Extranjero—Becas Chile—CONICYT" Program (to F.M.).

REFERENCES

Ahlrichs, R., Bär, M., Häser, M., Horn, H., & Kölme, C. (1989). Electronic structure calculations on workstation computers: The program system turbomole. *Chemical Physics Letters*, *162*, 165–169.

Ardèvol, A., & Rovira, C. (2011). The molecular mechanism of enzymatic glycosyl transfer with retention of configuration: Evidence for a short-lived oxocarbenium-like species. *Angewandte Chemie*, *50*, 1–6.

Baker, J. (1986). An algorithm for the location of transition states. *Journal of Computational Chemistry*, *7*, 385–395, Retrieved from.

Banerjee, A., Adams, N., Simons, J., & Shepard, R. (1985). Search for stationary points on surfaces. *The Journal of Physical Chemistry*, *89*, 52–57.

Becke, A. (1988). Density-functional exchange-energy approximation with correct asymptotic behavior. *Physical Review A*, *38*, 3098–3100.

Bennett, E. P., Mandel, U., Clausen, H., Gerken, T. A., Fritz, T. A., & Tabak, L. A. (2012). Control of mucin-type O-glycosylation: A classification of the polypeptide GalNAc-transferase gene family. *Glycobiology*, *22*, 736–756.

Billeter, S. R., Turner, A. J., & Thiel, W. (2000). Linear scaling geometry optimisation and transition state search in hybrid delocalised internal coordinates. *Physical Chemistry Chemical Physics*, *2*, 2177–2186.

Bobovská, A., Tvaroška, I., & Kóňa, J. (2014). A theoretical study on the catalytic mechanism of the retaining α-1,2-mannosyltransferase Kre2p/Mnt1p: The impact of different metal ions on catalysis. *Organic & Biomolecular Chemistry*, *12*, 4201–4210.

Bobovská, A., Tvaroška, I., & Kóňa, J. (2015). Theoretical study of enzymatic catalysis explains why the trapped covalent intermediate in the E303C mutant of glycosyltransferase GTB was not detected in the wild-type enzyme. *Glycobiology*, *25*, 3–7.

Bock, C. W., Katz, A. K., Markham, G. D., & Glusker, J. P. (1999). Manganese as a replacement for magnesium and zinc: Functional comparison of the divalent ions. *Journal of the American Chemical Society*, *2*, 7360–7372.

Boix, E., Swaminathan, G. J., Zhang, Y., Natesh, R., Brew, K., & Acharya, K. R. (2001). Structure of UDP complex of UDP-galactose:beta-galactoside-alpha−1,3-galactosyltransferase at 1.53-A resolution reveals a conformational change in the catalytically important C terminus. *The Journal of Biological Chemistry*, *276*, 48608–48614.

Breton, C., Fournel-Gigleux, S., & Palcic, M. M. (2012). Recent structures, evolution and mechanisms of glycosyltransferases. *Current Opinion in Structural Biology*, *22*, 540–549.

Brooks, B. R., Iii, C. L. B., MacKerell, A. D., Nilsson, L., Petrella, R. J., Roux, B., et al. (2009). CHARMM: The biomolecular simulation program. *Journal of Computational Chemistry*, *30*, 1545–1614.

Campbell, J., Davies, G., Bulone, V., & Henrissat, B. (1997). A classification of nucleotide-diphospho-sugar glycosyltransferases based on amino acid sequence similarities. *The Biochemical Journal*, *326*, 929–942.

Chan, J., Tang, A., & Bennet, A. J. (2012). A stepwise solvent-promoted SNi reaction of α-D-glucopyranosyl fluoride: Mechanistic implications for retaining glycosyltransferases. *Journal of the American Chemical Society*, *134*, 1212–1220.

Csonka, G. I., French, A. D., Johnson, G. P., & Stortz, C. A. (2009). Evaluation of density functionals and basis sets for carbohydrates. *Journal of Chemical Theory and Computation*, *5*, 679–692.

De Vries, A. H., Sherwood, P., Collins, S. J., Rigby, A. M., Rigutto, M., & Kramer, G. J. (1999). Zeolite structure and reactivity by combined quantum-chemical-classical calculations. *The Journal of Physical Chemistry B*, *103*, 6133–6141.

Eichkorn, K., Treutler, O., Ohm, H., Marco, H., & Ahlrichs, R. (1995). Auxiliary basis sets to approximate Coulomb potentials. *Chemical Physics Letters*, *240*, 283–290.

Eichkorn, K., Weigend, F., Treutler, O., & Ahlrichs, R. (1997). Auxiliary basis sets for main row atoms and transition metals and their use to approximate Coulomb potentials. *Theoretical Chemistry Accounts*, *97*, 119–124.

Elstner, M., Porezag, D., Jungnickel, G., Elsner, J., Haugk, M., Suhai, S., et al. (1998). Self-consistent-charge density-functional tight-binding method for simulations of complex materials properties. *Physical Review B*, *58*, 7260–7268.

Errey, J. C., Lee, S. S., Gibson, R. P., Martinez Fleites, C., Barry, C. S., Jung, P. M. J., et al. (2010). Mechanistic insight into enzymatic glycosyl transfer with retention of configuration through analysis of glycomimetic inhibitors. *Angewandte Chemie*, *49*, 1234–1237.

Foster, J. P., & Weinhold, F. (1980). Natural hybrid orbitals. *Journal of the American Chemical Society*, *102*, 7211–7218.

Frauenheim, T., Seifert, G., Elstner, M., & Sternberg, M. (2002). Atomistic simulations of complex materials: Ground-state and excited-state properties. *Journal of Physics: Condensed Matter*, *14*, 3015–3047.

Frisch, M. J., Trucks, G. W., Schlegel, H. B., Scuseria, G. E., Robb, M. A., Cheeseman, J. R., et al. (2009). *Gaussian 09, Revision D.01*. Wallingford, CT: Gaussian, Inc.

Fritz, T. A., Hurley, J. H., Trinh, L. B., Shiloach, J., & Tabak, L. A. (2004). The beginnings of mucin biosynthesis: The crystal structure of UDP-GalNAc:polypeptide alpha-N-acetylgalactosaminyltransferase-T1. *Proceedings of the National Academy of Sciences of the United States of America*, *101*, 15307–15312.

Fritz, T. A., Raman, J., & Tabak, L. A. (2006). Dynamic association between the catalytic and lectin domains of human UDP-GalNAc:polypeptide alpha-N-acetylgalactosaminyltransferase-2. *The Journal of Biological Chemistry*, *281*, 8613–8619.

Galili, U., & Swanson, K. (1991). Gene sequences suggest inactivation of alpha-1,3-galactosyltransferase in catarrhines after the divergence of apes from monkeys. *Proceedings of the National Academy of Sciences of the United States of America*, *93*, 7401–7404.

Glendening, E. D., Reed, A. E., Carpenter, J. E., & Weinhold, F. (2009). *NBO 3.1 Program*. Wallingford, CT: Gaussian Inc.

Gómez, H., Lluch, J. M., & Masgrau, L. (2012). Essential role of glutamate 317 in galactosyl transfer by α3GalT: A computational study. *Carbohydrate Research*, *356*, 204–208.

Gómez, H., Lluch, J. M., & Masgrau, L. (2013). Substrate-assisted and nucleophilically assisted catalysis in bovine α1,3-galactosyltransferase. Mechanistic implications for retaining glycosyltransferases. *Journal of the American Chemical Society*, *135*, 7053–7063.

Gómez, H., Polyak, I., Thiel, W., Lluch, J. M., & Masgrau, L. (2012). Retaining glycosyltransferase mechanism studied by QM/MM methods: Lipopolysaccharyl-α-1,4-galactosyltransferase C transfers α-galactose via an oxocarbenium ion-like transition state. *Journal of the American Chemical Society*, *134*, 4743–4752.

Gómez, H., Rojas, R., Patel, D., Tabak, L. A., Lluch, J. M., & Masgrau, L. (2014). A computational and experimental study of O-glycosylation. Catalysis by human UDP-GalNAc polypeptide:GalNAc transferase-T2. *Organic & Biomolecular Chemistry*, *12*, 2645–2653.

Guvench, O., Greene, S. N., Kamath, G., Brady, J. W., Venable, R. M., Pastor, R. W., et al. (2008). Additive empirical force field for hexopyranose monosaccharides. *Journal of Computational Chemistry*, *29*, 2544–2564.

Guvench, O., Hatcher, E. R., Venable, R. M., Pastor, R. W., & MacKerell, A. D. (2009). CHARMM additive all-atom force field for glycosidic linkages between hexopyranoses. *Journal of Chemical Theory and Computation*, *5*, 2353–2370.

Hart, G. W., & Copeland, R. J. (2010). Glycomics hits the big time. *Cell, 143*, 672–676.
Hoover, W. G. (1985). Canonical dynamics: Equilibrium phase-space distributions. *Physical Review A, 31*, 1695–1697.
Jamaluddin, H., Tumbale, P., Withers, S. G., Acharya, K. R., & Brew, K. (2007). Conformational changes induced by binding UDP-2F-galactose to α-1,3 galactosyltransferase-implications for catalysis. *Journal of Molecular Biology, 369*, 1270–1281.
Kästner, J., & Thiel, W. (2005). Bridging the gap between thermodynamic integration and umbrella sampling provides a novel analysis method: "Umbrella integration". *The Journal of Chemical Physics, 123*, 144104.
Kästner, J., & Thiel, W. (2006). Analysis of the statistical error in umbrella sampling simulations by umbrella integration. *The Journal of Chemical Physics, 124*, 234106.
Kóňa, J., & Tvaroška, I. (2009). Comparative DFT study on the α-glycosidic bond in reactive species of galactosyl diphosphates. *Chemical Papers, 63*, 598–607.
Křen, V. (2008). Glycoside vs. aglycon "The role of glycosidic residue in biological activity" In B. O. Fraser-Reid, K. Tatsuta, J. Thiem, G. L. Coté, S. Flitsch, & Y. Ito, et al. (Eds.), *Glycoscience* (pp. 2591–2638). Berlin: Springer.
Kubota, T., Shiba, T., Sugioka, S., Furukawa, S., Sawaki, H., Kato, R., et al. (2006). Structural basis of carbohydrate transfer activity by human UDP-GalNAc: Polypeptide alpha-N-acetylgalactosaminyltransferase (pp-GalNAc-T10). *Journal of Molecular Biology, 359*, 708–727.
Kumar, S., Rosenberg, J. M., Bouzida, D., Swendsen, R. H., & Kollman, P. A. (1992). The weighted histogram analysis method for free-energy calculations on biomolecules. 1. The method. *Journal of Computational Chemistry, 135*, 40–57.
Lairson, L., Henrissat, B., Davies, G., & Withers, S. (2008). Glycosyltransferases: Structures, functions, and mechanisms. *Annual Review of Biochemistry, 77*, 521–555.
Lee, S. S., Hong, S. Y., Errey, J. C., Izumi, A., Davies, G. J., & Davis, B. G. (2011). Mechanistic evidence for a front-side, SNi-type reaction in a retaining glycosyltransferase. *Nature Chemical Biology, 7*, 631–638.
Li, H., Robertson, A. D., & Jensen, J. H. (2005). Very fast empirical prediction and rationalization of protein pKa values. *Proteins: Structure, Function, and Bioinformatics, 61*, 704–721.
Lira-Navarrete, E., Iglesias-Fernández, J., Zandberg, W. F., Compaçón, I., Kong, Y., Corzana, F., et al. (2014). Substrate-guided front-face reaction revealed by combined structural snapshots and metadynamics for the polypeptide *N*-acetylgalactosaminyltransferase 2. *Angewandte Chemie International Edition, 53*, 8206–8210.
Liu, D. C., & Nocedal, J. (1989). On the limited memory BFGS method for large scale optimization. *Mathematical Programming, 45*, 503–528.
Lonsdale, R., Harvey, J. N., & Mulholland, A. J. (2012). A practical guide to modelling enzyme-catalysed reactions. *Chemical Society Reviews, 41*, 3025–3038.
MacKerell, A. D., Bashford, D., Bellott, M., Dunbrack, R. L., Eva seck, J. D., Field, M. J., et al. (1998). All-atom empirical potential for molecular modeling and dynamics studies of proteins. *Journal of Physical Chemistry B, 102*, 3586–3617.
MacKerell, A. D., Feig, M., & Brooks, C. L. (2004). Improved treatment of the protein backbone in empirical force fields. *Journal of the American Chemical Society, 126*, 698–699.
Milac, A. L., Buchete, N. V., Fritz, T. A., Hummer, G., & Tabak, L. A. (2007). Substrate-induced conformational changes and dynamics of UDP-N-acetylgalactosamine: Polypeptide N-acetylgalactosaminyltransferase-2. *Journal of Molecular Biology, 373*, 439–451.
Monegal, A., & Planas, A. (2006). Chemical rescue of alpha3-galactosyltransferase. Implications in the mechanism of retaining glycosyltransferases. *Journal of the American Chemical Society, 128*, 16030–16031.

Nocedal, J. (1980). Updating quasi-Newton matrices with limited storage. *Mathematics of Computation, 35*, 773–782.
Nosé, S. (1984). A unified formulation of the constant temperature molecular dynamics methods. *The Journal of Chemical Physics, 81*, 511–519.
Ohtsubo, K., & Marth, J. D. (2006). Glycosylation in cellular mechanisms of health and disease. *Cell, 126*, 855–867.
Perdew, J. P. (1986). Density-functional approximation for the correlation energy of the inhomogeneous electron gas. *Physical Review B, 33*, 8822–8824.
Perdew, J. P., & Yue, W. (1986). Accurate and simple density functional for the electronic exchange energy: Generalized gradient approximation. *Physical Review B, 33*, 8800–8802.
Persson, K., Ly, H. D., Dieckelmann, M., Wakarchuk, W. W., Withers, S. G., & Strynadka, N. C. J. (2001). Crystal structure of the retaining galactosyltransferase LgtC from Neisseria meningitidis in complex with donor and acceptor sugar analogs. *Nature Structural Biology, 8*, 166–175.
Phillips, J. C., Braun, R., Wang, W., Gumbart, J., Tajkhorshid, E., Villa, E., et al. (2005). Scalable molecular dynamics with NAMD. *Journal of Computational Chemistry, 26*, 1781–1802.
Reed, A. E., Curtiss, L. A., & Weinhold, F. (1988). Intermolecular interactions from a natural bond orbital, donor-acceptor viewpoint. *Chemical Reviews, 88*, 899–926.
Reed, A. E., & Weinhold, F. (1983). Natural bond orbital analysis of near-Hartree–Fock water dimer. *The Journal of Chemical Physics, 78*, 4066.
Reed, A. E., Weinstock, R. B., & Weinhold, F. (1985). Natural population analysis. *The Journal of Chemical Physics, 83*, 735.
Rojas-Cervellera, V., Ardèvol, A., Boero, M., Planas, A., & Rovira, C. (2013). Formation of a covalent glycosyl–enzyme species in a retaining glycosyltransferase. *Chemistry A European Journal, 19*, 14018–14023.
Ryckaert, J.-P., Ciccotti, G., & Berendsen, H. J. (1977). Numerical integration of the cartesian equations of motion of a system with constraints: Molecular dynamics of n-alkanes. *Journal of Computational Physics, 23*(3), 327–341. http://dx.doi.org/10.1016/0021-9991(77)90098-5.
Schäfer, A., Horn, H., & Ahlrichs, R. (1992). Fully optimized contracted Gaussian basis sets for atoms Li to Kr. *The Journal of Chemical Physics, 97*, 2571–2577.
Schäfer, A., Huber, C., & Ahlrichs, R. (1994). Fully optimized contracted Gaussian basis sets of triple zeta valence quality for atoms Li to Kr. *The Journal of Chemical Physics, 100*, 5829–5835.
Senn, H. M., & Thiel, W. (2009). QM/MM methods for biomolecular systems. *Angewandte Chemie, 48*, 1198–1229.
Sherwood, P., De Vries, A. H., Collins, S. J., Greatbanks, S. P., Burton, N. A., Vincent, M. A., et al. (1997). Computer simulation of zeolite structure and reactivity using embedded cluster methods. *Faraday Discussions, 106*, 79–92.
Sherwood, P., De Vries, A. H., Guest, M. F., Schreckenbach, G., Catlow, C. R. A., French, S. A., et al. (2003). QUASI: A general purpose implementation of the QM/MM approach and its application to problems in catalysis. *Journal of Molecular Structure: THEOCHEM, 632*, 1–28.
Slater, J. C. (1951). A simplification of the Hartree-Fock method. *Physical Review, 81*, 385.
Smith, W., & Forester, T. R. (1996). DL POLY 2. 0: A general-purpose parallel molecular dynamics simulation package. *Journal of Molecular Graphics, 14*, 136–141.
Solá, R. J., Rodríguez-Martínez, J. A., & Griebenow, K. (2007). Modulation of protein biophysical properties by chemical glycosylation: Biochemical insights and biomedical implications. *Cellular and Molecular Life Sciences, 64*, 2133–2152.

Souaille, M., & Roux, B. (2001). Extension to the weighted histogram analysis method: Combining umbrella sampling with free energy calculations. *Computer Physics Communications*, *135*, 40–57.

Soya, N., Fang, Y., Palcic, M. M., & Klassen, J. S. (2011). Trapping and characterization of covalent intermediates of mutant retaining glycosyltransferases. *Glycobiology*, *21*, 547–552.

Ten Hagen, K. G., Fritz, T. A., & Tabak, L. A. (2003). All in the family: The UDP-GalNAc: polypeptide N-acetylgalactosaminyltransferases. *Glycobiology*, *13*, 1R–16R.

Thiel, W. (2005). *Program MNDO2009, Version 7.0.* Mülheim: Max-Planck-Institute für Kohlenforschung.

Trnka, T., Kozmon, S., Tsvaroska, I., & Koca, J. (2015). Stepwise catalytic mechanism via short-lived intermediate inferred from combined QM/MM MERP and PES calculations on retaining glycosyltransferase ppGalNAcT2. *PLoS Computational Biology*, *11*, 1–21.

Tzeng, Y. L., & Stephens, D. S. (2000). Epidemiology and pathogenesis of *Neisseria meningitidis*. *Microbes and Infection*, *2*, 687–700.

Vosko, S. H., Wilk, L., & Nusair, M. (1980). Accurate spin-dependent electron liquid correlation energies for local spin density calculations: A critical analysis. *Journal of Physics*, *58*, 1200–1211.

Zhang, C., Griffith, B. R., Qiang, F., Albermann, C., Fu, X., Lee, I.-K., et al. (2006). Exploiting the reversibility of natural product glycosyltransferase-catalyzed reactions. *Science*, *313*, 1291–1294.

Zhao, Y., Schultz, N., & Truhlar, D. (2006). Design of density functionals by combining the method of constraint satisfaction with parametrization for thermochemistry, thermochemical kinetics, and noncovalent interactions. *Journal of Chemical Theory and Computation*, *2*, 364–382.

CHAPTER NINE

Excited States and Photochemistry of Chromophores in the Photoactive Proteins Explored by the Combined Quantum Mechanical and Molecular Mechanical Calculations

Lihong Liu[1], Ganglong Cui, Wei-Hai Fang
Key Laboratory of Theoretical and Computational Photochemistry, Ministry of Education, College of Chemistry, Beijing Normal University, Beijing, China
[1]Corresponding author: e-mail address: lihong.liu@bnu.edu.cn

Contents

Abstract

A photoactive protein usually contains a unique chromophore that is responsible for the initial photoresponse and functions of the photoactive protein are determined by the interaction between the chromophore and its protein surroundings. The combined quantum mechanical and molecular mechanical (QM/MM) approach is demonstrated to be a very useful tool for exploring structures and functions of a photoactive protein with the chromophore and its protein surroundings treated by the QM and MM methods, respectively. In this review, we summarize the basic formulas of the QM/MM approach and emphasize its applications to excited states and photoreactions of chromophores in rhodopsin protein, photoactive yellow protein, and green fluorescent protein.

Advances in Protein Chemistry and Structural Biology, Volume 100
ISSN 1876-1623
http://dx.doi.org/10.1016/bs.apcsb.2015.07.002

1. INTRODUCTION

Light is recognized ubiquitously throughout nature and is essential to life as the primary source of energy. In order to gain information from the environment by light stimulus, a wide variety of photoreceptors (light-sensitive proteins) evolved in nature, such as light-sensitive proteins of rhodopsins in the animal kingdom and flavoprotein light sensors in higher plants. In most cases, these light-sensitive proteins contain a chromophore that enables them to perceive and respond to different wavelengths of light and its photoresponsive properties are regulated by the protein environment. The initial irradiation of the chromophore and the subsequent relaxation processes change the interaction between the chromophore and the protein environment, which further leads to conformational rearrangements and altered properties of the protein and the other components around the protein. As a result, extensive structural, spectroscopic, and biochemical studies have been brought to bear on the associated light-responsive mechanisms. Recent advances have led to unprecedented progress in the level of our understanding of photoinduced functional processes for biologically important photoactive proteins.

Meanwhile, a great leap forward has been witnessed in the ability to carry out quantum mechanical calculations and dynamics simulations of excited states and photochemical processes of chromophores in the photoactive proteins in the past a few decade (Ai, Liao, Chen, Luo, & Fang, 2010; Ai, Liao, Fang, & Luo, 2012; Ai, Zhang, Chen, Luo, & Fang, 2010; Ben-Nun & Martinez, 1998; Bochenkova & Andersen, 2013; Gozem, Krylov, & Olivucci, 2013; Gozem et al., 2012; Gozem, Melaccio, et al., 2013; Mori & Martinez, 2013; Navizet et al., 2010). The *ab initio* quantum mechanical methods have been well established for theoretical investigations of structures and properties of a molecular system in the ground state. It is currently not practical to describe a protein and its environment using purely *ab initio* methods due to the numerous increasing of computational cost. The combined quantum mechanical and molecular mechanical (QM/MM) approach was initially developed by Warshel (Warshel & Karplus, 1975; Levitt & Warshel, 1975; Warshel & Karplus, 1972) and Levitt (Levitt & Warshel, 1975; Warshel & Levitt, 1976) for biomolecular systems and its accuracy and reliability were confirmed by numerous QM/MM calculations for different biological systems. The basic idea of the QM/MM scheme is to describe the relatively small and important region of system quantum mechanically, whereas the remainder is treated by a classical force field. For instance, the chromophore in a photoactive protein should be treated quantum mechanically, while the surroundings that may be somewhat less important are treated molecular mechanically, as depicted in Scheme 1. The main issues involved in the QM/MM calculation are the partition of the investigated system into the QM and MM moieties, the choice of the suitable QM method, and the treatment of the QM–MM interaction in the QM/MM boundary regions. Currently, the QM/MM approach has become a valuable tool for investigating chemical reactions in explicit solvents and for modeling enzymatic reactions and other biological processes, which have been extensively reviewed in the previous publications (Bakowies & Thiel, 1996; de Visser, 2009; Lin & Truhlar, 2005; Monard & Merz, 1999; Ridder & Mulholland, 2003; Senn & Thiel, 2009; Zhang, Lin, & Truhlar, 2007). Herein, we will focus entirely on the QM/MM calculations of excited states and photochemical reactions for a few chromophores in photoactive proteins. We begin with the basic QM/MM formulas in Section 2 with an emphasis on how to choose the QM methodology for excited electronic states.

Perhaps, the best known biological photoreceptor is the rhodopsin protein with its retinal chromophore, which is responsible for the vision process in our eyes. It should be noted that only the 11-*cis*-retinal moiety is involved in virtually all visual pigments (animal rhodopsin). The binding pocket for the retinal is formed by seven transmembrane helices for all rhodopsin proteins, which exhibit the sequence homology in the retinal binding region. The photoinduced isomerization of the retinal C=C double bond takes place easily in the tight protein environment, which is comparable with that for the retinal in the free volume. It is evident the uniquely arranged binding

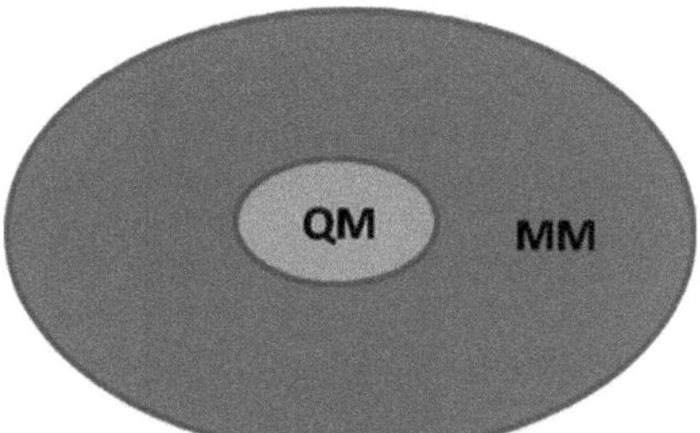

Scheme 1 The QM/MM scheme with the chromophore (C) treated quantum mechanically and the surroundings (S) treated molecular mechanically.

region provides the exceptionally high rate and efficiency of the retinal photodissociation. Since the photoinduced isomerization of the 11-*cis*-retinal into its all-*trans* form was discovered in the 1960s as the initial photochemical reaction in vision, the details of the mechanism for this important photochemical reaction has been the subject of experimental and theoretical investigations. In particular, numerous theoretical studies have tried to unravel the excited-state behavior and the key role of the conical intersection (CI) in the photoisomerization process in the past a few years, which is mainly elucidated in Section 3.

Unlike bacterial rhodopsins, photoactive yellow protein (PYP) is a highly soluble globular protein, which is found in purple photosynthetic bacteria as a water-soluble photosensor protein. After absorbing a photon in the visible region, the chromophore of PYP undergoes a fully reversible photocycle starting from a dark-state pG and then going through two early spectroscopic intermediates to form a red-shifted intermediate pR. The rearrangement of the hydrogen-bonding network triggers overall structural change of the protein moiety, and finally completes the whole photocycle. In this review, we will focus on the QM/MM investigation of the whole photocycle of the PYP photoresponse.

Green fluorescent protein (GFP) has been widely used as a fluorescent probe for *in vivo* biological imaging, which has brought about a revolution in the field of molecular biology. A broad range of GFP variants have been developed to improve the GFP stability, brightness, and fluorescent quantum yield for biological applications. Many structures of GFP variants (GFPs) have been determined by X-ray crystallography, in which 11 β-strands form an almost perfect cylinder, which is capped by short sections of α-helix. The *para*-hydroxybenzylidene-imidazolinone (*p*-HBDI) chromophore is totally encapsulated in the central of the cylinder, which is the main reason that high quantum yield of fluorescence can be observed in GFPs. The tremendous amount of work has devoted to GFPs and their applications, which have been the subject of a series of review articles. Herein, we will focus on recent progress toward understanding the excited-state behavior and photoreaction dynamics of the *p*-HBDI chromophore in GFPs from theoretical viewpoint. After a brief description of GFPs in Section 3, we conclude with a summary and perspective in Section 4. Finally, it should be pointed out that we only discuss a part of the QM/MM studies on GFP, PYP, and rhodopsin proteins in this review due to the tremendous amount of literature devoted to them.

2. METHOD

As mentioned in Section 1, the combined QM/MM scheme is to treat the region of chemical interest (e.g., chromophore) quantum mechanically and the rest of the system (e.g., surroundings) molecular mechanically. Theoretically, the QM/MM formalism can combine any of QM and MM methods. But for QM part in practical, the time-dependent density functional theory (Bartolotti, 1982; Dobson, Bunner, & Gross, 1997; Grabo, Petersilka, & Gross, 1999; Petersilka, Gossmann, & Gross, 1996) and the multireference methods (such as complete active space self-consistent field theory (CASSCF) (Knowles & Werner, 1985; Werner & Knowles, 1985), complete active space second perturbation theory (CASPT2) (Finley, Malmqvist, Roos, & Serrano-Andres, 1998), and multireference configuration interaction (MRCI) methods) are used in calculation of chormophore in the photoactive protein due to different accuracy and efficiency for specific system. Besides, semiempirical QM methods also have been used owing to its favorable computational efficiency, such as the OM2/MRCI (Fabiano, Keal, & Thiel, 2008; Hoffmann et al., 2006; Lan, Fabiano, & Thiel, 2009) and floating occupation molecular orbital (FOMO) method (Granucci & Toniolo, 2000; Slavicek & Martinez, 2010). As far as the choice of MM method is concerned, there are a number of biomolecular force fields available, such as Charmm and Amber force fields. The classical potential energy function of the MM region contains bonded terms (typically bond stretching, angle bending, torsions, and out-of-plane deformations or improper torsions), Lennard-Jones-type van der Waals terms, and Coulomb interaction terms between rigid point charges.

However, the total energy of the system cannot be simply written as the sum of the energies of the subsystems, the QM–MM coupling term have to be considered owning to the strong QM/MM interactions. Herein, we give a general review for the different strategies of QM–MM scheme in dealing with QM–MM coupling term.

First of all, we consider the basic aspect of the QM/MM, the subtractive and additive scheme. In the subtractive scheme, the total energy of the system is calculated as follows:

$$E_{\text{sub}}(T) = E_{\text{MM}}(T) + E_{\text{QM}}(C) - E_{\text{MM}}(C) \tag{1}$$

The $E_{\mathrm{sub}}(T)$ is the total energy calculated by the molecular mechanics, the $E_{\mathrm{QM}}(C)$ and $E_{\mathrm{MM}}(C)$ indicate the chromophore energy calculated by the quantum mechanics and molecular mechanics, respectively. The advantage and disadvantage of this subtractive QM/MM scheme are obvious. The advantage is simplicity and no explicit QM–MM coupling terms are needed, thus the standard QM and MM procedures can be used without any modifications, leading to a straightforward implementation. The disadvantage is that the electrostatic interactions are problematic since the coupling between the subsystems is handled entirely at the MM level. An alternative way to address this deficiency is the additive QM/MM scheme which is the most widely used now. The total energy of the system in this QM/MM scheme can be written as

$$E_{\mathrm{add}}(T) = E_{\mathrm{MM}}(S) + E_{\mathrm{QM}}(C) + E_{\mathrm{QM/MM}} \tag{2}$$

The $E_{\mathrm{MM}}(S)$ is the surrounding energy calculated by the molecular mechanics, while the $E_{\mathrm{QM}}(C)$ is the energy of the chromophore calculated by the quantum mechanics, and the $E_{\mathrm{QM/MM}}$ is the QM/MM coupling term. The QM/MM interaction can be further written as three different parts

$$E_{\mathrm{QM/MM}} = E^{b}_{\mathrm{QM/MM}} + E^{\mathrm{vwd}}_{\mathrm{QM/MM}} + E^{\mathrm{el}}_{\mathrm{QM/MM}} \tag{3}$$

The first term originates the covalent bonds that cross the QM–MM boundary. There are various approaches that have been devised to treat covalent bonds cut by the QM–MM boundary. The most well-known schemes are the link-atom scheme (Dapprich, Komaromi, Byun, Morokuma, & Frisch, 1999; de Vries et al., 1999; Eichinger, Tavan, Hutter, & Parrinello, 1999; Eurenius, Chatfield, Brooks, & Hodoscek, 1996; Field, Albe, Bret, Proust-De Martin, & Thomas, 2000; Field, Bash, & Karplus, 1990; Lyne, Hodoscek, & Karplus, 1999; Maseras & Morokuma, 1995; Ryde, 1996; Singh & Kollman, 1986; Swart, 2003; Woo, Cavallo, & Ziegler, 1998), the boundary-atom scheme (Antes & Thiel, 1999; DiLabio, Hurley, & Christiansen, 2002; DiLabio, Wolkow, & Johnson, 2005; Laio, VandeVondele, & Rothlisberger, 2002; Zhang, 2005, 2006; Zhang, Lee, & Yang, 1999), and the localized-orbital schemes (Hall, Hindle, Burton, & Hillier, 2000; Nicoll, Hindle, MacKenzie, Hillier, & Burton, 2001; Reuter, Dejaegere, Maigret, & Karplus, 2000; Rodriguez, Oliva, Gonzalez, van der Kamp, & Mulholland, 2007). This "QM–MM boundary" problem is particularly challenging when ground and excited states exist, especially the boundary involves electronic excitation. Another approach is the multicentered valence electron effective potentials (MC-VEEP) (Slavicek & Martinez, 2006), which ensures the correct behavior of the link atom for ground and excited states. By demanding that both occupied and virtual orbitals are described correctly on the MC-VEEP, we are able to ensure correct behavior even when the MC-VEEP borders an electronically excited region. The second term is the van der Waals interaction, it is typically described by a Lennard-Jones potential between atoms of chromophore and surroundings, just as the molecular mechanic does in surrounding molecules. As for the third term, the electrostatic coupling between QM and MM atoms, the mechanical-embedding (Bakowies & Thiel, 1996; Froese, Musaev, & Morokuma, 1998; Maseras & Morokuma, 1995), electrostatic-embedding (Cao, Liu, Fang, Xie, & Zhang, 2013; Ruckenbauer, Barbatti, Mueller, & Lischka, 2013; Schapiro et al., 2011), and polarizable-embedding (Schwabe, Beerepoot, Olsen, & Kongsted, 2015) strategies are used in the real implementation. Here, we restrict us in discussion of electrostatic embedding and refer readers to several literatures for the other two embedding schemes, since the electrostatic embedding is the most popular embedding scheme in use today. In the electrostatic embedding, the electronic interaction between chromophore and surroundings are collected into the Hamiltonian of the QM region via point charges of surroundings. The QM Hamiltonian (in atomic units) is thus written as follows:

$$\hat{H}^{\mathrm{el}}_{\mathrm{QM/MM}} = -\sum_{i}^{N}\sum_{J\in S}^{L}\frac{q_J}{|r_i - R_J|} + \sum_{\alpha}^{M}\sum_{J\in S}^{L}\frac{q_J Q_\alpha}{|R_\alpha - R_J|} \tag{4}$$

The symbols q_J are the MM point charges located at R_J; Q_α are the nuclear charges of the QM atoms at R_α; and ridesignate electron positions. The indices i, J, and α are run over the N electrons, L point charges, and M QM nuclei, respectively. Comparing with the mechanic embedding, the electrostatic embedding accounts for the interaction of the polarizable QM density with rigid MM charges, so that more sophisticated than mechanical embedding when dealing with electronic interaction between chromophore and its surroundings.

3. IMPLEMENTATION OF THE QM/MM APPROACH IN PROTEIN

Since the additive scheme of QM/MM formalism allows a partitioning of the total energy, and particularly of the QM–MM interaction energy, into various components, and thus offers the opportunity to analyze the effects of the protein environment. Herein, we review some of the implementation of the QM/MM approach in protein environment, mainly focus on the interactions between chromophore and surroundings. It can be expected that both electrostatic and steric effects of the protein surroundings will alter the geometries and energies of stationary points and CIs, so that leading to different photochemical reaction mechanisms.

3.1 Retinal Protonated Schiff Base

The retinal protonated Schiff base (RPSB) (Becker & Freedman, 1985; Becker, Freedman, Hutchinson, & Noe, 1985; Ben-Nun, Molnar, Schulten, & Martinez, 2002; Coccia, Varsano, & Guidoni, 2013; Doemer, Maurer, Campomanes, Tavernelli, & Rothlisberger, 2014; Du & Davidson, 1990; Froese, Komaromi, Byun, & Morokuma, 1997; Garavelli, Negri, & Olivucci, 1999; Kholmurodov, Fel'dman, & Ostrovsky, 2007; Liu et al., 2009; Menger & Kliger, 1976; Munoz-Losa, Martin, Galvan, Sanchez, & Aguilar, 2011; Rohrig, Guidoni, & Rothlisberger, 2005; Shriver, Mateescu, & Abrahamson, 1979; Sugihara, Buss, Entel, Elstner, & Frauenheim, 2002; Weingart, Schapiro, & Buss, 2007; Zhou, Liu, Han, & He, 2014), as the chromophore of the rhodopsin family of proteins, has been widely investigated (Cembran, Bernardi, Olivucci, & Garavelli, 2004, 2005; Chalfie, 2009; Chalfie, Tu, Euskirchen, Ward, & Prasher, 1994; Garavelli, 2006; Jung et al., 2013; Khoroshyy, Der, & Zimanyi, 2013; Kim et al., 2012; Martinez, 2006; Nienhaus, 2008; Shimomura, 2009; Shriver et al., 1979; Sovdat et al., 2012; Tsien, 2009; Wei, Wang, Chen, Fang, & Wang, 2014; Yen, Hayden, Dreaden, Szymanski, & El-Sayed, 2011) for its ability to convert light into mechanical energy. When triggered by light, the RPSB chromophore undergoes ultrafast *cis–trans* isomerization through particular double bond, converting light into mechanical energy efficiently. For instance, RPSB isomerizes at C11=C12 double bond in rhodopsin (Rh), and C13=C14 double bond in bacteriorhodopsin (bR) and halorhodopsin (hR). In addition to the bond selectivity of isomerization in the different protein environment, the lifetime and the quantum yield of the isomerization are also different. Ultrafast spectroscopy experiments have shown that the isomerization of RPSB in bR (0.5 ps) occurs much faster than in hR (several ps), with the efficiency of isomerization in bR and hR are 64% and $\sim$30%, respectively. To elucidate the differences in lifetimes and quantum yields in bR, hR, Martinez et al. (Punwong, Martinez, & Hannongbua,

2014) simulated the excited state dynamics of bR and hR using *ab initio* multiple spawning and a hybrid QM/MM description of the potential energy surfaces and nonadiabatic coupling matrix elements. They conclude that the origin of this difference presumably lies in the specific arrangements of key residues around the chromophore. The stabilization of the positive charge of RPSB on S_1 state plays a key part in the different isomerization rates of RPSB in hR and bR. Since the distance between the complex counterion and the chromophore in bR is larger than that in hR, the stabilization of the positive charge on the Schiff base in bR is weaker comparing to that in hR. The relatively weaker stabilization leads to the delocalization of the positive charge to the β-ionone ring side of the chromophore, which facilitates torsion about the C13=C14 bond. That is, the electrostatic effects due to the complex counterion are at least partially responsible for the different excited state lifetimes in bR and hR.

The hydrogen-out-of-plane (HOOP) mechanism is widely accepted for the RPSB photoisomerization in protein, which is firstly supposed by Mathies et al., based on the analysis of Raman spectra (Migani, Robb, & Olivucci, 2003). Olivucci et al. (Schapiro et al., 2011) conclude that the ultrafast photoisomerizations of RPSB in rhodopsin and bathorhodopsin are modulated by bond length alternation (BLA) and HOOP-driven electronic effects, using QM/MM models based on a hydrogen link-atom and electrostatic-embedding schemes with the frontier placed at the C_δ–C_ε bond of the Lys296 side chain to investigate the mechanics of the photoisomerization. The chromophore in rhodopsin and bathorhodopsin is 11-*cis* (PSB11) and all-*trans* protonated Schiff base, respectively, they are treated by the *ab initio* multiconfigurational second-order perturbation theory with active space comprising the full π-system of PSB11 (12 electrons in 12 π-type orbitals), and others treated by molecular mechanics with the AMBER charges account for S_0 polarization effects in a mean-field way. It confirmed that the phase and amplitude of the HOOP mode controls the stereochemical outcome of the photoisomerization. The amplitude of the BLA and HOOP oscillation can affect the ratio between successful and unsuccessful events. A lower or higher amplitude shall change the dynamics at the point of decay and during the initial evolution on the S_0 energy surface owing to the fact that in these regions the motion is highly anharmonic. These results give a new insight of the RPSB photoisomerization mechanism. However, their steric and electrostatic effects are still unclear since the protein environments are so complex with many kinds of protein residues. In addition, the arrangement of the key residues also plays a crucial role in the photoisomerization. That is, the complex environment of retina makes hard to get the key point of the environment influences. The latest investigation of RPSB by Martinez et al. (Punwong, Owens, & Martinez, 2015) compares the dynamics *in vacuo* and in solvent after photo excitation to S_1 state, gives a side view of the RPSB isomerization mechanism in protein. As mentioned previously, the photoisomerization process of RPSB in protein environment is highly selective and efficient. But in solvated environments such as methanol, several photoisomerization products are possible, including isomerization about the C9=C10, C11=C12, and C13=C14 bonds, the quantum yield of isomerization is only 10−30% in methanol. Moreover, the timescale of 2−10 ps for isomerization in methanol is much slower than that in proteins, where isomerization occurs within hundreds of femtoseconds. The protein environment increases the selectivity, quantum yield, and rate of photoinduced isomerization in RPSB. Martinez et al. ran nonadiabatic dynamic simulations of RBSB based on the "full quantum" full multiple spawning method. The condensed-phase simulation is carried out using a hybrid quantum mechanics/molecular mechanics (QM/MM) scheme for the potential energy surfaces and their couplings. The chromophore of RPSB was treated with a multireference reparameterized semiempirical FOMO-CI method, in which the self-consistent field problem is solved using a FOMO method, refer to this as FOMO-CASCI (6/6). Since some treatment of dynamic electron correlation is implicit in the method, and one could hope for accuracy intermediate between CASSCF and CASPT2 with computational cost significantly decreasing. The QM region in their simulations includes all isomerizable double bonds in the RPSB chromophore excluding the β-ionone ring with butylamine linker at the two ends of the full RPSB chromophore, as depicted in Scheme 2. The MM region of methanol solvent molecules is described using the OPLS–AA force field.

The dynamic simulations reproduce the lifetime and quantum yield in experiments. The possible reasons for the differences are steric and

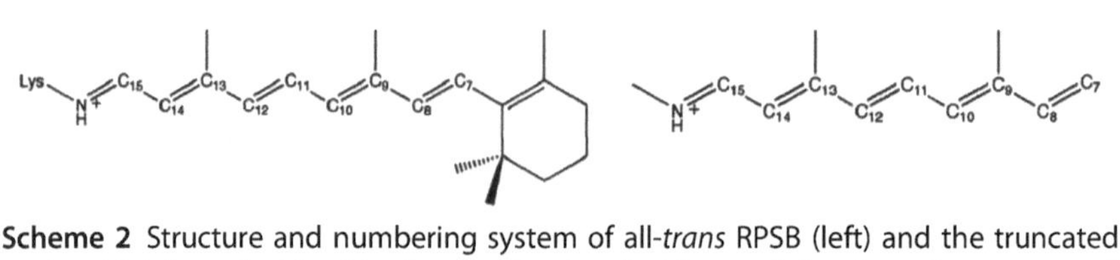

Scheme 2 Structure and numbering system of all-*trans* RPSB (left) and the truncated analog (right) studied in this chapter. *Reprinted with permission from Punwong et al. (2015). Copyright@2015 American Chemical Society.*

electrostatic effects of the solvent. They firstly make the rigid torsion cuts of RPSB in methanol and in gas phase with all of the heavy atoms (C and N) on the RPSB backbone fixed and others relaxed on S_1 potential energy surface, mainly focus on the torsion around C9=C10, C11=C12, and C13=C14 bonds, shown in Fig. 1. *In vacuo*, the torsion about the C11=C12 bond is barrierless (starting from the all-*trans* structure on the right-hand side of the figure), while torsion about the C9=C10 or C13=C14 bonds exhibits very small (<0.05 eV) barriers. These results explain the observation that isomerization in isolated RPSB is ultrafast and most likely to occur about the C11=C12 bond. The situation changes rather dramatically for RPSB in MeOH with significant (~0.2 eV) barriers to torsion about any of the C9=C10, C11=C12, and C13=C14 bonds. Among these barriers, the smallest one is the torsion around the C11=C12 bond, which is consistent with the observation that the primary isomerization channel is torsion about the C11=C12 bond in MeOH. The torsional barriers presence explains the difference in isomerization timescales for isolated and solvated RPSB. To identify the reason of the differences of the PES, they also make the rigid torsion cuts of RPSB in MeOH without electrostatic interactions through setting the atom charges in MeOH to zero, shown with dashed line in Fig. 1. The minimum energy profiles (MEPs) are nearly parallel to the results from RPSB in isolation, and the barriers are significantly reduced or completely removed. Thus, they conclude that the electrostatic interactions with MeOH are the primary cause of the torsional barriers. In order to further explore the origin of the electrostatic solvation-induced barrier, they have calculated the atomic charges on the RPSB molecule along the torsional pathways. The molecule has been divided into those atoms that are to the left of the twisting bond (including the nitrogen atom) and the other atoms which are to the right. The charges are then summed over each side of the molecule and displayed in Fig. 2. As the molecule twisting around any of the three double bonds about 130°, the positive charge localizes on one side of the molecule while delocalized over the RPSB backbone in Franck–Condon (FC) region. Since the S_1 barrier in MeOH occurs between 120° and 130° torsion, as shown in Fig. 1. These results suggest that the observed barrier originates from the solvent reorientation as the charge distribution changes during torsion about one of the double bonds. These findings provide further evidence that electrostatic effects play a more crucial role than steric effects of the different behavior of RPSB in isolation and MeOH.

Their direct comparison of the excited-state dynamics in vacuum and neat solvent environments provides insight into the solvent effects on the

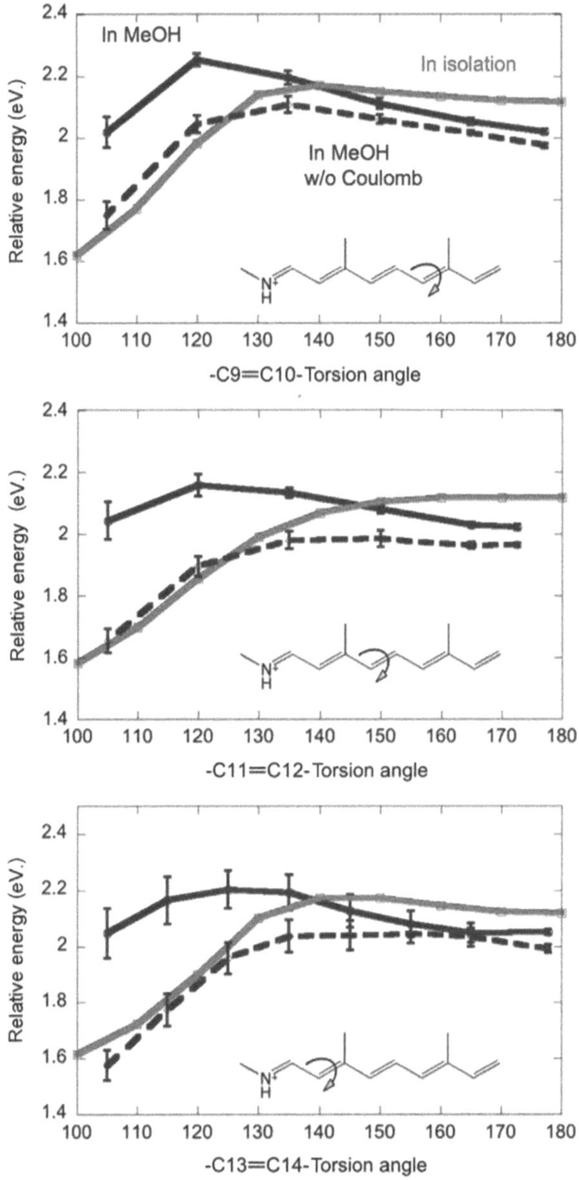

Figure 1 Rigid torsion cuts of RPSB in methanol (solid blue (dark gray in the print version) lines), in the gas phase (solid red (light gray in the print version) lines), and in methanol without electrostatic interaction (dashed blue (dark gray in the print version) lines). The curves for torsion about C9=C10, C11=C12, and C13=C14 are the top, middle, and bottom panels, respectively. In all cases, the zero of energy is taken to be the all-*trans* configuration of RPSB on S_0. Results in solvent (blue (dark gray in the print version)) are obtained by averaging over nine different solvent configurations, and depicted error bars are calculated by bootstrapping. *Reprinted with permission from Punwong et al. (2015). Copyright@2015 American Chemical Society.*

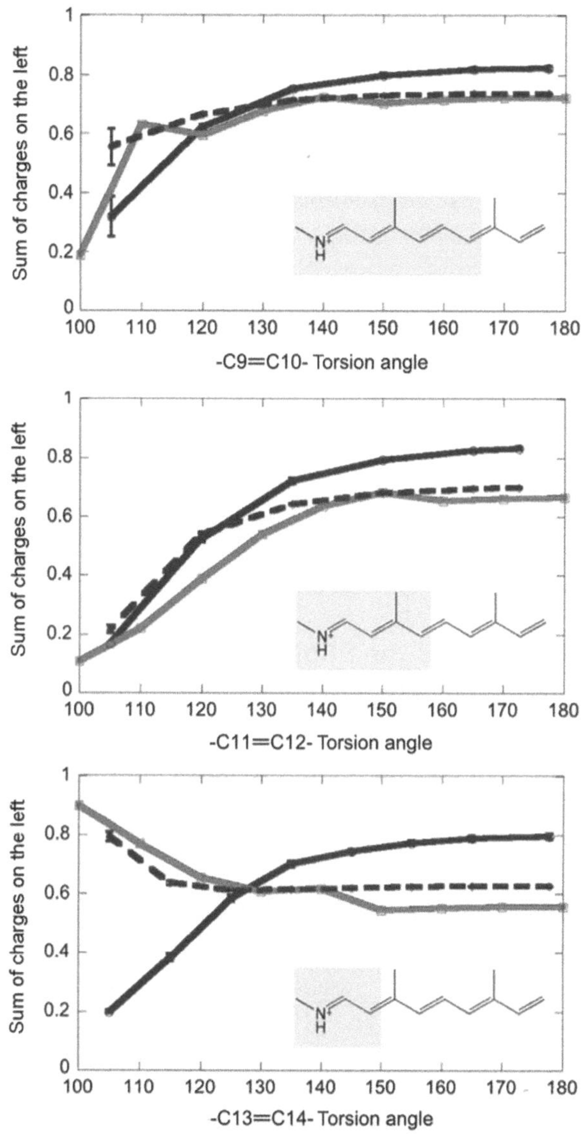

Figure 2 Sum of atomic charges on the left (Schiff base) side of the twisting bond at various torsion angles. Molecular geometries correspond to those used in Fig. 1. Charges are shown for RPSB in isolation (solid red (light gray in the print version)), MeOH (solid blue (dark gray in the print version)), and MeOH without electrostatic interactions (dashed blue (dark gray in the print version)). As in Fig. 1, solvated results correspond to an average for nine distinct solvent configurations, and error bars for this case are from bootstrapping (note that most of the error bars are within the line width). *Reprinted with permission from Punwong et al. (2015). Copyright@2015 American Chemical Society.*

photochemical mechanism and bond selectivity of RPSB isomerization. They also compare the counterion effects through run simulations including a Cl^- ion. The inclusion and the position of a Cl^- counterion in the methanol solvent has little effect on the photoisomerization timescale and the bond selectivity of RPSB, in contrast to previous studies (Cembran et al., 2004, 2005; Garavelli, 2006; Martinez, 2006).

From the rigid cut calculations, the slow photoisomerization in methanol environment is due to the emergence of torsional barriers which mainly originate from the electrostatic interactions between RPSB and the solvent. Since the intramolecular charge transfer occurs along with the torsion of RPSB, they suspect that the protein pocket of Rh shield the electrostatic effects of the counterion that partially recovers the ideal *in vacuo* condition, which leads to the ultrafast, unique, and efficient conversion. They emphasize that efficient and selective isomerization in protein environments is due to both the particular arrangement of protein residues and the low dielectric constant found in these environments.

3.2 The Photocycle of the Photoactive Yellow Protein

The PYP, which was first discovered in the cytoplasm of *Halorhodospira halophila* bacterium, is a cytoplasmic blue light photoreceptor initiates the negative phototaxis pathway. Upon photoirradiation, the chromophore of PYP undergoes a fully reversible photocycle starting from a dark-state pG and then going through two early spectroscopic intermediates I_0 and $I_0^{\ddagger}$ that may be equivalent to IT and I_{CP}, respectively. Following the initial relaxation, wt-PYP decays quickly to a red-shifted intermediate pR. However, the precise mechanism for generating pR is still unclear. Ihee et al. (Jung et al., 2013; Kim et al., 2012) proposed that IT may simultaneously decay to the two intermediates pR_1 and pR_2 (via I_{CP}) through a parallel isomerization reaction, while Zimányi et al. (Khoroshyy et al., 2013) proposed different mechanism that pR_2 is produced after pR_1 via a sequential reaction. In both mechanisms the intermediate pR_2 undergoes a protonation reaction to capture one proton from the Glu46 residue, leading to a blue-shifted intermediate pB′ and an energetically unstable, charged $Glu46^-$. The $Glu46^-$ triggers a large conformational change in the protein, relaxes to a putative signaling state pB which forms a new blue-shifted intermediate. Ihee et al. (Kim et al., 2012) also published another possibility; the pB may be reached directly from pR_1 because they exhibit similarities in the chromophore orientation and the surrounding hydrogen-bonding networks.

Eventually, the photocycle is completed via a series of deprotonation and reisomerization processes. A structural recovery of PYP is achieved through a new intermediate pG′ with a deprotonated chromophore that facilitates the occurrence of reisomerization. Identify all possible intermediates in the photocycle of PYP is significant challenges for both experimental and theoretical investigations. Since there are so many debated issues, such as the parallel versus sequential kinetic pathways, the assignment of various intermediates, and the identity of the donor and acceptor for proton transfer in the multistep protonation–deprotonation reaction, Chen and Fang et al. (Wei et al., 2014) carry out a comprehensive computational study to address the overall process using a combined QM/MM approach, at the level of theory of CASPT2/CASSCF/AMBER. In the computation of the constrained MEPs along the physically motivated reaction coordinates modeling all possible photoisomerization and proton transfer processes, the pCA chromophore, and its adjacent residues (complete or partial) as well as the crystal water molecules are included in QM subsystem, while the remaining residues, water molecules, and counterions were treated using MM. Different QM/MM partitions were adopted to account for various steps in the photocycle of PYP. The first step of photoisomerization is described with QM1, which includes the pCA chromophore and a portion of the Glu46 and Cys69 residues. The latter isomerization steps $I_{CP} \rightarrow pR_2$ and $pR_1 \rightarrow pR_2$, as well as subsequent processes, which involve two typical structural deformations via the simultaneous torsion along two nonadjacent or adjacent bonds, are described by QM2 with whole Cys69 residue included. To compute the MEPs of the protonation–deprotonation steps, additional one and three crystal water molecules have been added to the QM3 and QM4 subsystems, respectively. For the QM3, part of Tyr42 was also added to account for its role of proton transfer relay.

Their study provides a comprehensive picture of the overall photocycle of the isomerization and protonation reactions upon the photoexcitation of the wild-type PYP. As illustrated in Scheme 3, after photoirradiation to $S_1(^1\pi\pi^*)$ state, the pG isomerizes along b bond through one-bond flip (OBF) to CI (S_0/S_1), followed by "hula twist" (HT), "bicycle pedal"(BP), and OBF isomerization reactions to form *cis* isomers pR_1, I_{CP}, and I_{CT}, respectively. The pB could be achieved by either protonation of pR_1 or $I_{CP} \rightarrow pB$ transition. Starting from I_{CP}, a barrier of 16.2 kcal/mol bicycle pedal isomerization needs to over to form an intermediate pR_2, with a fast protonation reaction (6.1 kcal/mol) to get pB′. There is another high barrier of 28.4 kcal/mol that needs to surmount through bicycle

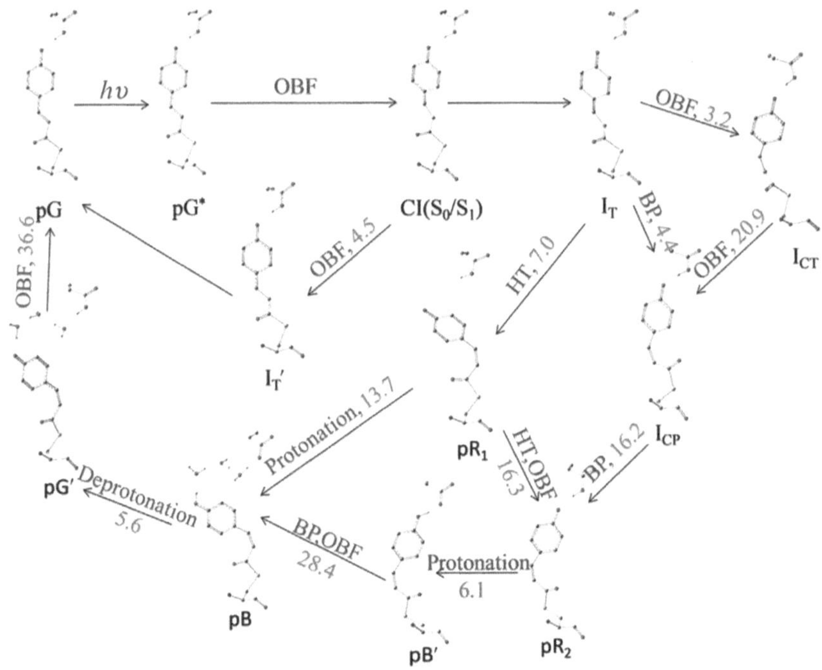

Scheme 3 Mechanistic illustration of the overall PYP photocycle: the protonation–deprotonation and isomerization reactions of the hula twist (HT), bicycle pedal (BP), and one-bond flip (OBF) are shown in blue (light gray in the print version) along the special one or two bonds and the related barriers (kcal/mol) are also highlighted in red (light gray in the print version). *From Wei et al. (2014), reproduced by permission of the PCCP Owner Societies.*

pedal isomerization and a reverse OBF isomerization, which leads to a hydrophobic–hydrophilic transformation and the formation of pB. After that, the exposed COO^- group of Glu46 attracts water solvent into the binding pocket and triggers the fast deprotonation reaction of pCA. The photocycle is finally completed by the recovery of the phenoxy ring repelling water molecules out of the binding pocket through an OBF isomerization with a large barrier of 36.6 kcal/mol. Their studies unveil that the parallel mechanism through $pR_1 \rightarrow pB$ transition is a favorable channel but coexists with the sequential model via the $pR_1 \rightarrow pR_2$ transformation.

3.3 The Green Fluorescent Protein and Its S65T/H148D Double Mutant

The wild-type GFP (wt-GFP) has been first isolated from the jellyfish *Aequorea victoria* and has been widely investigated (Baranov et al., 2012;

Chalfie, 2009; Chalfie et al., 1994; Chen et al., 2007; Cubitt et al., 1995; Elsliger, Wachter, Hanson, Kallio, & Remington, 1999; Nienhaus, 2008; Shimomura, 2009; Stoner-Ma et al., 2005, 2008; Tsien, 1998; Tsien, 2009; Zimmer, 2002) by experimental and theoretical scientists because of the reporter function in cell and molecular biology expression through eliminating bright green fluorescence when exposed to light in the blue to ultraviolet range. Due to the potential for widespread usage of GFT family, many different mutants of GFP have been designed and engineered for satisfying bioimaging needs. It is thus desirable to study the photophysics and photochemistry of the GFP mutant at the atomistic level to enhance the understanding of the GFP fluorescence emitting state and its deactivation process. The fluorescence is to originate from the anionic 4-(4-hydroxybenzylidene)-1,2-dimethyl-1H-imidazol-5(4H)-one (*p*-HBDI$^-$) through a multistep excited-state proton transfer of neutral chromophore *p*-HBDI, along the intermolecular hydrogen-bonding network that consists of a water molecule (W22), Ser205, and Glu222 (Brejc et al., 1997; Fang, Frontiera, Tran, & Mathies, 2009; Meech, 2009), upon photoexcitation of wt-GFP. The protein environment of GFP plays a crucial role in the energy and the character of excited states. For instance, gas-phase action spectroscopy experiments on the GFP chromophore have reported an absorption maximum blue shift of 0.06 eV relative to the protein environment, which is consistent with the 0.02 eV blue shift by the QM/MM calculations with the CASPT2 approach in the QM part. While the protein environment has a negligible effect on the excitation energy of the bright state for neutral GFP, the electrostatic effect is obvious in different environments for the ionized states of chromophore. Krylov et al. investigated the effect of the protein environment on the electronic properties of the GFP chromophore (Bravaya, Khrenova, Grigorenko, Nemukhin, & Krylov, 2011). While the protein environment has a very small effect on the character and excitation energy of the singlet $\pi\pi^*$ state of GFP, its effect on the VDE is more significant (the VDE increases from 2.5 eV in the gas phase to 5.0 eV in the solvated protein). They also suspect that VDE of the bare chromophore in aqueous solution would be even larger owing to more efficient stabilization of the anionic species in polar solvents. They also investigated the possible existence of the CTTS states; they found that a tightly packed GFP does not present cavities suitable to accommodate a CTTS electron other than the GFP mutants with larger water-filled cavities in the vicinity of the chromophore. The protein can also affect the photocycle of GFP, the excited-state lifetime drops by four orders of magnitude, i.e., from nanoseconds in the protein to subpicosecond in solutions (Meech, 2009) due to the radiationless relaxation to S_0 state. Jonasson and Demachy (Jonasson et al., 2011) have determined a coupled torsional potential $V(\tau,\varphi)$ for the S_1 state of the anionic chromophore of the GFP, by fitting a grid of SA2-CAS(4,3)-PT2 energy values. They implemented this potential in the parallel module PMEMD of the AMBER package, and ran 24 long trajectories (15-ns-long each) of the brightly fluorescent S65T-GFP using this new program. They observed a mean first passage time at the φ twisted geometry of 4 ns, which corresponds to a fluorescence lifetime around 2.2 ns and a quantum yield around 0.5, which is qualitatively consistent with the experimental data. The twist movement is very fast once the chromophore leaves the fluorescent state valley, suggesting that it happens when collective motions of the protein allow the chromophore to move in its pocket. Their work is the first theoretical estimate of the fluorescence lifetime. In 2013, Krylov et al. focus on the QM/MM computed structures and spectra of the model systems to mimic the chromophore-containing domains of wt-GFP and the S65T mutant (Grigorenko, Nemukhin, Polyakov, Morozov, & Krylov, 2013). They arrange different structures with different energies along the proton transfer route $A \rightarrow I \rightarrow B$, characterized by first-principles calculations. They found that the system with the neutral chromophore (form A) is the lowest in energy for wt-GFP, whereas the systems with the anionic chromophore that are assigned to the B- and I-forms are about 1 kcal/mol higher in energy. The respective population ratio between A and B (5:1) is in excellent agreement with the experimentally derived one (6:1) (Chattoraj, King, Bublitz, & Boxer, 1996). In the S65T mutant, the systems with the anionic chromophore are significantly lower in energy (about 18 kcal/mol) relative to the neutral chromophore structure. They reexamined the role of the nearby amino acid residues in the chromophore-containing pocket. The calculations reveal that the structural differences between the I- and B-forms (the former has a slightly red-shifted absorption relative to the latter) are based on the Glu222 position, other than the Thr203 orientation. They conclude that the hydrogen bond between the chromophore and the His148 residue stabilizes the structures with the deprotonated phenolic ring in the I- and B-forms for wt-GFP, while both of the His148 and Thr203 residues are responsible for a considerable energy gap between the lowest energy structure of the B type with the anionic chromophore from other structures for the S65T mutant.

Another interesting behavior of GFP analogs is that the S65T mutant lost the fluorescence emission of GFP, but could partially recovered in the

S65T/H148D double mutant (Kondo, Heisler, Stoner-Ma, Tonge, & Meech, 2010; Shi et al., 2007; Shu et al., 2007; Stoner-Ma et al., 2008). Zhang and Fang et al. employ a combined quantum mechanics/molecular mechanics (QM/MM) method to investigate the deactivation mechanism of S65T/H148D double mutant, the *para*-chromophore (*p*-1) and Asp148 are included in QM subsystem, while the remaining amino acid residues, counterions, and water molecules are collected into MM subsystem, shown in Scheme 4. The hydrogen link atoms are used to saturate the broken covalent bonds between the boundary of QM and MM regions. To highlight the QM/MM interaction effect, the QM only (*p*-HBDI chromophore) is also studied in this section.

They start with the MEPs on S_1 state along with O9–H10 distance which leads to the formation of keto tautomer fluorescent state. As shown in Fig. 3, the deprotonation of the *p*-HBDI chromophore is barrierless both *in vacuo* and in protein with energy sharply decreasing from ~70 kcal/mol at the FC point to ~45 kcal/mol at the S_1 keto tautomer. The H10 proton rapidly migrates from O9 to O11 along the downhill relaxation pathway, leading to the ultrafast excited-state proton transfer which is fully consistent with

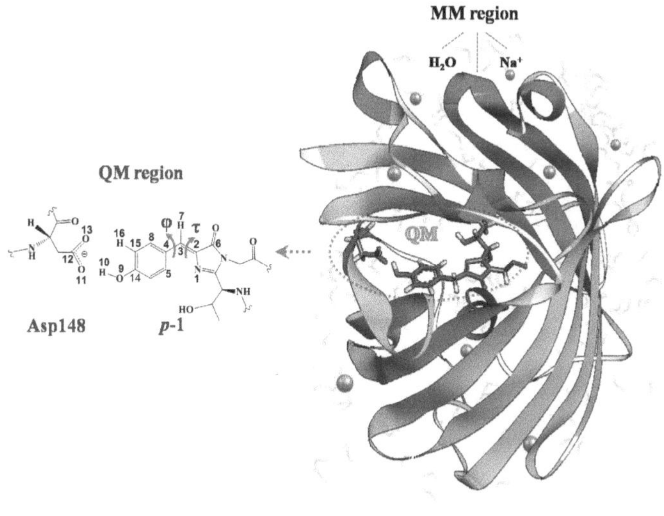

Scheme 4 QM/MM computational protocol: The QM subsystem (left) includes the *para*-chromophore (*p*-1) and Asp148. The MM subsystem comprises the remaining amino acid residues, counterions, and water molecules. The key rotations around the P-bond (f) and the I-bond (t) are indicated in red (light gray in the print version). Broken covalent bonds separating the QM and MM regions are saturated using hydrogen link atoms (wavy lines).

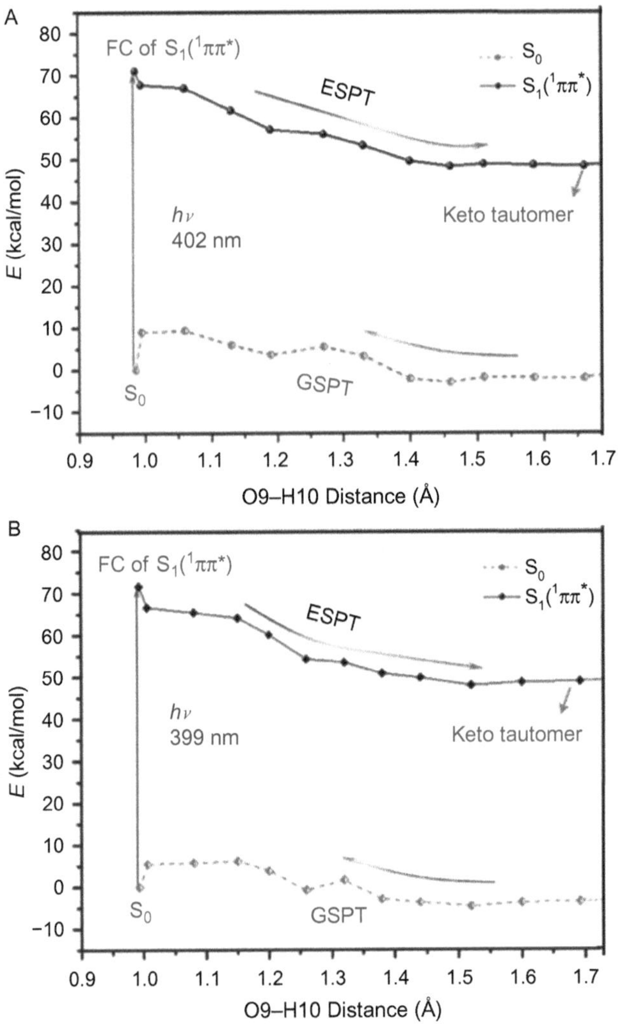

Figure 3 Minimum-energy profiles (relaxed S_1 state and unrelaxed S_0 state) along the O9–H10 proton transfer reaction coordinates in the S65T/H148D mutant (A) and *in vacuo* (B). FC, Franck–Condon point; ESPT, excited-state proton transfer; GSPT, ground-state proton transfer.

the sub-175 femtosecond timescale observed experimentally (Kondo et al., 2010; Morise, Shimomur, Johnson, & Winant, 1974; Shi et al., 2007; Shu et al., 2007). Since the ESPT process in protein is nearly the same with that *in vacuo*, the deprotonated Asp148 provides the only driving force for the ultrafast proton transfer observed in the S65T/H148D mutant.

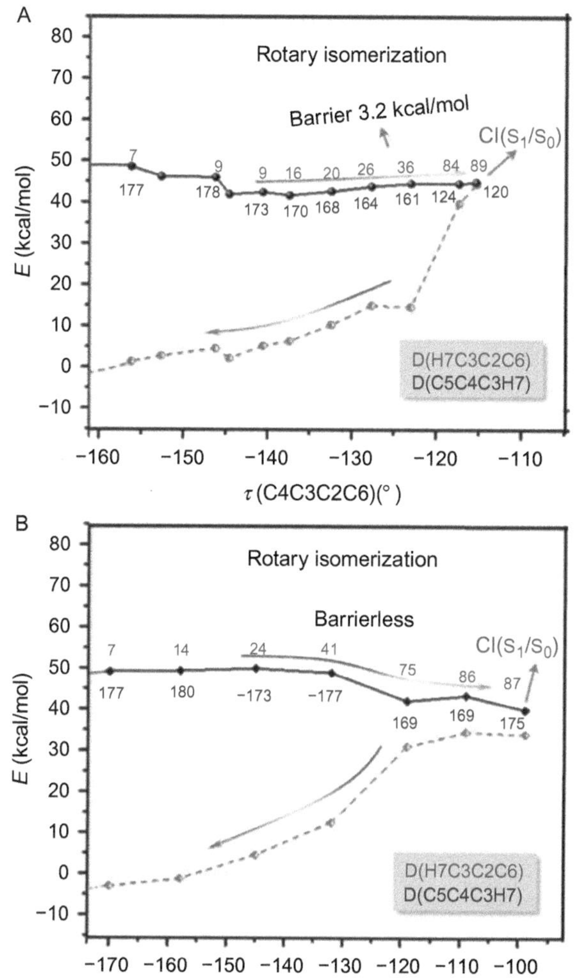

Figure 4 Minimum-energy profiles (relaxed S_1 state and unrelaxed S_0 state) along the C4C3C2C6 torsion reaction coordinates in the S65T/H148D mutant (A) and *in vacuo* (B). CI, conical intersection.

After that, the MEPs of *p*-HBDI chromophore in protein and *in vacuo* are shown in Fig. 4, in function of C4C3C2C6 dihedral angle. A barrierless process *in vacuo* while a small barrier of 3.2 kcal/mol is found in protein. *In vacuo*, the S_1 deactivation occurs by the classical OBF isomerization mechanism, mainly through nearly synchronous rotation around the C4C3C2C6 (from −175° to −99°) and H7C3C2C6 (from 7° to 87°) dihedral angles. The C5C4C3H7 dihedral angle remains close to 180° during the whole excited-state relaxation process, and a conical intersection region of $CI(S_1/S_0)$ is reached at the end of the isomerization process, with S_1–S_0 energy gap is ~6 kcal/mol at the CASPT2/CASSCF level. Thus, OBF deactivation mechanism can be easily accessed *in vacuo*, which is consistent with the ultralow fluorescence quantum yield that is observed in experiment. In contrast, a concerted asynchronous HT mechanism is proposed in S65T/H148D GFP mutant. Besides the twist around C4C3C2C6 and H7C3C2C6 dihedral angles, the C5C4C3H7 dihedral angle decreases to 120° with the C3–C4/C2–C3 bond length increase/decrease from 1.40/1.49 to 1.45/1.46 Å. The small barrier of 3.2 kcal/mol on the minimum energy path could be easily accessed since it is ~28 kcal/mol lower than the $S_1(^1\pi\pi^*)$ state at the FC point. The S_1/S_0 CI region is also reached at the end of the isomerization process, which leads to the ultrafast $S_1 \rightarrow S_0$ nonadiabatic transition. However, the S_1 deactivation of S65T/H148D GFP mutant should be less efficient than *in vacuo* since there is a small isomerization barrier of ~3 kcal/mol (actually will less than this value) in the protein environment which is qualitatively compatible with the observed fluorescence quantum yield of 0.21 (Shu et al., 2007). They emphasize that the HOOP motion (H7 atom) plays an important role in isomerization process in the rigid protein surrounding, which is necessitated by the rigid cavity near the GFP chromophore to avoid the strong steric interactions with adjacent residues in the double mutant. Since fluorescence quantum yield of *p*-HBDI chromophore observed *in vacuo* and in solvent is similar, they suspect that the steric interactions between QM and MM predominates the deactivation mechanism. Contrast to the OBF photoisomerization *in vacuo*, the volume-conserving HT mechanism combined with HOOP motion in the central CH moiety are preferred in protein.

4. CONCLUSION AND PERSPECTIVE

In this review, we summarized the general methodological aspects of the QM/MM approach, and the applications in bimolecular systems mainly focus on the electronic and steric interactions between chromophore and surroundings. As documented in our review, the QM/MM calculations provide detailed mechanistic insight into the biochemical reactions, such as the retinal protonated Schiff base, the PYP, and the GFP.

The QM/MM studies of the retinal protonated Schiff base in rhodopsin show that the S_1 energy surfaces decide the dynamic simulation behaviors, such as the lifetime and the selectivity of the torsion bonds. The barriers in the solvent environment mainly originate from the electrostatic interactions between RPSB and the solvent, because of the redistribution of the solvent molecules as the charge transfer during torsion about one of the double bonds. The efficient and selective isomerization in protein environments is due to both the particular arrangement of protein residues and the low dielectric constant found in these environments. The specific isomerization around C11=C12 bond in the cavity of the protein surroundings is due to the HOOP motion.

The PYP photocycle has been investigated by combined QM/MM approach, the overall photocycle can be achieved by competitive parallel/sequential reactions, followed by a series of slow volume-conserving BP/HT and OBF isomerization reactions, as well as fast protonation–deprotonation processes and the hydrophobic–hydrophilic state transformation.

The GFP family, as invaluable fluorescent makers in cell biology, has a huge number of mutants that have been investigated. S65T/H148D double mutant recovers the fluorescence-emitting through proton transfer forming S_1 keto state, which is long-lived enough to give a fluorescence quantum yield of 0.21, less than in wt-GFP (0.8) but much higher than in the S65T mutant (0.02). These differences can be attributed to different topologies of the S_1 potential energy surfaces. They found that Asp148 provides the exclusive driving force for the ultrafast sub-175 fs formation of the fluorescent state, which is little affected by the protein surrounding. The environment completely changes the deactivation path of the fluorescent state (S_1 keto tautomer), from a one-bond-flip mechanism *in vacuo* to a concerted asynchronous HT mechanism in the double mutant. Closer analysis of MEPs shows that the favored HT pathway involves concerted asynchronous motion and strongly benefits from simultaneous HOOP motion in the final stage of the photoisomerization. These insights provide a detailed understanding of the HT photoisomerization mechanism and its specific features in GFP.

It is clear that the QM/MM methods are suitable for treating many other complex systems, with the chemical reactions localizing in an active site (typically of the order of 100 atoms) under the influence of an interacting larger environment. Up to now, the QM/MM scheme is a good compromise between computational cost and accuracy. However, the QM/MM approach still needs to be improved in order to become a highly reliable method that can be applied to a wide range of problems, such as the continued improvements in the QM methods and MM representation, the embedding scheme, the boundary issues, and so on.

REFERENCES

Ai, Y.-J., Liao, R.-z., Chen, S.-f, Luo, Y., & Fang, W.-H. (2010). Theoretical studies on photoisomerizations of (6-4) and dewar photolesions in DNA. *Journal of Physical Chemistry B, 114*, 14096–14102.

Ai, Y.-J., Liao, R.-Z., Fang, W.-H., & Luo, Y. (2012). Theoretical studies on the isomerization mechanism of the ortho-green fluorescent protein chromophore. *Physical Chemistry Chemical Physics, 14*, 13409–13414.

Ai, Y.-J., Zhang, F., Chen, S.-F., Luo, Y., & Fang, W.-H. (2010). Importance of the intramolecular hydrogen bond on the photochemistry of anionic hydroquinone (fadh(-)) in DNA photolyase. *Journal of Physical Chemistry Letters, 1*, 743–747.

Antes, I., & Thiel, W. (1999). Adjusted connection atoms for combined quantum mechanical and molecular mechanical methods. *Journal of Physical Chemistry A, 103*, 9290–9295.

Bakowies, D., & Thiel, W. (1996). Hybrid models for combined quantum mechanical and molecular mechanical approaches. *The Journal of Physical Chemistry, 100*, 10580–10594.

Baranov, M. S., Lukyanov, K. A., Borissova, A. O., Shamir, J., Kosenkov, D., Slipchenko, L. V., et al. (2012). Conformationally locked chromophores as models of excited-state proton transfer in fluorescent proteins. *Journal of the American Chemical Society, 134*, 6025–6032.

Bartolotti, L. J. (1982). Time-dependent Kohn-sham density-functional theory. *Physical Review A, 26*, 2243–2244.

Becker, R. S., & Freedman, K. (1985). A comprehensive investigation of the mechanism and photophysics of isomerization of a protonated and unprotonated Schiff-base of 11-cis-retinal. *Journal of the American Chemical Society, 107*, 1477–1485.

Becker, R. S., Freedman, K., Hutchinson, J. A., & Noe, L. J. (1985). Kinetic-study of the photoisomerization of a protonated Schiff-base of 11-cis-retinal over the picosecond-to-second time regimes. *Journal of the American Chemical Society, 107*, 3942–3944.

Ben-Nun, M., & Martinez, T. J. (1998). Electronic energy funnels in cis-trans photoisomerization of retinal protonated Schiff base. *Journal of Physical Chemistry A, 102*, 9607–9617.

Ben-Nun, M., Molnar, F., Schulten, K., & Martinez, T. J. (2002). The role of intersection topography in bond selectivity of cis-trans photoisomerization. *Proceedings of the National Academy of Sciences of the United States of America, 99*, 1769–1773.

Bochenkova, A. V., & Andersen, L. H. (2013). Ultrafast dual photoresponse of isolated biological chromophores: Link to the photoinduced mode-specific non-adiabatic dynamics in proteins. *Faraday Discussions, 163*, 297–319.

Bravaya, K. B., Khrenova, M. G., Grigorenko, B. L., Nemukhin, A. V., & Krylov, A. I. (2011). Effect of protein environment on electronically excited and ionized states of the green fluorescent protein chromophore. *The Journal of Physical Chemistry B, 115*, 8296–8303.

Brejc, K., Sixma, T. K., Kitts, P. A., Kain, S. R., Tsien, R. Y., Ormo, M., et al. (1997). Structural basis for dual excitation and photoisomerization of the Aequorea victoria green fluorescent protein. *Proceedings of the National Academy of Sciences of the United States of America, 94*, 2306–2311.

Cao, J., Liu, L.-H., Fang, W.-H., Xie, Z.-Z., & Zhang, Y. (2013). Photo-induced isomerization of ethylene-bridged azobenzene explored by *ab initio* based non-adiabatic dynamics simulation: A comparative investigation of the isomerization in the gas and solution phases. *The Journal of Chemical Physics, 138*, 134306.

Cembran, A., Bernardi, F., Olivucci, M., & Garavelli, M. (2004). Counterion controlled photoisomerization of retinal chromophore models: A computational investigation. *Journal of the American Chemical Society, 126*, 16018–16037.

Cembran, A., Bernardi, F., Olivucci, M., & Garavelli, M. (2005). The Retinal chromophore/chloride ion pair: Structure of the photo isomerization path and interplay of charge transfer and covalent states. *Proceedings of the National Academy of Sciences of the United States of America, 102*, 6255–6260.

Chalfie, M. (2009). Gfp: Lighting up life (Nobel lecture). *Angewandte Chemie (International Edition in English), 48*, 5603–5611.

Chalfie, M., Tu, Y., Euskirchen, G., Ward, W. W., & Prasher, D. C. (1994). Green fluorescent protein as a marker for gene-expression. *Science, 263*, 802–805.

Chattoraj, M., King, B. A., Bublitz, G. U., & Boxer, S. G. (1996). Ultra-fast excited state dynamics in green fluorescent protein: Multiple states and proton transfer. *Proceedings of the National Academy of Sciences of the United States of America, 93*, 8362–8367.

Chen, K.-Y., Cheng, Y.-M., Lai, C.-H., Hsu, C.-C., Ho, M.-L., Lee, G.-H., et al. (2007). Ortho green fluorescence protein synthetic chromophore; excited-state intramolecular proton transfer via a seven-membered-ring hydrogen-bonding system. *Journal of the American Chemical Society, 129*, 4534–4535.

Coccia, E., Varsano, D., & Guidoni, L. (2013). Protein field effect on the dark state of 11-cis retinal in rhodopsin by quantum Monte Carlo/molecular mechanics. *Journal of Chemical Theory and Computation, 9*, 8–12.

Cubitt, A. B., Heim, R., Adams, S. R., Boyd, A. E., Gross, L. A., & Tsien, R. Y. (1995). Understanding, improving and using green fluorescent proteins. *Trends In Biochemical Sciences, 20*, 448–455.

Dapprich, S., Komaromi, I., Byun, K. S., Morokuma, K., & Frisch, M. J. (1999). A new oniom implementation in Gaussian98. Part I. The calculation of energies, gradients, vibrational frequencies and electric field derivatives. *THEOCHEM-Journal of Molecular Structure, 461*, 1–21.

de Visser, S. P. (2009). Density functional theory (Dft) and combined quantum mechanical/molecular mechanics (Qm/Mm) studies on the oxygen activation step in nitric oxide synthase enzymes. *Biochemical Society Transactions, 37*, 373–377.

de Vries, A. H., Sherwood, P., Collins, S. J., Rigby, A. M., Rigutto, M., & Kramer, G. J. (1999). Zeolite structure and reactivity by combined quantum-chemical-classical calculations. *The Journal of Physical Chemistry B, 103*, 6133–6141.

DiLabio, G. A., Hurley, M. M., & Christiansen, P. A. (2002). Simple one-electron quantum capping potentials for use in hybrid Qm/Mm studies of biological molecules. *The Journal of Chemical Physics, 116*, 9578–9584.

DiLabio, G. A., Wolkow, R. A., & Johnson, E. R. (2005). Efficient silicon surface and cluster modeling using quantum capping potentials. *The Journal of Chemical Physics, 122*, 044708.

Dobson, J. F., Bunner, M. J., & Gross, E. K. U. (1997). Time-dependent density functional theory beyond linear response: An exchange-correlation potential with memory. *Physical Review Letters, 79*, 1905–1908.

Doemer, M., Maurer, P., Campomanes, P., Tavernelli, I., & Rothlisberger, U. (2014). Generalized Qm/Mm force matching approach applied to the 11-cis protonated Schiff base chromophore of rhodopsin. *Journal of Chemical Theory and Computation, 10*, 412–422.

Du, P., & Davidson, E. R. (1990). Abinitio study on the excitation-energies of the protonated Schiff-base of 11-cis-retinal. *Journal of Physical Chemistry, 94*, 7013–7020.

Eichinger, M., Tavan, P., Hutter, J., & Parrinello, M. (1999). A hybrid method for solutes in complex solvents: Density functional theory combined with empirical force fields. *The Journal of Chemical Physics, 110*, 10452–10467.

Elsliger, M. A., Wachter, R. M., Hanson, G. T., Kallio, K., & Remington, S. J. (1999). Structural and spectral response of green fluorescent protein variants to changes in pH. *Biochemistry, 38*, 5296–5301.

Eurenius, K. P., Chatfield, D. C., Brooks, B. R., & Hodoscek, M. (1996). Enzyme mechanisms with hybrid quantum and molecular mechanical potentials .1. Theoretical considerations. *International Journal of Quantum Chemistry, 60*, 1189–1200.

Fabiano, E., Keal, T. W., & Thiel, W. (2008). Implementation of surface hopping molecular dynamics using semiempirical methods. *Chemical Physics*, *349*, 334–347.

Fang, C., Frontiera, R. R., Tran, R., & Mathies, R. A. (2009). Mapping Gfp structure evolution during proton transfer with femtosecond Raman spectroscopy. *Nature*, *462*, U200–U274.

Field, M. J., Albe, M., Bret, C., Proust-De Martin, F., & Thomas, A. (2000). The dynamo library for molecular simulations using hybrid quantum mechanical and molecular mechanical potentials. *Journal of Computational Chemistry*, *21*, 1088–1100.

Field, M. J., Bash, P. A., & Karplus, M. (1990). A combined quantum-mechanical and molecular mechanical potential for molecular-dynamics simulations. *Journal of Computational Chemistry*, *11*, 700–733.

Finley, J., Malmqvist, P. A., Roos, B. O., & Serrano-Andres, L. (1998). The multi-state Caspt2 method. *Chemical Physics Letters*, *288*, 299–306.

Froese, R. D. J., Komaromi, I., Byun, K. S., & Morokuma, K. (1997). Theoretical studies of protonated and non-protonated Schiff bases of retinal: Ground state structures and energies of the all-trans, 13-cis, 11-cis, 9-cis, 6,7-cis, and 6-S-cis isomers. *Chemical Physics Letters*, *272*, 335–340.

Froese, R. D. J., Musaev, D. G., & Morokuma, K. (1998). Theoretical study of substituent effects in the diimine-M(Ii) catalyzed ethylene polymerization reaction using the Imomm method. *Journal of the American Chemical Society*, *120*, 1581–1587.

Garavelli, M. (2006). Computational organic photochemistry: Strategy achievements and perspectives. *Theoretical Chemistry Accounts*, *116*, 87–105.

Garavelli, M., Negri, F., & Olivucci, M. (1999). Initial excited-state relaxation of the isolated 11-cis protonated Schiff base of retinal: Evidence for in-plane motion from *ab initio* quantum chemical simulation of the resonance Raman spectrum. *Journal of the American Chemical Society*, *121*, 1023–1029.

Gozem, S., Huntress, M., Schapiro, I., Lindh, R., Granovsky, A. A., Angeli, C., et al. (2012). Dynamic electron correlation effects on the ground state potential energy surface of a retinal chromophore model. *Journal of Chemical Theory and Computation*, *8*, 4069–4080.

Gozem, S., Krylov, A. I., & Olivucci, M. (2013). Conical intersection and potential energy surface features of a model retinal chromophore: comparison of Eom-Cc and multireference methods. *Journal of Chemical Theory and Computation*, *9*, 284–292.

Gozem, S., Melaccio, F., Lindh, R., Krylov, A. I., Granovsky, A. A., Angeli, C., et al. (2013). Mapping the excited state potential energy surface of a retinal chromophore model with multireference and equation-of-motion coupled-cluster methods. *Journal of Chemical Theory and Computation*, *9*, 4495–4506.

Grabo, T., Petersilka, M., & Gross, E. K. U. (1999). Molecular excitation energies from time-dependent density functional theory. *Journal of Molecular Structure-Theochem*, *501–502*, 353–367.

Granucci, G., & Toniolo, A. (2000). Molecular gradients for semiempirical Ci wavefunctions with floating occupation molecular orbitals. *Chemical Physics Letters*, *325*, 79–85.

Grigorenko, B. L., Nemukhin, A. V., Polyakov, I. V., Morozov, D. I., & Krylov, A. I. (2013). First-principles characterization of the energy landscape and optical spectra of green fluorescent protein along the a ->I ->B proton transfer route. *Journal of the American Chemical Society*, *135*, 11541–11549.

Hall, R. J., Hindle, S. A., Burton, N. A., & Hillier, I. H. (2000). Aspects of hybrid Qm/Mm calculations: The treatment of the Qm/Mm interface region and geometry optimization with an application to chorismate mutase. *Journal of Computational Chemistry*, *21*, 1433–1441.

Hoffmann, M., Wanko, M., Strodel, P., Konig, P. H., Frauenheim, T., Schulten, K., et al. (2006). Color tuning in rhodopsins: The mechanism for the spectral shift between

bacteriorhodopsin and sensory Rhodopsin Ii. *Journal of the American Chemical Society, 128*, 10808–10818.
Jonasson, G., Teuler, J.-M., Vallverdu, G., Merola, F., Ridard, J., Levy, B., et al. (2011). Excited state dynamics of the green fluorescent protein on the nanosecond time scale. *Journal of Chemical Theory and Computation*, 7, 1990–1997.
Jung, Y. O., Lee, J. H., Kim, J., Schmidt, M., Moffat, K., Srajer, V., et al. (2013). Volume-conserving trans-cis isomerization pathways in photoactive yellow protein visualized by picosecond X-ray crystallography. *Nature Chemistry, 5*, 212–220.
Kholmurodov, K. T., Fel'dman, T. B., & Ostrovsky, M. A. (2007). Interaction of chromophore, 11-cis-retinal, with amino acid residues of the visual pigment rhodopsin in the region of protonated Schiff base: A molecular dynamics study. *Russian Chemical Bulletin, 56*, 20–27.
Khoroshyy, P., Der, A., & Zimanyi, L. (2013). Effect of hofmeister cosolutes on the photocycle of photoactive yellow protein at moderately alkaline pH. *Journal of Photochemistry and Photobiology B-Biology, 120*, 111–119.
Kim, T. W., Lee, J. H., Choi, J., Kim, K. H., van Wilderen, L. J., Guerin, L., et al. (2012). Protein structural dynamics of photoactive yellow protein in solution revealed by pump-probe X-ray solution scattering. *Journal of the American Chemical Society, 134*, 3145–3153.
Knowles, P. J., & Werner, H. J. (1985). An efficient 2nd-order Mc Scf method for long configuration expansions. *Chemical Physics Letters, 115*, 259–267.
Kondo, M., Heisler, I. A., Stoner-Ma, D., Tonge, P. J., & Meech, S. R. (2010). Ultrafast dynamics of protein proton transfer on short hydrogen bond potential energy surfaces: S65t/H148d Gfp. *Journal of the American Chemical Society, 132*, 1452–1453.
Laio, A., VandeVondele, J., & Rothlisberger, U. (2002). A hamiltonian electrostatic coupling scheme for hybrid Car-Parrinello molecular dynamics simulations. *The Journal of Chemical Physics, 116*, 6941–6947.
Lan, Z., Fabiano, E., & Thiel, W. (2009). Photoinduced nonadiabatic dynamics of pyrimidine nucleobases: On-the-fly surface-hopping study with semiempirical methods. *The Journal of Physical Chemistry B, 113*, 3548–3555.
Levitt, M., & Warshel, A. (1975). Computer-simulation of protein folding. *Nature, 253*, 694–698.
Lin, H., & Truhlar, D. G. (2005). Redistributed charge and dipole schemes for combined quantum mechanical and molecular mechanical calculations. *Journal of Physical Chemistry A, 109*, 3991–4004.
Liu, J., Liu, M. Y., Nguyen, J. B., Bhagat, A., Mooney, V., & Yan, E. C. Y. (2009). Thermal decay of rhodopsin: Role of hydrogen bonds in thermal isomerization of 11-cis retinal in the binding site and hydrolysis of protonated Schiff base. *Journal of the American Chemical Society, 131*, 8750–8751.
Lyne, P. D., Hodoscek, M., & Karplus, M. (1999). A hybrid Qm-Mm potential employing hartree-fock or density functional methods in the quantum region. *Journal of Physical Chemistry A, 103*, 3462–3471.
Martinez, T. J. (2006). Insights for light-driven molecular devices from *ab initio* multiple spawning excited-state dynamics of organic and biological chromophores. *Accounts of Chemical Research, 39*, 119–126.
Maseras, F., & Morokuma, K. (1995). Imomm—A new integrated *ab-initio* plus molecular mechanics geometry optimization scheme of equilibrium structures and transition-states. *Journal of Computational Chemistry, 16*, 1170–1179.
Meech, S. R. (2009). Excited state reactions in fluorescent proteins. *Chemical Society Reviews, 38*, 2922–2934.
Menger, E. L., & Kliger, D. S. (1976). Photoisomerization kinetics of 11-cis-retinal, its Schiff-base, and its protonated Schiff-base. *Journal of the American Chemical Society, 98*, 3975–3979.

Migani, A., Robb, M. A., & Olivucci, M. (2003). Relationship between photoisomerization path and intersection space in a retinal chromophore model. *Journal of the American Chemical Society, 125*, 2804–2808.

Monard, G., & Merz, K. M. (1999). Combined quantum mechanical/molecular mechanical methodologies applied to biomolecular systems. *Accounts of Chemical Research, 32*, 904–911.

Mori, T., & Martinez, T. J. (2013). Exploring the conical intersection seam: The seam space nudged elastic band method. *Journal of Chemical Theory and Computation, 9*, 1155–1163.

Morise, H., Shimomur, O., Johnson, F. H., & Winant, J. (1974). Intermolecular energy-transfer in bioluminescent system of Aequorea. *Biochemistry, 13*, 2656–2662.

Munoz-Losa, A., Martin, M. E., Galvan, I. F., Sanchez, M. L., & Aguilar, M. A. (2011). Solvent effects on the radiative and nonradiative decay of a model of the rhodopsin chromophore. *Journal of Chemical Theory and Computation, 7*, 4050–4059.

Navizet, I., Liu, Y.-J., Ferre, N., Xiao, H.-Y., Fang, W.-H., & Lindh, R. (2010). Color-tuning mechanism of firefly investigated by multi-configurational perturbation method. *Journal of the American Chemical Society, 132*, 706–712.

Nicoll, R. M., Hindle, S. A., MacKenzie, G., Hillier, I. H., & Burton, N. A. (2001). Quantum mechanical/molecular mechanical methods and the study of kinetic isotope effects: Modelling the covalent junction region and application to the enzyme xylose isomerase. *Theoretical Chemistry Accounts, 106*, 105–112.

Nienhaus, G. U. (2008). The green fluorescent protein: A key tool to study chemical processes in living cells. *Angewandte Chemie (International Edition in English), 47*, 8992–8994.

Petersilka, M., Gossmann, U. J., & Gross, E. K. U. (1996). Excitation energies from time-dependent density-functional theory. *Physical Review Letters, 76*, 1212–1215.

Punwong, C., Martinez, T. J., & Hannongbua, S. (2014). Direct Qm/Mm simulation of photoexcitation dynamics in bacteriorhodopsin and halorhodopsin. *Chemical Physics Letters, 610*, 213–218.

Punwong, C., Owens, J., & Martinez, T. J. (2015). Direct Qm/Mm excited-state dynamics of retinal protonated Schiff base in isolation and methanol solution. *The Journal of Physical Chemistry B, 119*, 704–714.

Reuter, N., Dejaegere, A., Maigret, B., & Karplus, M. (2000). Frontier bonds in Qm/Mm methods: A comparison of different approaches. *Journal of Physical Chemistry A, 104*, 1720–1735.

Ridder, L., & Mulholland, A. J. (2003). Modeling biotransformation reactions by combined quantum mechanical/molecular mechanical approaches: From structure to activity. *Current Topics in Medicinal Chemistry, 3*, 1241–1256.

Rodriguez, A., Oliva, C., Gonzalez, M., van der Kamp, M., & Mulholland, A. J. (2007). Comparison of different quantum mechanical/molecular mechanics boundary treatments in the reaction of the hepatitis C virus Ns3 protease with the Ns5a/5b substrate. *The Journal of Physical Chemistry B, 111*, 12909–12915.

Rohrig, U. F., Guidoni, L., & Rothlisberger, U. (2005). Solvent and protein effects on the structure and dynamics of the rhodopsin chromophore. *Chemphyschem, 6*, 1836–1847.

Ruckenbauer, M., Barbatti, M., Mueller, T., & Lischka, H. (2013). Nonadiabatic photodynamics of a retinal model in polar and nonpolar environment. *Journal of Physical Chemistry A, 117*, 2790–2799.

Ryde, U. (1996). The coordination of the catalytic zinc ion in alcohol dehydrogenase studied by combined quantum-chemical and molecular mechanics calculations. *Journal of Computer-Aided Molecular Design, 10*, 153–164.

Schapiro, I., Ryazantsev, M. N., Manuel Frutos, L., Ferre, N., Lindh, R., & Olivucci, M. (2011). The ultrafast photoisomerizations of rhodopsin and bathorhodopsin are modulated by bond length alternation and hoop driven electronic effects. *Journal of the American Chemical Society, 133*, 3354–3364.

Schwabe, T., Beerepoot, M. T. P., Olsen, J. M. H., & Kongsted, J. (2015). Analysis of computational models for an accurate study of electronic excitations in Gfp. *Physical Chemistry Chemical Physics*, *17*, 2582–2588.

Senn, H. M., & Thiel, W. (2009). Qm/Mm methods for biomolecular systems. *Angewandte Chemie (International Edition in English)*, *48*, 1198–1229.

Shi, X., Abbyad, P., Shu, X., Kallio, K., Kanchanawong, P., Childs, W., et al. (2007). Ultrafast excited-state dynamics in the green fluorescent protein variant S65t/H148d. 2. Unusual photophysical properties. *Biochemistry*, *46*, 12014–12025.

Shimomura, O. (2009). Discovery of green fluorescent protein (Gfp) (Nobel lecture). *Angewandte Chemie (International Edition in English)*, *48*, 5590–5602.

Shriver, J. W., Mateescu, G. D., & Abrahamson, E. W. (1979). Proton and C-13 nuclear magnetic-resonance spectroscopy study of the conformation of a protonated 11-cis-retinal Schiff-base. *Biochemistry*, *18*, 4785–4792.

Shu, X., Kallio, K., Shi, X., Abbyad, P., Kanchanawong, P., Childs, W., et al. (2007). Ultrafast excited-state dynamics in the green fluorescent protein variant S65t/H148d. 1. Mutagenesis and structural studies. *Biochemistry*, *46*, 12005–12013.

Singh, U. C., & Kollman, P. A. (1986). A combined ab initio quantum-mechanical and molecular mechanical method for carrying out simulations on complex molecular-systems—Applications to the $CH_3Cl + Cl$- exchange-reaction and gas-phase protonation of polyethers. *Journal of Computational Chemistry*, *7*, 718–730.

Slavicek, P., & Martinez, T. J. (2006). Multicentered valence electron effective potentials: A solution to the link atom problem for ground and excited electronic states. *The Journal of Chemical Physics*, *124*, 084107.

Slavicek, P., & Martinez, T. J. (2010). *Ab initio* floating occupation molecular orbital-complete active space configuration interaction: An efficient approximation to CASSCF. *The Journal of Chemical Physics*, *132*, 234102.

Sovdat, T., Bassolino, G., Liebel, M., Schnedermann, C., Fletcher, S. P., & Kukura, P. (2012). Backbone modification of retinal induces protein-like excited state dynamics in solution. *Journal of the American Chemical Society*, *134*, 8318–8320.

Stoner-Ma, D., Jaye, A. A., Matousek, P., Towrie, M., Meech, S. R., & Tonge, P. J. (2005). Observation of excited-state proton transfer in green fluorescent protein using ultrafast vibrational spectroscopy. *Journal of the American Chemical Society*, *127*, 2864–2865.

Stoner-Ma, D., Jaye, A. A., Ronayne, K. L., Nappa, J., Meech, S. R., & Tonge, P. J. (2008). An alternate proton acceptor for excited-state proton transfer in green fluorescent protein: Rewiring Gfp. *Journal of the American Chemical Society*, *130*, 1227–1235.

Sugihara, M., Buss, V., Entel, P., Elstner, M., & Frauenheim, T. (2002). 11-Cis-retinal protonated Schiff base: Influence of the protein environment on the geometry of the rhodopsin chromophore. *Biochemistry*, *41*, 15259–15266.

Swart, M. (2003). Addremove: A new link model for use in Qm/Mm studies. *International Journal of Quantum Chemistry*, *91*, 177–183.

Tsien, R. Y. (1998). The green fluorescent protein. *Annual Review of Biochemistry*, *67*, 509–544.

Tsien, R. Y. (2009). Constructing and exploiting the fluorescent protein paintbox (Nobel lecture). *Angewandte Chemie (International Edition in English)*, *48*, 5612–5626.

Warshel, A., & Karplus, M. (1972). Calculation of ground and excited-state potential surfaces of conjugated molecules.1. Formulation and parametrization. *Journal of the American Chemical Society*, *94*, 5612–5625.

Warshel, A., & Karplus, M. (1975). Semiclassical trajectory approach to photoisomerization. *Chemical Physics Letters*, *32*, 11–17.

Warshel, A., & Levitt, M. (1976). Theoretical studies of enzymic reactions—Dielectric, electrostatic and steric stabilization of carbonium-ion in reaction of lysozyme. *Journal of Molecular Biology*, *103*, 227–249.

Wei, L., Wang, H., Chen, X., Fang, W., & Wang, H. (2014). A comprehensive study of isomerization and protonation reactions in the photocycle of the photoactive yellow protein. *Physical Chemistry Chemical Physics*, *16*, 25263–25272.

Weingart, O., Schapiro, I., & Buss, V. (2007). Photochemistry of visual pigment chromophore models by *ab initio* molecular dynamics. *The Journal of Physical Chemistry B*, *111*, 3782–3788.

Werner, H. J., & Knowles, P. J. (1985). A 2nd order multiconfiguration Scf procedure with optimum convergence. *The Journal of Chemical Physics*, *82*, 5053–5063.

Woo, T. K., Cavallo, L., & Ziegler, T. (1998). Implementation of the Imomm methodology for performing combined Qm/Mm molecular dynamics simulations and frequency calculations. *Theoretical Chemistry Accounts*, *100*, 307–313.

Yen, C. W., Hayden, S. C., Dreaden, E. C., Szymanski, P., & El-Sayed, M. A. (2011). Tailoring plasmonic and electrostatic field effects to maximize solar energy conversion by bacteriorhodopsin, the other natural photosynthetic system. *Nano Letters*, *11*, 3821–3826.

Zhang, Y. K. (2005). Improved pseudobonds for combined *ab initio* quantum mechanical/ molecular mechanical methods. *The Journal of Chemical Physics*, *122*, 024114.

Zhang, Y. (2006). Pseudobond *ab initio* Qm/Mm approach and its applications to enzyme reactions. *Theoretical Chemistry Accounts*, *116*, 43–50.

Zhang, Y. K., Lee, T. S., & Yang, W. T. (1999). A pseudobond approach to combining quantum mechanical and molecular mechanical methods. *The Journal of Chemical Physics*, *110*, 46–54.

Zhang, Y., Lin, H., & Truhlar, D. G. (2007). Self-consistent polarization of the boundary in the redistributed charge and dipole scheme for combined quantum-mechanical and molecular-mechanical calculations. *Journal of Chemical Theory and Computation*, *3*, 1378–1398.

Zhou, P., Liu, J., Han, K., & He, G. (2014). The photoisomerization of 11-cis-retinal protonated Schiff base in gas phase: Insight from spin-flip density functional theory. *Journal of Computational Chemistry*, *35*, 109–120.

Zimmer, M. (2002). Green fluorescent protein (GFP): Applications, structure, and related photophysical behavior. *Chemical Reviews*, *102*, 759–781.

AUTHOR INDEX

Note: Page numbers followed by "*f*" indicate figures, "*t*" indicate tables and "*s*" indicate schemes.

C

D

G

H

I

J

K

L

M

N

O

R

S

U

V

W

SUBJECT INDEX

Note: Page numbers followed by "*f*" indicate figures, "*t*" indicate tables and "*s*" indicate schemes.

I

J

K

L

M